Logic Circuit Design

Shimon P. Vingron

Logic Circuit Design

Selected Methods

Shimon P. Vingron
Bärenkogelweg 21
2371 Hinterbrühl
Austria

ISBN 978-3-642-27656-9 e-ISBN 978-3-642-27657-6
DOI 10.1007/978-3-642-27657-6
Springer Heidelberg Dordrecht London New York

Library of Congress Control Number: 2012935652

Printed on acid-free paper

Springer is part of Springer Science+Business Media (www.springer.com)

For my wife, Dora,
the spirit of the family,
the companion and love of my life.

Preface

This book introduces you to the **design of logic circuits**. In these times of the scientific journal it has become customary for a scientific book, in the main, to contain well-established knowledge while new findings are presented in scientific journals, preferably ones peer reviewed. As J.R.R. Tolkien says of his Hobbits in the prologue to 'The Fellowship of the Ring', '...they liked to have books filled with things that they already knew, set out fair and square with no contradictions'. But this is not a Hobbitian book: While I certainly have tried to set things out fairly and squarely and hope to have avoided contradictions, major parts of this book contain new findings. These, when put together and presented in context, draw a totally new picture of sequential circuits. As the whole is more than just the sum of its parts, I thought it advisable to give a picture as complete as possible, thus the book form, and not to split the material into small parts adequate for journals, but lacking in meaning if not seen in context.

The subject matter is divided into three divisions, the first covering circuits that have no memorising ability, the **combinational circuits**, the second presents pure memory circuits, the **latches**, the third investigates circuits which have a memorising ability to various degrees, the **sequential circuits**. The presentation is not theoretical, in that most proofs have been omitted, hopefully making the text more readable. But there is still enough algebraic content to warrant using paper and pencil parallel to reading the book.

Part I, on **Combinational Circuits**, draws completely upon the first three divisions of Vingron (2004); their approximately 260 pages have not only been excerpted to the present 95, but a number of improvements have been put in place (Sects. 1.1–1.3, and 2.2, Chaps. 5, 6, and 8). Especially Chap. 8, on the composition of combinational circuits, points to a new and important design technique. With a heavy heart I have refrained from all proofs in connection with normal forms (Chap. 4) and have omitted chapters on *nand* and *nor* design techniques, and on *hazards*. Combinational circuits are quite easily recognised as being describable by the laws of logic. In a rather roundabout way this insight was first achieved by analysing circuits built of electric relays, but is, of course, independent of the technology by which a circuit is realised.

Part II is on **Latches**, these being the elementary form of sequential circuits, and allowing us to study memorisation in its most basic form. Memorisation is conventionally taken to be a time-dependent phenomenon, a notion that has always stood in the way of describing latches (and subsequently sequential circuits) by the the laws of logic (as was so successfully done for combinational circuits). The major achievement, presented in Chap. 9, is to find a time-*independent* description of memory making it possible to use logic in its conventional form (not, temporal logic) in working with latches. Despite a superfluous similarity with the division on latches of Vingron (2004), the present division on latches has been completely reworked and rewritten (especially Chaps. 9 and 10) putting the theory of latches on a new footing.

Part III is on ***asynchronous*** **sequential circuits**. Their theory was initiated by Huffman (1954) landmark paper '*The Synthesis of Sequential Switching Circuits*'. In this wonderfully readable paper (no maths!) he actually didn't present a theory—he put forth verbally expressed procedures saying how to develop asynchronous circuits. His presentation and arguments were so persuasive that all subsequent theory (it came to be known as automata theory) seemed to be focussed on providing a mathematical background and justification. The quantized time-dependent mathematical models that automata theory developed led to the introduction of **synchronous** sequential circuits. They are usually explained as being clock driven meaning that all switching activity takes place during brief and periodically recurring time intervals. Asynchronous circuits (Huffman's original sequential circuits) are not clocked. They employ continuously present signals, and the outputs can change whenever an input changes. Automata theory had practically no success in advancing the theory of asynchronous circuits.

Part III breaks with the standard synthesis procedure for asynchronous circuits as initiated by Huffman. For one thing, the three representations used in specifying an asynchronous circuit—the events graph, the word-recognition tree and the flow table (a slightly advanced version of Huffman's famous flow table)—are time-independent idealisations. This allows us to build on the theory of latches, as put forth in Part II, and to freely use logic in the same way as for combinational circuits and for latches. The basic problem confronting the standard synthesis procedure is finding a binary encoding for the flow table. Chapter 15 presents an algorithmic solution for the encoding problem. Another hitherto unsolved problem is that of verifying a sequential design. A solution to this problem is given in Chap. 18. Many of the methods and procedures of Part III are greatly improved and expanded versions found in Vingron (2004). I have refrained from a discussion of standard theory as there are excellent books on the subject (my favourites being Krieger (1969), and Dietmeyer (1971)).

Sections and chapters containing relevantly new material are marked by an asterisk ($*$). Much of the information conveyed is contained in the figures which is why I have designed and drawn them very carefully. In general I would suggest not to just fly over them. The index contains not only page numbers, rather it also refers to figures (as in: *see* Fig. 13.5) or equations (as in: *see* Eq. (13.5)). You will find the chapter number (e.g., Chap. 13) in the page headers. I always appreciate comments, and invite you to write me at ***vingron@kabsi.at***.

Hinterbrühl
Austria

Shimon P. Vingron

Contents

Part I Combinational Circuits

Part I
Combinational Circuits

Combinational circuits are the simplest binary switching-circuits as they have no memorising ability. Because they occur in their own right, and are also used in all circuits which do have a memorising ability, they are basic to all applications and thus fundamentally important. Nevertheless, it took roughly one-hundred years from their advent in 1835, when the Scottish-American scientist J. Henry invented the electro-magnetic relay, to Shannon's (1938) influential paper 'A Symbolic Analysis of Relays and Switching Circuits', *the* paper that founded switching algebra. From this late start onward, switching theory has remained technology driven. Even before Veitch (1952) and Karnaugh (1953) had put forth the so-called Karnaugh map for simplifying combinational circuits, the transistor had been invented (in 1948), and we were already working with the third computer generation (electro-magnetic relays, electronic radio valves, transistors). At this point in the development we still had no idea how to calculate simple memory devices, let alone general circuits with a memorising ability, so-called sequential circuits. All such circuits were developed intuitively.

The topics chosen for presentation in this division are a balanced extract of problems commonly encountered when working with combinational circuits. It is easy to list many further areas that are of basic interest, but being reticent therein allows us to focus our sights. The material of this division is taken from Vingron (2004) but much of it has been reworked quite basically, especially Chaps. 5, 6 and 8.

Chapter 1
Logic Variables and Events

Switching circuits process binary input signals logically, the result itself being a binary output signal. The logical operations used in processing the input signals require the binary signals to be transformed into logic variables. Giving substance to the previous two sentences is the subject of this chapter.

The first three sections concentrate on the *specification* of circuits, here, Sect. 1.3 introducing two tools of importance: the *events graph* and the *events table*. The prime result of this chapter is the introduction of *logic variables* and *logic formulas* in Sect. 1.4.

1.1 Specifying a Circuit in Plain Prose

Let us employ a running example of a switching circuit to introduce some basic concepts. As formal methods for describing a circuit's behaviour are, as yet, lacking, we shall do the obvious, and use plain prose to specify a **running example**:

> Imagine a windowless room with two doors and a light switch at each door. Every time someone passes through a door (to become the sole occupant of the room, or to leave it empty) he brings the switch next to that door from its current to its alternative position, thereby switching the lights on if they were off, and off if they were on. Correspondingly, the room's lights go on if they were off, and vice versa, go off if they were on. Thus, the basic idea is: whenever someone enters the room he turns the electric lights on, and then turns them off when he leaves the room by the same, or by the other door.
>
> Our task is to design a circuit that takes as its two input signals the current (or voltage) from the two switches, and produces as output signal the lighting condition in the room.

There are a few things to note in the above specification. Firstly, care was taken to state the number of inputs and outputs to the circuit. Secondly, each of

S.P. Vingron, *Logic Circuit Design*, DOI 10.1007/978-3-642-27657-6_1,

the circuit's signals occurs in either of two distinctly different states: The current, voltage, or lights are either on or off. Signals with this property are called **binary signals**. Usually the two alternative states of a binary signal are referred to by the integers 0 and 1. The 0 is most frequently associated with the lower energy level of a binary signal, the 1 with the higher energy level (If the energy level of a signal's state is irrelevant, one is free to associate 0 and 1 with the signal's states in any way convenient). Less obvious aspects of the above specification are: The wording of the specification is **redundant**. For human beings redundancy is helpful in grasping things more quickly, and often more thoroughly. Formal systems (on the other hand) need no redundancy (whatsoever). Finally, the above *specification is complete*, meaning that it contains all the information needed to design the circuit; you will soon see how this statement can be verified.

1.2 Analogue and Binary Timing Diagrams

Casting the usually cumbersome verbal specification into a technical form is frequently done via a **timing diagram**, a diagram which depicts the time-dependent behaviour of a signal. A timing diagram for our running example is shown in Fig. 1.1. This being our first acquaintance with a timing diagram, we ponder it and its relationship to the *running example* quite conscientiously.

t_0: At an arbitrarily chosen starting time, t_0, when you open a door to enter the room, assume that the contacts of both switches are open letting no current flow. Accordingly, we interpret both input signals to the circuit to be 0. The lights will be off, and we again choose the integer 0 to describe this state.

t_1: Entering the room, you activate, say, switch 1. The current that can now flow over the closed contacts of the switch is referred to (when flowing freely) by the integer 1. But, current cannot *jump* from a zero value to its maximum value: It first rises slowly, then ever faster until it almost reaches its maximum value from when on it grows ever more slowly towards its maximum of 1. Only at time t_1, when the current has reached a certain height (say, 80% of the

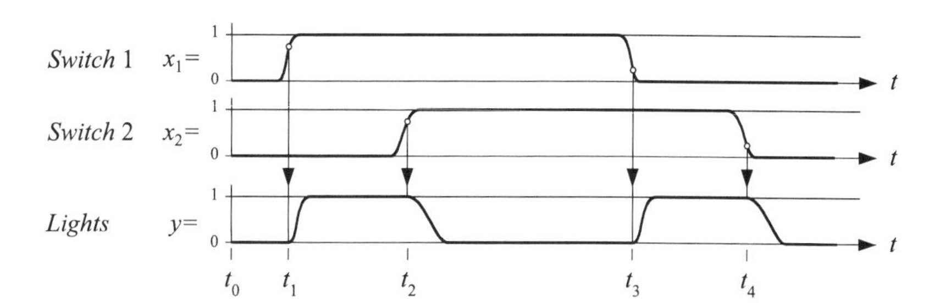

Fig. 1.1 A timing diagram of the running example

maximum value), will the circuit register that the input signal has become 1, and will *start* turning the lights on. As with the current, it takes time for the lights to obtain their full brightness, as the form of the post-t_1 output signal depicts.

t_2: You walk through the room leaving it via the opposite door and bringing switch 2 into its alternative state (a value of 1 for switch 2) to turn the lights off as you leave (letting the value for the lights drop to zero). Again, note the gradual rise of the value for x_2, and the gradual decline of the output value y.

t_3: To turn the lights on, a person entering the room tips the switch at hand (say switch 1). Please take a few moments to consider the relationship between the declining signal (x_1) and the rising signal (y).

t_4: When leaving the room by the opposite door, switch 2 will be brought to its alternative state to turn the lights off.

We now take a closer look at the concept of a signal, and the way analogue signals are transformed into binary signals. A **signal** is the information carried by, or contained in, a quantifiable physical entity (such as voltage, speed, force, etc.). Colloquially, the physical entity itself (and not the information it carries) is frequently called the signal. To be independent of any physical entity and its units of measurement (e.g., Volt, m/s, Newton) we let any physical entity being considered vary *continuously* between 0 and 1. Signals that can change their values continuously are called **analogue signals**. An example of an analogue signal is given in Fig. 1.2a. For all practical purposes, physical entities in a real-world environment

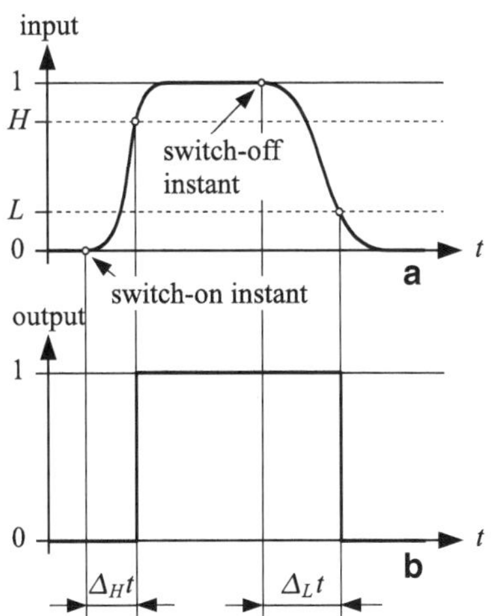

Once the input signal exceeds the high value *H*, the output jumps to 1 and remains 1 until the input signal drops below the low value *L*.

Once the input signal drops below the low value *L*, the output jumps to 0 and remains 0 until the input rises above the high level *H*.

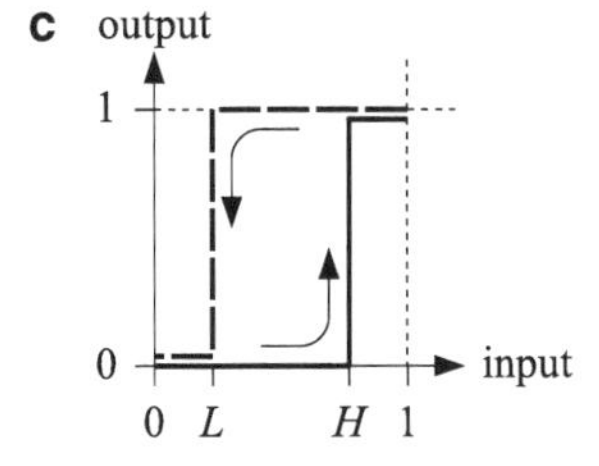

Fig. 1.2 Transforming an analogue signal (**a**) into a binary signal (**b**) with the help of an ideal hysteresis (**c**)

change their values continuously. But switching circuits use **binary signals**, signals whose values are either 0 or 1, and thus are supposed to *jump instantaneously* from one value to another. Clearly, it is quite a challenge to transform real-world analogue signals into binary signals. This is done with the help of an ideal **hysteresis**, a function with an input for the analogue signal, and an output for the binary signal. The input-output behaviour of the hysteresis is shown in Fig. 1.2c and explained in the text of the figure. Electronically, the ideal hysteresis is well approximated by a circuit known as a Schmitt trigger.

Let us apply the hysteresis function to the analogue signal of Fig. 1.2a. The instant the analogue signal (which is fed into the hysteresis) is switched on, it starts rising, first slowly and then with ever increasing speed. Only when it reaches a certain high level H does the output of the hysteresis jump from 0 to 1. Note the time lag $\Delta_H t$ from the instant the analogue signal is switched on to the moment when the binary output signal becomes 1. The analogue signal continues to rise until it reaches its 1-value. When it is switched off, it starts declining. Only when it has dropped to low level L does the output signal of the hysteresis jump from 1 to 0. Here too, we register a time delay, $\Delta_L t$, from when the analogue signal is switched off to the moment the binary signal drops to 0.

We are now well equipped to transform our *analogue* timing diagram, repeated in Fig. 1.3a, into its *binary* equivalent of Fig. 1.3b. As the instant when an *input* signal jumps from 0 to 1, we choose the moment when the analogue signal reaches the H-level. As the instant when an *input* signal jumps from 1 to 0, we choose the moment when the analogue signal drops to the L-level. The instantaneous

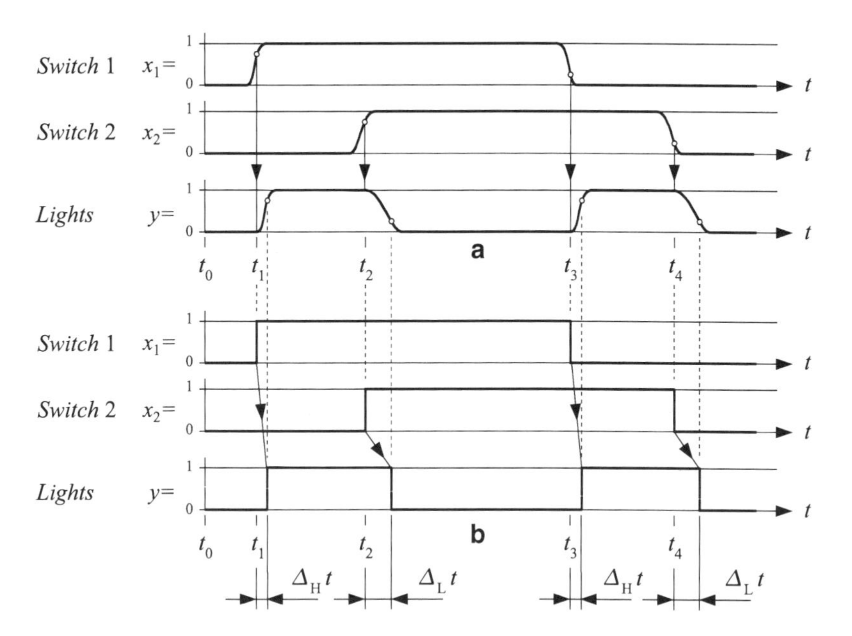

Fig. 1.3 Developing the **binary timing diagram (b)** from an **analogue timing diagram (a)**

zero-to-one transitions, and the one-to-zero transitions of the *output* signal are governed by the concept of an ideal hysteresis.

The **binary timing diagram** of Fig. 1.3b implies a causal dependence of the output signal y on the input signals x_1 and x_2 that is quite misleading. The implication is that the **rising edge** of x_1 at t_1 causes the output y to jump (instantaneously) from 0 to 1 after the time interval $\Delta_H t$ has elapsed. On the other hand, the rising edge of y at $t_3 + \Delta_H t$ seems to be caused by the **falling edge** of x_1 at t_3. Do note that, in a like manner, both the rising *and* falling edges of x_2 seem to cause the falling edges of y. If we wanted to design circuits along these lines, we would quickly find ourselves in rather deep water. To avoid this, we introduce an idealisation that allows a simpler conceptual and mathematical approach.

1.3 Events Graph and Events Table*

Mathematically, the problem with the binary timing-diagram of Fig. 1.3b is that the output variable y is *not* a function of the ordered pair (x_1, x_2) of input variables x_1 and x_2: Every occurring input pair—$(0,0)$, $(0,1)$, $(1,0)$, and $(1,1)$—maps to 0 *as well as* to 1. Let it suffice to point this out for only one input pair, for $(x_1, x_2) = (1,1)$ which is present from t_2 to t_3. From t_2 to $t_2 + \Delta_L t$ the output is 1, while from $t_2 + \Delta_L t$ to t_3 the output is 0. As can be seen from the diagram of Fig. 1.3b, the problem with two output values per input pair stems from the presence of the time lags $\Delta_L t$ and $\Delta_H t$.

To remedy this situation we postulate circuits to be **well-behaved**, that is, the output may only change when the value of at least one input variable changes, which does not mean that the output must change when an input variable changes its value. In general not every change of an input value necessitates a change in the output value. As desired, postulating that a circuit be well behaved reduces the time delays in Fig. 1.3 to zero. We call the *binary timing-diagram of a well behaved circuit* an **events graph**. The events graph for our running example is shown in Fig. 1.4.

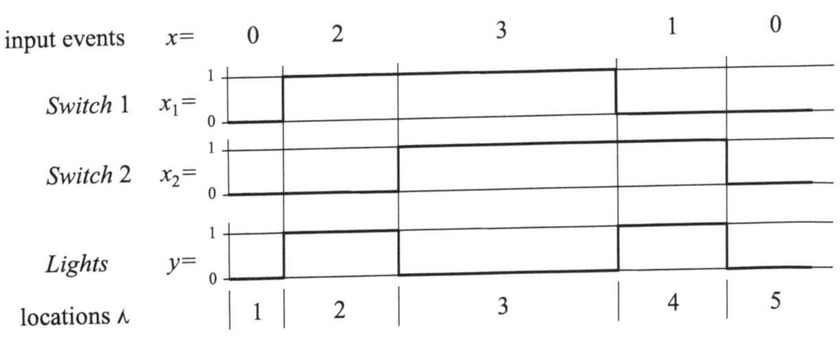

Fig. 1.4 An **events graph** for the running example

Superficially, the events graph seems to differ insignificantly from the binary timing diagram. But the basic and very important difference is that the events graph is *not* a time dependent representation of signals.

The events graph consists of a sequence of *input events* and associated output values, an **input event** being an ordered n-tuple $(e_1, e_2, \ldots, e_n)$ of arbitrary but fixed values $e_1, e_2, \ldots, e_n$ of the input variables $x_1, x_2, \ldots, x_n$, respectively. One draws the input events of an events graph successively from left to right in a sequence one finds appropriate in describing the problem at hand, parting two neighbouring input events by vertical lines,[1] and then numbering the sequence of input events successively from left to right; each of these numbers is called a **location** λ, and gives the position of an input event x (together with the associated output value) within the chosen sequence of input events.

The output value in a location is constant because we postulate our circuits to be well behaved. As no signal, neither input nor output signal, changes its value between the vertical lines it is irrelevant how far apart they are drawn. Frequently, one draws them in equal distances.

The events graph for our running example has the important property of expressing the output value of the circuit as a *function* of the input values. A **function** always maps the input event, an ordered n-tuple $(e_1, e_2, \ldots, e_n)$ of individual input values $e_1, e_2, \ldots, e_n$, to one and the same output value, and does so for each input event. For instance, at locations 1 and 5 the input event is $(0, 0)$ which in both cases is mapped to the output value 0. If, at location $\lambda = 5$, the output were 1, the output would not be a *function* of the input events—it would be a *relation*. A **relation** will map at least one input event to more than one output value. An events graph which depicts the output values as a *function* of the input events specifies a **combinational circuit**; in the case of a *relation* it specifies a **sequential circuit**. Sequential circuits are the topic of Parts II and III.

Each individual value e_i of an input event $(e_1, e_2, \ldots, e_n)$ is, in the case of binary circuits, either 0 or 1 so that each such event represents an n-digit binary number. Most frequently we express these binary numbers by their decimal equivalents

$$e = \sum_{i=1}^{n} e_i \cdot 2^{n-i}.$$

It is characteristic for an events graph of a *combinational* circuit that whenever a certain input event e is depicted, it initiates the same output value. This possible redundancy in the representation is avoided when using an *events table* instead of an events graph. The **events table** has one column for each input event, the bottom row containing the output values (unique for each input event). The events table for our running example is shown in Fig. 1.5a.

---

[1] In other words: We draw a vertical line whenever at least one input variable changes its value.

a

λ_5 λ_1	λ_4	λ_2	λ_3
$x_1=0$	$x_1=0$	$x_1=1$	$x_1=1$
$x_2=0$	$x_2=1$	$x_2=0$	$x_2=1$
$y=0$	$y=1$	$y=1$	$y=0$

Explicit form

b

$x =$	0	1	2	3
$x_1 =$	0	0	1	1
$x_2 =$	0	1	0	1
$y =$	0	1	1	0

Brief form

Fig. 1.5 Events table corresponding to the events graph of Fig. 1.4

For clarity (i.e., to concentrate better on the information—the 0s and 1s—of the table), as well as to reduce the amount of writing necessary, we usually use the events table in its **brief form**, shown in Fig. 1.5b, in which the variables have been '*extracted*' from the columns of each row. From the top to the penultimate row, the succession of 0s and 1s of a single column is its input event e. An events table whose variables are not extracted is called the **explicit form** of the table.

An advantage of the events table is that it enables us to recognise when a combinational circuit is **completely specified**; this is the case when an *output value*, either a 1 or a 0, is specified for every input event, i.e., for every column. If a circuit specification does not state what output value is initiated by each of the input events, it is said to be **incomplete**. Most technical problems—thankfully—lead to incompletely specified circuits allowing us to choose output values which lead to simpler realisations. Our running example, as mentioned, is completely specified.

In the next section we turn to the interpretation of the events table, this leading us directly into the field of logic.

1.4 Logic Variables and Logic Formulas*

The events table is a tool of major importance: Not only does it enable an *unambiguous specification* of a combinational circuit, its analysis also leads directly to the introduction of *logic variables* and to the *logical evaluation of functions*. Let us start with an informal, yet stringent, interpretation of the events table of our running example, the table repeated below. It states when the output y is 0 and when it is 1. This is expressed in a semi-formal manner in the sentence next to the table.

$x_1=0$	$x_1=0$	$x_1=1$	$x_1=1$
$x_2=0$	$x_2=1$	$x_2=0$	$x_2=1$
$y=0$	$y=1$	$y=1$	$y=0$

The output variable y is 1
if and only if
x_1 is 0 AND x_2 is 1
OR
x_1 is 1 AND x_2 is 0

Employing the symbols

$\Leftrightarrow$	...	for *logically equivalent* or *if and only if* (***iff***),
$\wedge$	...	for AND, and
$\vee$	...	for OR

the sentence to the right of the table can be written as:

$$(y = 1) \Leftrightarrow \big((x_1 = 0) \wedge (x_2 = 1)\big) \vee \big((x_1 = 1) \wedge (x_2 = 0)\big). \tag{1.1}$$

Before continuing, take a few moments to ponder the above formula, more precisely, the expressions $y = 1$, $x_i = 0$, and $x_i = 1$ in the inner parentheses. Stating that a certain input variable x_i is 0 does *not* mean that x_i really is 0. In fact, we have to *measure* the value of x_i to know its value. Let us refer to a measured value of x_i as an **instance** of x_i. We could then substitute the measured value, its instance, for x_i in $x_i = 0$ obtaining either $0 = 0$ or $1 = 0$, as the case may be. These expressions are *propositions*. A **proposition** is a statement that is either **true** (which we denote as **1**) or **false** (denoted as **0**). The term $x_i = 0$, on the other hand, has only the *form* of a proposition, and is thus called a **propositional form**; it acts as a *logic variable*, a variable whose value is either **1** or **0**. Specifying each input x_i to have a certain value, and stating the associated value of the output y, provides us with a single column of the events table.

The rather cumbersome expression (1.1) is greatly simplified by introducing the following **logic variables** X_i, $\overline{X}_i$, Y, and $\overline{Y}$ as a shorthand notation for the expressions written within the inner parentheses, i.e. defining

$$\boxed{\begin{array}{ll} X_i :\Leftrightarrow x_i = 1, & \overline{X}_i :\Leftrightarrow x_i = 0 \\ Y :\Leftrightarrow y = 1, & \overline{Y} :\Leftrightarrow y = 0 \end{array}} \tag{1.2}$$

Rewriting (1.1) with the help of the logic variables X_i, $\overline{X}_i$, Y, and $\overline{Y}$ leads to what is called a **logic formula** or **Boolean formula**[2]:

$$Y \Leftrightarrow (\overline{X}_1 \wedge X_2) \vee (X_1 \wedge \overline{X}_2). \tag{1.3}$$

Colloquially, the symbol $\Leftrightarrow$, called **logical equivalence**, refers to its left and right sides being equal from the point of view of logic, allowing us—among other things—to use the left-hand and right-hand expressions interchangeably in logic formulas (substituting one by the other). For a detailed discussion of logic equivalence see Chap. 8. The symbol $:\Leftrightarrow$ takes a (single) logic variable on its left, this being used as a shorthand notation for a logic expression on the right of the symbol $:\Leftrightarrow$. The symbol $:\Leftrightarrow$ can be read as '*... is defined to be logically equivalent to ...*'.

---

[2]For a stringent explanation of what a logic formula is, please refer to Sect. 8.1.

Alternatively to stating when y is 1, we could have used the events table to state precisely when the output variable y is 0, writing this in analogy to (1.1) as

$$(y = 0) \Leftrightarrow \big((x_1 = 0) \wedge (x_2 = 0)\big) \vee \big((x_1 = 1) \wedge (x_2 = 1)\big), \qquad (1.4)$$

which, with (1.2), may be written as the logic formula

$$\overline{Y} \Leftrightarrow (\overline{X}_1 \wedge \overline{X}_2) \vee (X_1 \wedge X_2). \qquad (1.5)$$

The formulas (1.3) and (1.5) are said to be **complementary**.

1.5 Drawing the Logic Circuit

A logic circuit is comprised of **gate**, these being graphic symbols representing logic connectives. The gates for AND, OR and NOT (INVERSION) are shown in Fig. 1.6. To invert an input signal you draw a small circle where the input signal enters the gate. If you want to invert an output signal, you draw a small circle where the output signal leaves the gate.

Logic formulas, as (1.3) and (1.5), state which gates to use and how to connect them.

Connecting gates to create a logic circuit. To transform a logic formula into a circuit consisting of gates, you draw an AND gate for each $\wedge$-connective of the formula, and an OR-gate for each $\vee$-connective. The arguments of the $\wedge$- and $\vee$-connectives in the logic formula are taken as inputs to the associated gates, this ensuring the correct connection between the gates.

Some people like to draw a circuit starting with the innermost parentheses of a formula, thus drawing the circuit from the inputs to the output. Others prefer to start with the outermost parentheses, thus drawing the circuit from the output to the inputs. Use whichever method you feel more comfortable with.

The logic circuits of Fig. 1.7 all represent the same formula, i.e. (1.3). Figure 1.7b demonstrates that you can stretch a gate to a desired size, while Fig. 1.7c shows that

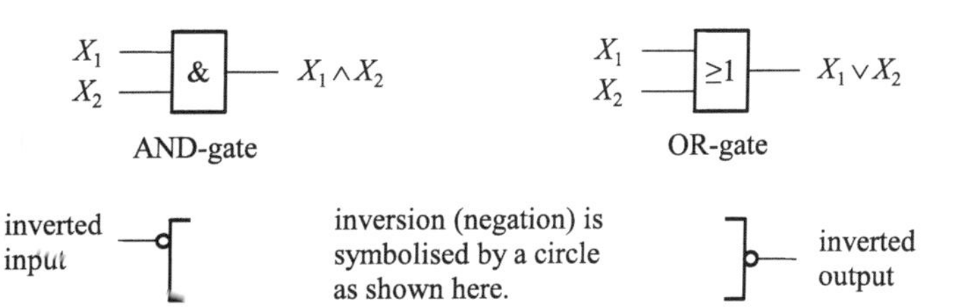

Fig. 1.6 Gates as graphical symbols for logical connectives

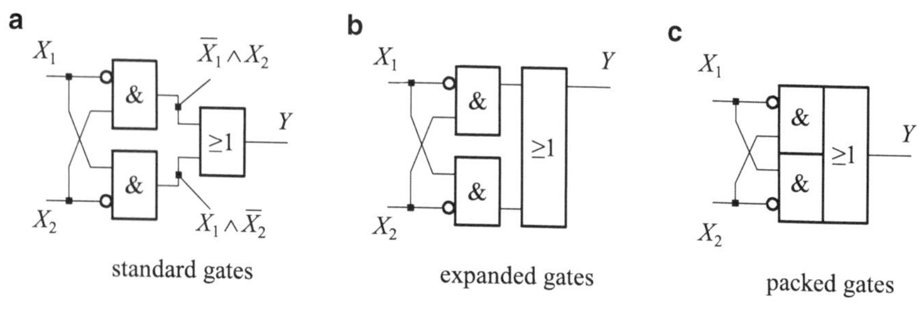

Fig. 1.7 Drawing the circuit for $Y \Leftrightarrow (\overline{X}_1 \wedge X_2) \vee (X_1 \wedge \overline{X}_2)$

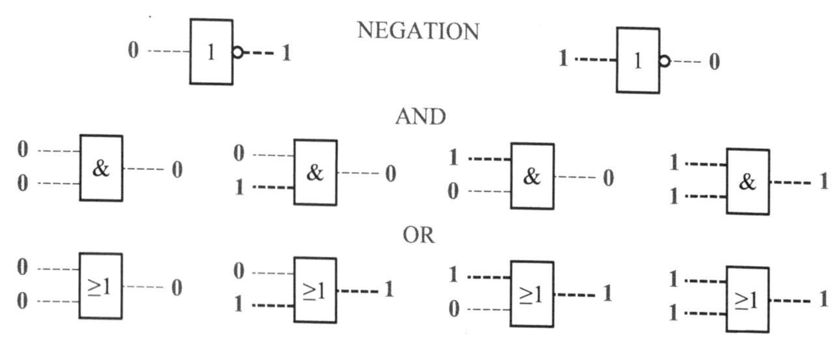

Fig. 1.8 Explaining the behaviour of NEGATION, AND, and OR

gates may be densely packed. The output signal, by the way, need not be positioned in the middle of a gate.

To fully understand these gates, we need to know precisely how their output signals react to the input signals. This is shown in Fig. 1.8 for the NEGATION, AND gate, and OR gate. Thick input or output lines represent TRUTH (i.e. **1**), thin lines indicate FALSITY (i.e. **0**).

Chapter 2
Switching Devices

Logic gates, as introduced in the previous chapter, represent the connectives used in logic formulas. Assume you have drawn a circuit diagram using logic gates, and you now want to build the circuit using switching devices of a given technology (pneumatic, electric, or electronic). To be able to do so you need to know how to translate each logic symbol into a symbol or a collection of symbols of the technology you intend to employ. This chapter is restricted to discussing symbols used for pneumatic valves, electric relays, and CMOS transistors, and to showing how they correspond to logic gates.

In general, any quantifiable physical entity (such as voltage, pressure, force, etc.) can be used as a signal. To be independent of the actual physical entity of any *binary* signal (and its unit of measurement, such as Volt, psi, Newton, etc.) it is convenient and customary to map its two values to the *integers* 0 and 1. If nothing is said to the contrary, the smaller numerical value of a signal is mapped to 0, while the larger numerical value is mapped to 1, and these two integers are not assigned any unit of measurement. In this book the lower case letters x and y (most often with indices) shall always be used as variables whose range is $\{0, 1\}$ (denoted as $x, y \in \{0, 1\}$, and meaning that the value of x and that of y is—in any given moment—either 0 or 1), and referred to as a binary **numeric variables**. These variables, x and y, are used to denote the input and output signals of actual switching devices and their symbols.

2.1 Pneumatic Valves

One of the most common pneumatic valves is the *spool-and-sleeve* design, the principle of which is shown in Fig. 2.1. A *zero* signal refers to atmospheric pressure, a *one* signal to a pressure of say 90 psi (pounds per square inch). If the pilot pressure is 0 the valve is said to be non-actuated and a spring presses the spool in the leftmost position. Actuating the valve by applying a 1-signal to the pilot chamber causes the

S.P. Vingron, *Logic Circuit Design*, DOI 10.1007/978-3-642-27657-6_2,

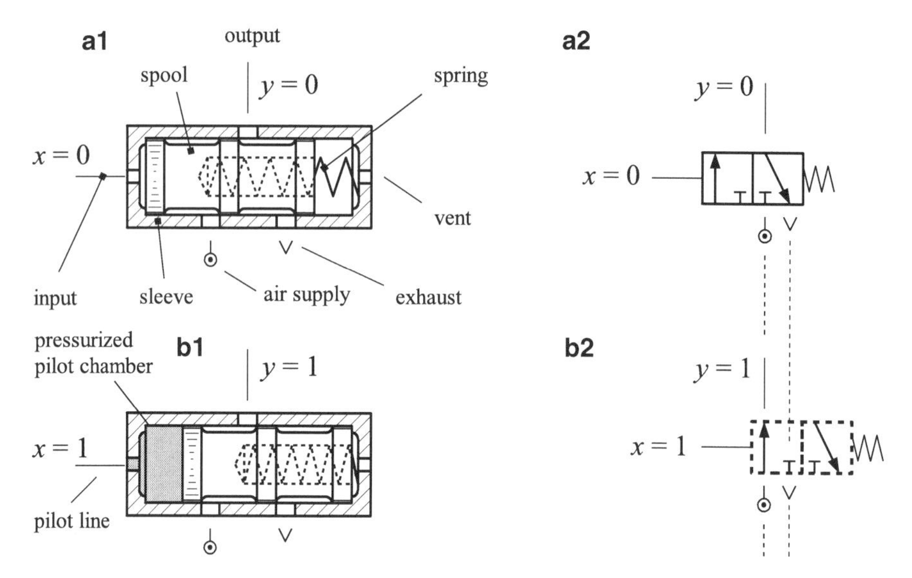

Fig. 2.1 A normally open valve

spool to be pressed to the right (against the force of the spring). In the non-actuated valve you will notice that the output port and the exhaust port are connected while the air supply is blocked. In the actuated valve the output is connected to the air supply and the exhaust is blocked. Let us refer to this valve as a **normally open valve** (or **NO-valve**, for short), this naming convention being borrowed from the realm of electric relays. *Normal* refers to the pilot pressure x being 0 (the valve is not actuated), whereas *open* refers to the path between the air supply and the output being *interrupted*. The standard symbol for the NO-valve is that of Fig. 2.1a2. The rectangle to which the spring is attached (in our case, the right rectangle) shows the port connections in the non-actuated (the *normal*) case, the left rectangle depicts the port connections in the actuated case.

I took the liberty of drawing the *controlled lines*—exhaust, air supply, and output—as being detached from the symbol's rectangles to better show how the symbol is to be understood in the actuated case: Think of the controlled lines as being static and the rectangles being pushed to the right by the pilot pressure, as indicated in Fig. 2.1b2. When the valve is not actuated, the spring pushes the rectangles back into the position shown in Fig. 2.1a2. In actual use, the controlled lines are always drawn attached to the rectangle (the one with the spring), and *the symbol is always drawn in the* non-actuated *mode*. It is *never* drawn in the actuated mode indicated in Fig. 2.1b2. By **convention**, all device symbols in a circuit are drawn in their **normal state**—i.e., their non-actuated mode—which means that they are drawn as if their input signals were zero.

An alternate behaviour to the NO-valve is obtained if the air supply and the exhaust are interchanged as shown in Fig. 2.2a. When the valve is *not* actuated the

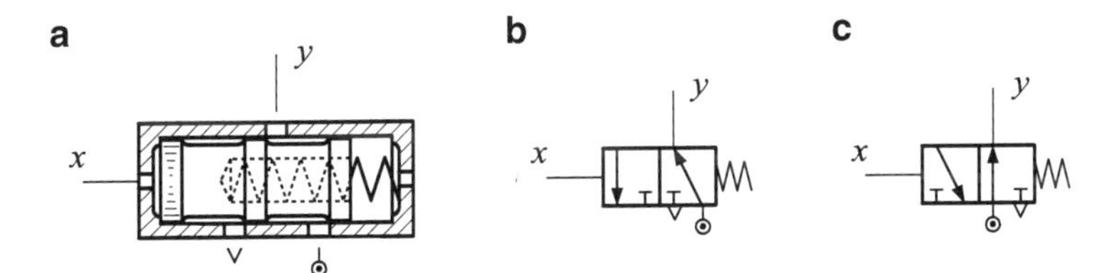

Fig. 2.2 A normally closed valve

output is connected to the air supply while actuating the valve connects the output to the exhaust. This behaviour is described by either of the symbols of Fig. 2.2b, c, and the valve is called a **normally closed valve** or **NC-valve**.

To analyse the behaviour of a spool-and-sleeve valve (Fig. 2.3) it suffices to work with the valve symbol. The input-output behaviour of such a valve is best documented in events tables as shown in Fig. 2.3a2, b2.

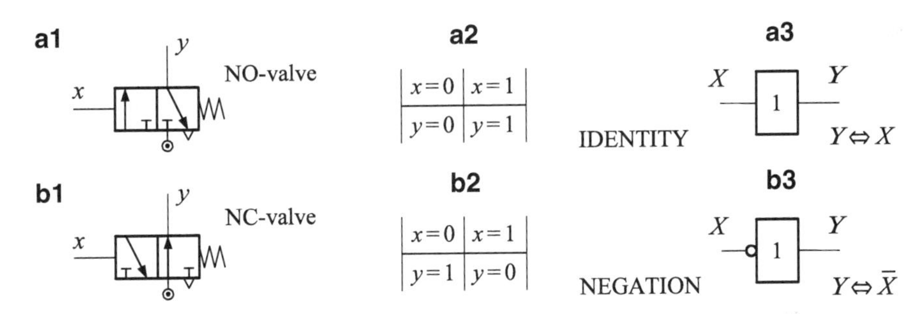

Fig. 2.3 Spool-and-sleeve valves as power amplifiers

According to the ATE of Fig. 2.3a2, 'the output y is 1 iff the input x is 1', or (in falling back on Sect. 1.4)

$$y = 1 \Leftrightarrow x = 1,$$

$$Y \Leftrightarrow X,$$

this function being called the IDENTITY, as the output Y is always identical to the input X. The symbol for the gate is shown in Fig. 2.3a3. You will probably not be alone in asking what good the IDENTITY function is, as its output is always the same as its input. Frankly, *the IDENTITY function serves no logical purpose.* But, it is of real technical importance as a *power amplifier*. In pneumatics, power is determined as the product of flow times pressure. A low-power pilot-signal can control a high-power air-supply to the output.

The behaviour of the NC-valve, on the other hand, is fully documented in the events table of Fig. 2.3b2, and from this follows

$$y = 1 \Leftrightarrow x = 0,$$

$$Y \Leftrightarrow \overline{X},$$

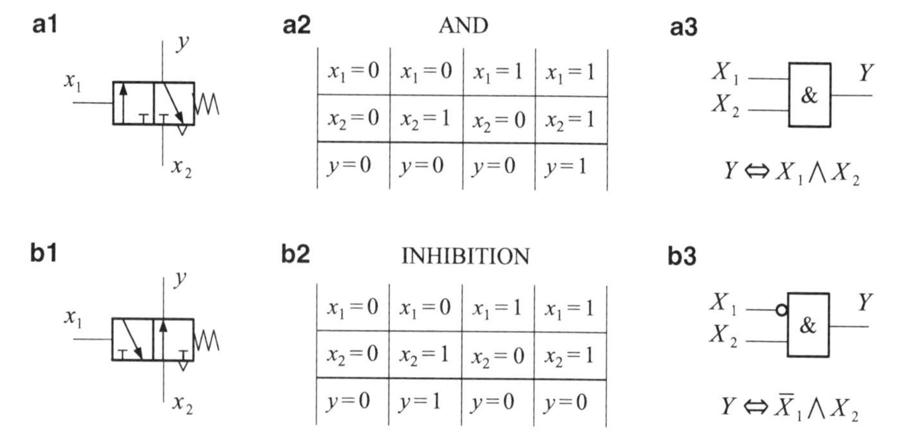

Fig. 2.4 Spool-and-sleeve valves as logic switches

The above function is called a NEGATION, the output always having the complementary value to the input.

The air supply of a spool-and-sleeve valve can be replaced by an input *signal*, as indicated in Fig. 2.4. The difference is that the air supply is *always* 1 while a signal can *vary*, meaning it can be either 0 or 1.

You will hopefully be able to recognise the valves' behaviour as being documented in the two event tables. The functions associated with these event tables are called AND and INHIBITION. The latter name stems from the fact that x_1 *inhibits* y, meaning that, when $x_1 = 1$, the output y is blocked. It is probably routine to you by now to express the events tables of Fig. 2.4a2, b2 in the following formal manner:

$$y = 1 \Leftrightarrow x_1 = 1 \text{ AND } x_2 = 1, \qquad y = 1 \Leftrightarrow x_1 = 0 \text{ AND } x_2 = 1,$$

$$Y \Leftrightarrow X_1 \wedge X_2, \qquad Y \Leftrightarrow \overline{X}_1 \wedge X_2.$$

In the above usages the spool-and-sleeve valve cannot realise the logical OR function, a function no design engineer would like to do without. A cheap and effective realisation of the OR function is the shuttle valve of Fig. 2.5a. The ball in the design shown will always seal the input of lower pressure, connecting the output with the input of higher pressure. This valve is called a **passive** valve as the air of the output signal is derived from an input line, not from an air supply. This is a drawback, as is the fact that the valve does not have an exhaust of its own; the output must be exhausted via an input line. The standard symbol for this valve (Fig. 2.5b) is obviously inspired by the design shown in Fig. 2.5a. The events table describing the valve's behaviour marks the valve as a realisation of the OR function.

To see that the events table of Fig. 2.5c does describe the OR function, read the table row-wise (and not column-wise, as we have been doing up till now). First note that 'if $x_1 = 1$ then $y = 1$' (this covering columns 2 and 3), and then that 'if $x_2 = 1$

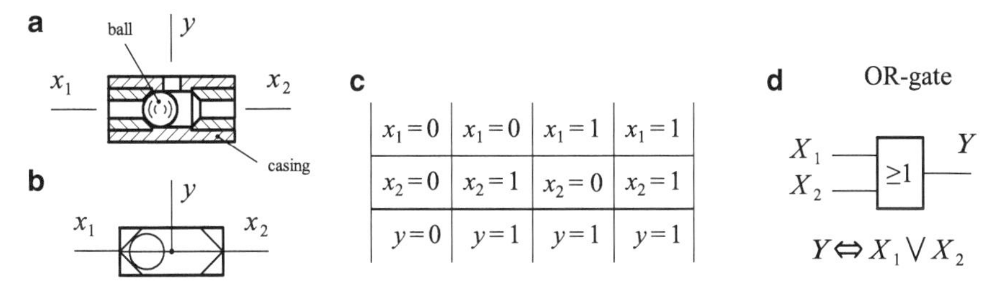

Fig. 2.5 A shuttle valve to realise the OR function

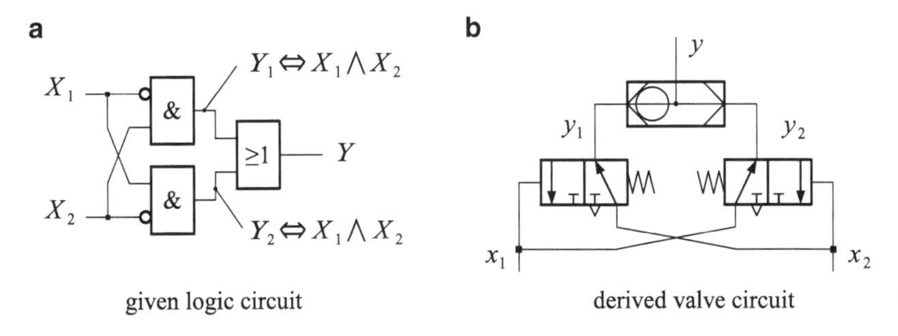

Fig. 2.6 A valve realisation of a logic circuit

then $y = 1$' (which covers columns 1 and 3). Thus we reason that '$y = 1$ iff $x_1 = 1$ OR $x_2 = 1$', and write this as

$$y = 1 \Leftrightarrow x_1 = 1 \text{ OR } x_2 = 1,$$
$$Y \Leftrightarrow X_1 \vee X_2.$$

We are now in a good position to transform a given logic circuit into one consisting of valve symbols. Such a transition is shown in Fig. 2.6 for the running example of Chap. 1. Of course this example is only used to demonstrate a technique, and has nothing to do with the original problem of switching lights on or off. To replace the INHIBITIONs of Fig. 2.6a by NC-valves, note the correspondence laid down in Fig. 2.4.

2.2 Electric Relays

In its simplest form, an electric relay (see Fig. 2.7) consists of a coil, a plunger, and a contact that is actuated by the plunger. When currant flows through the coil its magnetic field attracts the plunger, thereby opening a *normally closed contact* (Fig. 2.7a) and closing a *normally open contact* (Fig. 2.7d). Interrupting the currant

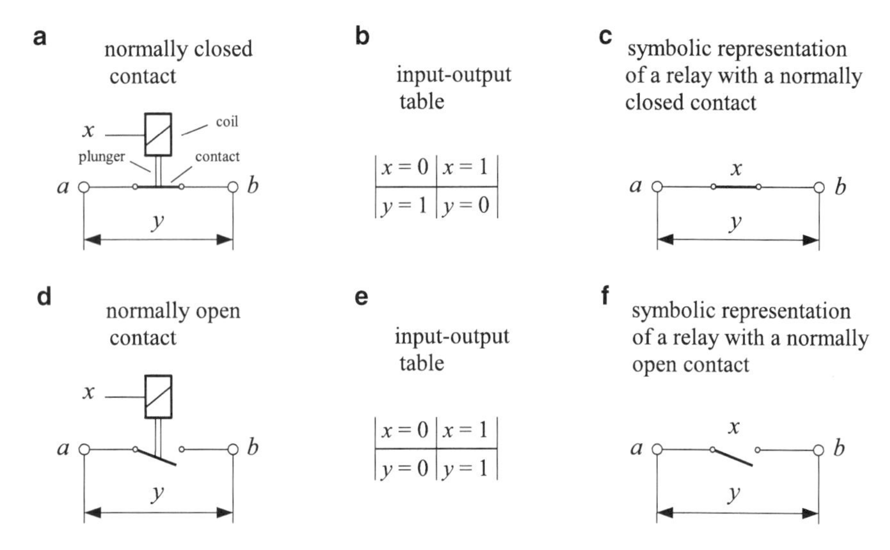

Fig. 2.7 The electric relay: principle operation, IO-table, symbol

to the coil allows a spring to return the plunger, and thus the contacts, to their initial positions.

The **input signal** to a relay is the voltage applied to the relay's coil, this voltage being represented by the binary variable $x \in \{0, 1\}$ of Fig. 2.7. The **output signal** $y \in \{0, 1\}$ stands for the **transmission** of the relay's contact, i.e., its ability to pass or transmit currant. The transmission y of a closed contact is taken to be 1, that of an open contact is 0.

Relays (as those of Fig. 2.7a, d) are *always* drawn under the assumption that their input signals are 0, i.e., that zero voltage is applied to their coils—this being referred to as the **normal state** of the relays. Thus, a *normally open contact*—a so-called NO-contact—is open ($y = 0$) when the relay's coil is not energised (when $x = 0$), and closed ($y = 1$) when $x = 1$. A *normally closed contact*—a NC-contact—is closed ($y = 1$) when the relay's coil is not energised ($x = 0$), and open ($y = 0$) when $x = 1$. This behaviour is expressed in the input-output tables (the events tables) of Fig. 2.7b, e. Stating when the outputs y of these tables are 1 is expressed formally in the logic formulas

$$y = 1 \Leftrightarrow x = 0, \qquad y = 1 \Leftrightarrow x = 1,$$

$$Y \Leftrightarrow \overline{X}, \qquad Y \Leftrightarrow X,$$

which show clearly that the relay with a normally closed contact is a NEGATION, while the relay with a normally open contact is an IDENTITY.

The **symbolic representation of a relay** (see Fig. 2.7c, f) consists of only two things: (a) a contact drawn in its *normal state*, its transmission y standing for the output of the relay, and (b) an input variable—in our example, x—which is the input signal to the coil activating the contact. The symbolic representation of a relay is an anachronism in that it has neither an input lead, nor a clearly defined output lead (such as the *logic symbols* of Sect. 1.5, or the *pneumatic symbols* of Sect. 2.1 have).

As stated for pneumatic valves, by **convention**, all relays in a circuit are drawn in their **normal state** which means that they are drawn as if their input signals were zero.

The coil of a relay usually activates multiple contacts, some normally open, the others normally closed. A single contact, or two or more interconnected contacts that transmit current to a *single load* comprise a **relay network**; for example, see the serial network of Fig. 2.8b or the parallel network of Fig. 2.9b. The interconnected contacts of a relay network are always drawn *in their normal state*, assigning to each contact the input variable of the coil activating it (in our cases, the variables x, x_1, x_2, etc.). **The way the contacts are connected** unequivocally determines when the network's transmission y is 1. The network's transmission y is usually assigned to the network's load. A **load** is any device (such as a lamp, an electric motor, a relay coil, etc.) that receives and dissipates electric energy. The load is always needed to avoid the network from short circuiting. The only relay networks to be touched on here are the serial-parallel networks, well knowing that this does not do the possibilities of relay networks justice.

Figure 2.8a shows two relays in series, with a lamp as load. As explained in the above paragraph, the *relay network* is drawn as in Fig. 2.8b. The analysis of the input-output behaviour of Fig. 2.8a (or Fig. 2.8b, if you prefer) is undertaken in the events table of Fig. 2.8c, column 3 of this table stating when y is 1:

$$y = 1 \Leftrightarrow x_1 = 1 \text{ AND } x_2 = 0,$$

$$Y \Leftrightarrow X_1 \wedge \overline{X}_2.$$

This result lets us **conjecture** that the serial connection of two contacts is expressed by the logic AND ($\wedge$), and that, vice versa, the logic AND ($\wedge$) is realised by connecting two contacts in series. For our example, this conjecture allows us to

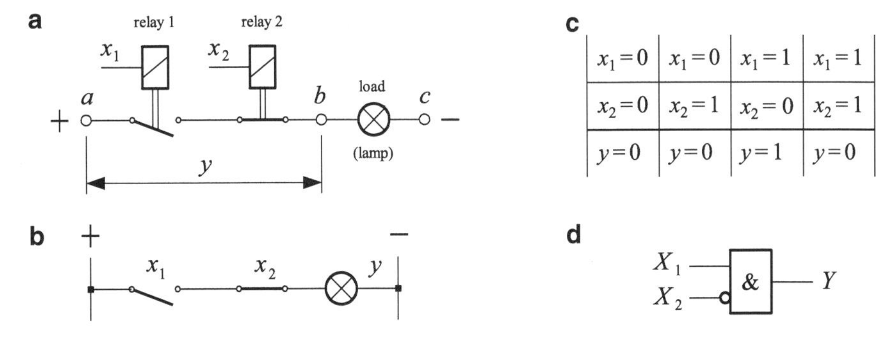

$x_1=0$	$x_1=0$	$x_1=1$	$x_1=1$
$x_2=0$	$x_2=1$	$x_2=0$	$x_2=1$
$y=0$	$y=0$	$y=1$	$y=0$

Fig. 2.8 Principle of a serial network

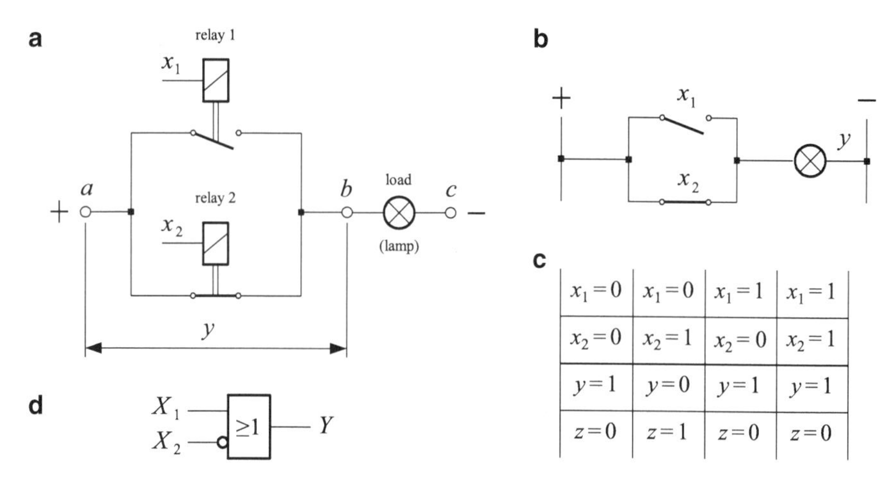

$x_1=0$	$x_1=0$	$x_1=1$	$x_1=1$
$x_2=0$	$x_2=1$	$x_2=0$	$x_2=1$
$y=1$	$y=0$	$y=1$	$y=1$
$z=0$	$z=1$	$z=0$	$z=0$

Fig. 2.9 Principle of a parallel network

state that the behaviour of the circuits of the Figs. 2.8a, b is equivalent to that of the gate of sub-figure (d).

Figure 2.9a depicts relays connected in parallel, again with a lamp as a load; the symbolic relay network is drawn in Fig. 2.9b. From either of these representations we deduce that the circuit's transmission y is 1 when the transmission of at least one of the contacts is 1; this is expressed in the events table of sub-figure (c) when concentrating only on the transmission y (i.e., for the moment, not considering the variable z).

As with the shuttle valve of Fig. 2.5, we read the events table of Fig. 2.9c row-wise arguing that $y = 1$ *iff* $x_1 = 1$ (hereby describing columns 3 and 4) OR $x_2 = 0$ (which covers column 1 and again column 3),

$$y = 1 \Leftrightarrow x_1 = 1 \;\text{ OR }\; x_2 = 0,$$

$$Y \Leftrightarrow X_1 \vee \overline{X}_2.$$

This result lets us **conjecture** that the parallel connection of two contacts is expressed by the logic OR ($\vee$), and that, vice versa, the logic OR ($\vee$) is realised by connecting two contacts in parallel. For the present example, this conjecture allows us to state that the behaviour of the circuits of the Fig. 2.9a, b is equivalent to that of the gate of sub-figure (d).

Inverting a Relay Circuit

There are two ways to invert a circuit. The simplest and most direct way is to invert the circuit's output. For instance, suppose we want to invert the circuit of

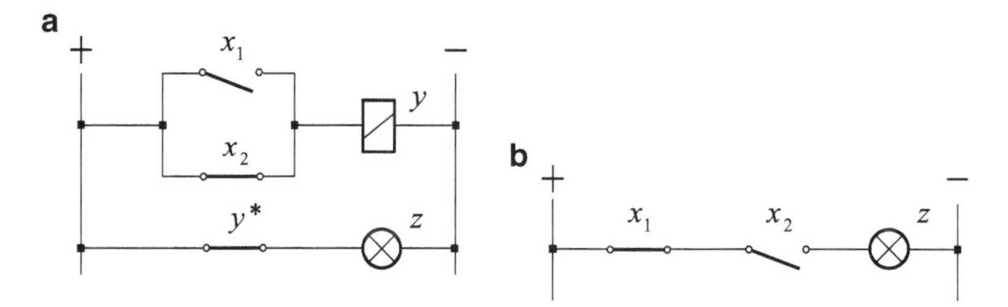

Fig. 2.10 Expressing the OR-function using double negation

Fig. 2.9b. This is done by replacing the lamp—which is the circuit's output load—by an inverter, that is, by a normally closed relay as shown in Fig. 2.10a. The relay's coil, y, is now the load for the original circuit, while the lamp is the load, z, for the inverter. The values of z are shown in the bottom row of Fig. 2.9c—they are the complement of those for y.

The second way to invert a circuit is to calculate the inverted circuit directly from the inverted output z of Fig. 2.9c, obtaining

$$Z \Leftrightarrow \overline{X}_1 \wedge X_2,$$

which leads to the circuit of Fig. 2.10b.

Realising Feedback

Feedback plays a dominating role in the design of so-called latches, these being elementary memory circuits. In fact they are so important that a whole Part of this book is devoted to them. In this early stage, we only want to see how to transform a logic feedback circuit, say, that of Fig. 2.11a, into a relay circuit. The output OR-gate of sub-figure (a) requires our contact network to consist of two parallel paths, one of which contains the NO-contact x_2. The AND-gate of sub-figure (a) leads to the serial circuit consisting of the NC-contact x_1 and the NO-contact named y^*. The load of the relay network is the coil y of a relay. To realise the *feedback* of the

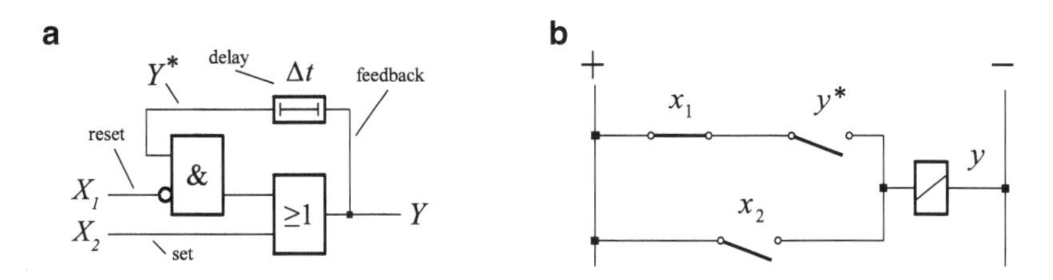

Fig. 2.11 Transforming the logic circuit of a latch into a relay network

latch pictured in sub-figure (a) we define the NO-contact y^* of sub-figure (b) to be actuated by the coil y.

The delay Δt in the logic circuit of sub-figure (a) expresses the fact that the NO-contact y^* follows the coil's state of activation with the time delay of Δt. The logic circuit is thus modelled after the relay circuit. But it must be noted that the *theory* of feedback latches does not require the existence of a delay in the feedback loop.

2.3 CMOS Transistors

The transistor was invented in 1948 by John Bardeen, Walter Brattain and William Shockley and soon replaced the vacuum tube as a cheap amplifier. We shall take a look at only one transistor device, of a large proliferation, namely the CMOS technology. The transistors used are **insulated-gate field-effect transistors (IG-FET)**—Fig. 2.12—more commonly called **MOS** for the sequence of the layers of material used in their design—**metal-oxide-silicon**. There are two types of these transistors, according to the polarity (**n**egative or **p**ositive) of the electric charges used to create a conducting channel between the *drain* and the *source* regions: *n-channel MOS* (**NMOS**) and *p-channel MOS* (**PMOS**). Digital circuits that employ both these transistor types are called **complementary MOS (CMOS)**, and have the advantage of needing only a single level of supply voltage.

For current to be able to flow between the drain and the source region, you need to *create a conducting channel* in the silicon between these regions. This is achieved by applying the proper voltage between the gate and the substrate. In an n-channel MOS you obtain a conducting channel by applying a more positive potential to the gate than to the substrate (the voltage from gate to substrate is positive), thus attracting negative charges to the gate while repulsing positive charges. In a p-channel MOS the conducting channel is obtained by making the voltage from substrate to gate positive, thus attracting positive charges to the gate while repulsing negative charges from the gate. *In actual usage, the source is always connected*

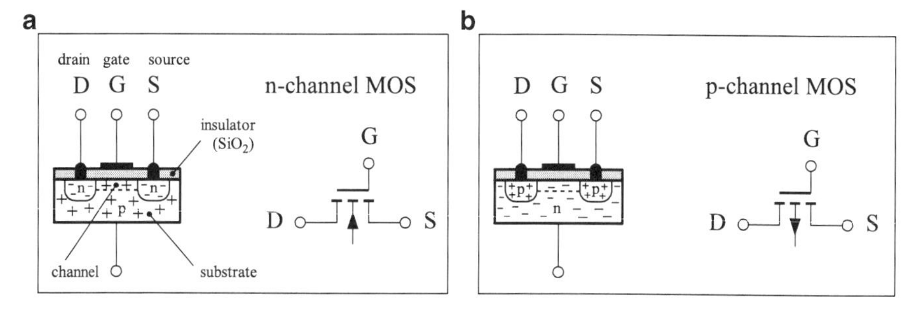

Fig. 2.12 Principle of the n-channel and p-channel MOS

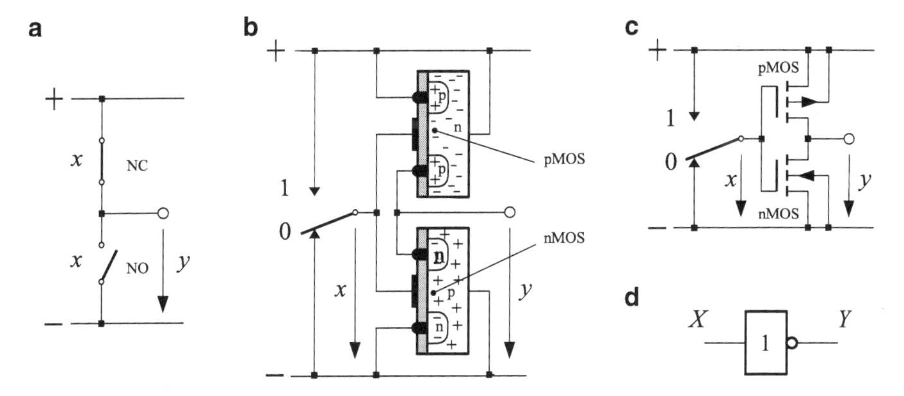

Fig. 2.13 The CMOS inverter

to the substrate. To obtain a channel, the density of source-type charges must be enhanced giving these transistors the name **enhancement type transistors**.

The simplest CMOS circuit is the **inverter** of Fig. 2.13. Its principle of operation is basic to CMOS technology. If you understand how the inverter works, you almost automatically understand the next two gates.The idea behind the inverter is shown in the relay circuit of Fig. 2.13a: The two serial contacts are always in a state of alternate transmission—when one is closed, the other is open. When the top switch, the NC-contact, is closed and the bottom switch, the NO-contact, open, the full supply voltage is fed through to the output y. *Vice versa*, the output y is zero when the top switch is open and the bottom one closed.

The relay circuit translates directly to the transistor circuit of Fig. 2.13b, the NC-contact being replaced by a PMOS, the NO-contact by an NMOS. In the circuit as drawn, the gates of both transistors are at ground potential dew to the way the switch governing the gate potential is set. As the substrate of the PMOS is connected to high potential a conducting p-channel is created. The NMOS transistor, on the other hand, is non-conducting as it has 0 voltage between gate and substrate, both being connected to ground. The voltage measurable at the output y is thus high. If the switch is brought to the alternate position, in which the potential at the gates of both transistors is high, the PMOS becomes non-conducting while the NMOS becomes conducting. The output voltage y is thus brought to zero. The actual circuit, of course, is always drawn using MOS symbols, giving you Fig. 2.13c. As the input and output of this circuit are always complementary, it is an inverter or a negation, the symbol for which is shown in Fig. 2.13d.

There are two successful ways, of four trivial ones, to generalise the principle inverter circuit of Fig. 2.13a: One of these is shown in Fig. 2.14a, the other in Fig. 2.15a, and both lead to circuits superbly adapted to CMOS technology.

The first generalisation of the inverter of Fig. 2.13a is achieved by substituting the single NC-contact by two serial NC-contacts, and substituting the single NO-contact by two parallel NO-contacts—this leading to Fig. 2.14a. The equivalent

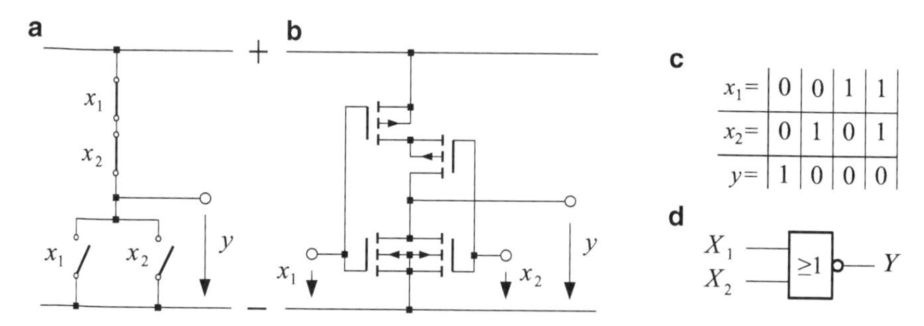

Fig. 2.14 The CMOS NOR-circuit

CMOS circuit is obtained by substituting PMOS transistors for NC-contacts, and NMOS transistors for NO-contacts. As with the inverter of Fig. 2.13c, note that in Fig. 2.14b the sources and substrates of the PMOS transistors are connected to high potential, or at least higher potential, while the sources and substrates of the NMOS transistors are connected to low (or lower) potential.

The simplest way to analyse the circuit of Fig. 2.14 is to develop the events table shown in Fig. 2.14c. You start out with the table of input events, leaving the bottom row, the one for the output y, empty. Then, for each input event, you deduce from the behaviour of the circuit the value for the output y, entering the outputs into the bottom row. Considering the result, you will notice that inverting each output value leads to the **or** function. In other words, the events table of Fig. 2.14c is that of a **not-or** function, usually referred to as the **nor** function. The logic symbol used for this function is shown in Fig. 2.14d; it is simply an OR gate with an inverted output.

The second generalisation of the inverter leads to Fig. 2.15a and that to Fig. 2.15b. The analysis needed to obtain the events table of Fig. 2.15c is easy so that I again leave the details to you. The result, as you will have noticed, is the inversion of the **and** function, i.e. is a **not-and** function, and is thus referred to as the **nand** function, the logic symbol for which is shown in Fig. 2.15d.

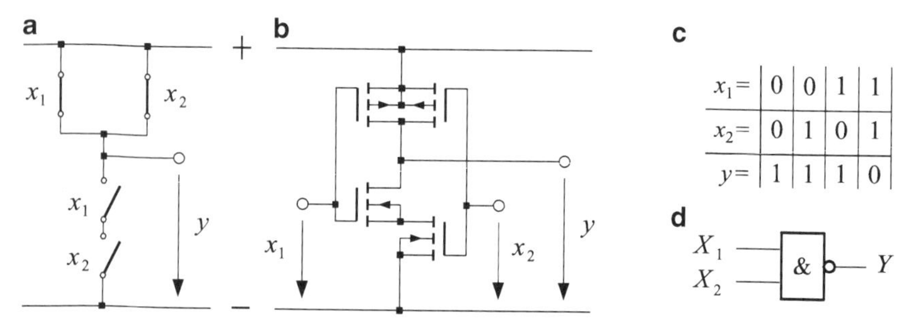

Fig. 2.15 The CMOS NAND-circuit

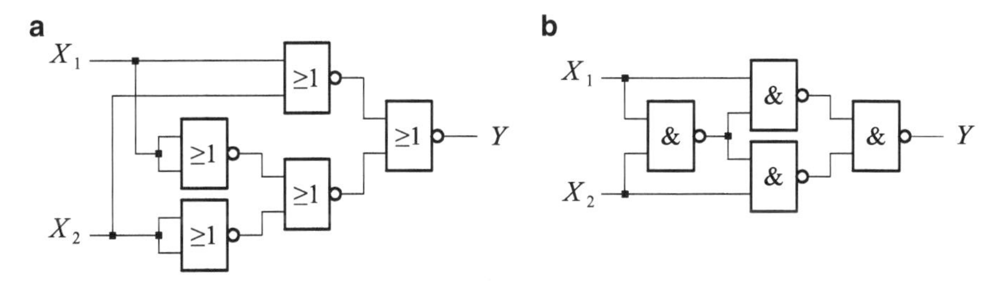

Fig. 2.16 NOR and NAND versions of our running example

Of course the CMOS circuits discussed here are only a small excerpt of the circuits used. But they are basic and absolutely essential to the understanding of the techniques employed. One of the most important omissions is a discussion of circuits with feedback, circuits that lead to the realisation of memory devices—so-called *latches* and *flip-flops*. These circuits are discussed, together with their theory, in later chapters.

The logic building blocks of CMOS technology are NAND-, NOR- and NEGATION-gates. These, therefore, one wishes to use when realising logic circuits in CMOS technology. To get an impression of pure NOR- and NAND-design, take a look at Fig. 2.16. These circuits are equivalent to that of Fig. 1.5 of our running example of Chap. 1. Transforming larger AND, OR and NOT based circuits into NOR or NAND circuits can be quite demanding and is discussed in a later chapter.

Chapter 3
Elementary Logic Functions

Elementary logic functions are the building blocks of all logic functions however complicated they may be, and are thus of prime importance. In the following pages we discuss all elementary logic functions and how they interact. Much of this material, such as the summaries of theorems, is only intended as reference to be at hand when needed (e.g., in calculations, or for theoretical considerations).

3.1 Logic Functions

An **elementary logic function** F maps a pair of ordered *truth values*, FALSITY (for which we write $\mathbf{0}$) and TRUTH (denoted as $\mathbf{1}$), into the set $\{\mathbf{0}, \mathbf{1}\}$ of truth values

$$F : \{\mathbf{0}, \mathbf{1}\}^2 \mapsto \{\mathbf{0}, \mathbf{1}\}$$

These functions have, at most, two input variables and are summarised in the table of Fig. 3.1. The middle column of this table shows function symbols commonly used in switching algebra. These function symbols—$\wedge, >, <, \oplus, \vee, \tilde{\vee}, \leftrightarrow, \leftarrow, \rightarrow, \tilde{\wedge}$; $\neg$—are called **logic connectives**. Those that take two arguments are **dyadic logic connectives**. The only one that takes a single argument, the NEGATION symbol $\neg$, is referred to as a **monadic logic connective**.

Actually, the symbols $>$ and $<$ for the INHIBITIONs, $\tilde{\vee}$ for NOR, and $\tilde{\wedge}$ for NAND are not common—these symbols were introduced by **Iverson** for APL. Iverson's notation for the INHIBITIONs was chosen because it is mnemonic and because there is no generally accepted notation for these functions. In the case of NOR and NAND Iverson's notation is mnemonically clearly superior to the respective notations of **Sheffer** ($X_1 \downarrow X_2 :\Leftrightarrow X_1 \tilde{\vee} X_2$) and **Nicod** ($X_1 | X_2 :\Leftrightarrow X_1 \tilde{\wedge} X_2$) sometimes used. In switching algebra the negation of a variable X is commonly denoted as $\overline{X}$ whereas in logic the negation would preferably be written as $\neg X$.

S.P. Vingron, *Logic Circuit Design*, DOI 10.1007/978-3-642-27657-6_3,

Definition $X_1 \leftrightarrow \mathbf{0\,0\,1\,1}$ $X_2 \leftrightarrow \mathbf{0\,1\,0\,1}$	Logic notation	Name of the logic function
$Y_0 \leftrightarrow \mathbf{0\,0\,0\,0}$	$Y_0 \Leftrightarrow \mathbf{0}$	CONTRADICTION, FALSITY
$Y_1 \leftrightarrow \mathbf{0\,0\,0\,1}$	$Y_1 \Leftrightarrow X_1 \wedge X_2$	CONJUNCTION, AND
$Y_2 \leftrightarrow \mathbf{0\,0\,1\,0}$	$Y_2 \Leftrightarrow X_1 > X_2$	INHIBITION
$Y_3 \leftrightarrow \mathbf{0\,0\,1\,1}$	$Y_3 \Leftrightarrow X_1$	IDENTITY
$Y_4 \leftrightarrow \mathbf{0\,1\,0\,0}$	$Y_4 \Leftrightarrow X_1 < X_2$	*(transposed)* INHIBITION
$Y_5 \leftrightarrow \mathbf{0\,1\,0\,1}$	$Y_5 \Leftrightarrow X_2$	IDENTITY
$Y_6 \leftrightarrow \mathbf{0\,1\,1\,0}$	$Y_6 \Leftrightarrow X_1 \oplus X_2$	ANTIVALENCE, XOR
$Y_7 \leftrightarrow \mathbf{0\,1\,1\,1}$	$Y_7 \Leftrightarrow X_1 \vee X_2$	DISJUNCTION, OR
$Y_8 \leftrightarrow \mathbf{1\,0\,0\,0}$	$Y_8 \Leftrightarrow X_1 \tilde{\vee} X_2$	NOR, Sheffer function
$Y_9 \leftrightarrow \mathbf{1\,0\,0\,1}$	$Y_9 \Leftrightarrow X_1 \leftrightarrow X_2$	(material) EQUIVALENCE
$Y_{10} \leftrightarrow \mathbf{1\,0\,1\,0}$	$Y_{10} \Leftrightarrow \overline{X_2}$	NEGATION, NOT
$Y_{11} \leftrightarrow \mathbf{1\,0\,1\,1}$	$Y_{11} \Leftrightarrow X_1 \leftarrow X_2$	*(transposed)* IMPLICATION
$Y_{12} \leftrightarrow \mathbf{1\,1\,0\,0}$	$Y_{12} \Leftrightarrow \overline{X_1}$	NEGATION, NOT
$Y_{13} \leftrightarrow \mathbf{1\,1\,0\,1}$	$Y_{13} \Leftrightarrow X_1 \to X_2$	(material) IMPLICATION
$Y_{14} \leftrightarrow \mathbf{1\,1\,1\,0}$	$Y_{14} \Leftrightarrow X_1 \tilde{\wedge} X_2$	NAND, Nicod function
$Y_{15} \leftrightarrow \mathbf{1\,1\,1\,1}$	$Y_{15} \Leftrightarrow \mathbf{1}$	TAUTOLOGY, TRUTH

Fig. 3.1 Table of the elementary logic-functions

Happily, it suffices to use sub-sets of the set of elementary connectives, $\{\neg, \wedge, >, <, \oplus, \vee, \tilde{\vee}, \leftrightarrow, \leftarrow, \rightarrow, \tilde{\wedge}\}$, to be able to write the logic expression of a general logic function. These *sets of necessary and sufficient connectives* are frequently referred to as **functionally complete sets** (of connectives). They arc compilcd in the leftmost and rightmost columns of Fig. 3.2. Without proof at this point, let it be said that all general functions can be realised with the AND, OR, and NOT

set	$\neg A \Leftrightarrow$	$A \wedge B \Leftrightarrow$	$A \vee B \Leftrightarrow$	dual set
$\{\tilde{\wedge}\}$	$A \tilde{\wedge} A$	$(A \tilde{\wedge} B) \tilde{\wedge} (A \tilde{\wedge} B)$	$(A \tilde{\wedge} A) \tilde{\wedge} (B \tilde{\wedge} B)$	$\{\tilde{\vee}\}$
$\{\wedge, \neg\}$	$\neg A$	$A \wedge B$	$\neg(\neg A \wedge \neg B)$	$\{\vee, \neg\}$
$\{\rightarrow, \mathbf{0}\}$	$A \rightarrow \mathbf{0}$	$\big(A \rightarrow (B \rightarrow \mathbf{0})\big) \rightarrow \mathbf{0}$	$(A \rightarrow B) \rightarrow B$	$\{>, \mathbf{1}\}$
$\{\rightarrow, \neg\}$	$\neg A$	$\neg(A \rightarrow \neg B)$	$\neg A \rightarrow B$	$\{>, \neg\}$
$\{\oplus, \wedge, \mathbf{1}\}$	$A \oplus \mathbf{1}$	$A \wedge B$	$A \oplus B \oplus (A \wedge B)$	$\{\leftrightarrow, \vee, \mathbf{0}\}$
$\{\oplus, \vee, \mathbf{1}\}$	$A \oplus \mathbf{1}$	$A \oplus B \oplus (A \vee B)$	$A \vee B$	$\{\leftrightarrow, \wedge, \mathbf{0}\}$

Fig. 3.2 Functionally complete sets of connectives

original	$\mathbf{0}$	$\wedge$	$>$	$<$	$\oplus$	$\vee$	$\tilde{\vee}$	$\leftrightarrow$	$\neg$	$\leftarrow$	$\rightarrow$	$\tilde{\wedge}$	$\mathbf{1}$
dual	$\mathbf{1}$	$\vee$	$\leftarrow$	$\rightarrow$	$\leftrightarrow$	$\wedge$	$\tilde{\wedge}$	$\oplus$	$\neg$	$>$	$<$	$\tilde{\vee}$	$\mathbf{0}$

Fig. 3.3 Dual connectives

functions, i.e., $\{\wedge, \vee, \neg\}$ is a functionally complete set. The formulas in the three central columns of a given row state how to realise the functions NOT, AND, and OR using the connectives stated in the functionally complete set of the leftmost column. The functionally complete sets in the rightmost column are **dual** to the sets in the leftmost column. This refers to the fact that replacing truth values and connectives by their duals, according to Fig. 3.3, transforms one correct formula into another one which is then also correct. The formulas are called (mutually) dual.

The sequence in which the logic connectives are acted upon in a given formula is determined solely by parentheses. But this causes their number to grow quite rapidly for lengthy formulas. To counteract this, one uses a *convention of adhesion* that introduces an artificial priority at least between the most commonly used logical connectives.

Convention of adhesion: A given connective of $(\neg, \wedge, \vee, \rightarrow, \leftrightarrow)$ has a stronger adhesion to its arguments than any connective of the quintuple listed to its right. In other words, any connectives listed to the left of another is acted upon before acting upon this other one.

This convention allows us to rewrite the following formula in the indicated ways:

$$\Big((A \wedge B) \vee ((\neg A) \wedge B)\Big) \leftrightarrow B \Leftrightarrow$$
$$(A \wedge B) \vee (\neg A \wedge B) \leftrightarrow B \Leftrightarrow$$
$$A \wedge B \vee \neg A \wedge B \leftrightarrow B$$

When $\wedge$- and $\vee$-connectives occur alternatively, one is prone to confuse them. It is thus common to adhere to the

Convention of omitting the $\wedge$-connective: One may write PQ instead of $P \wedge Q$ when it seems appropriate.

The last of the previous formulas may then be written $AB \vee \neg AB \leftrightarrow B$ which is quite easy to read, as you will hopefully agree. In contrast to the above procedure, it is sometimes advantageous not to **bind** arguments to (dyadic) connectives but to **separate** the arguments from the connectives. This is done in the

Convention of separation: *Dots* attached to dyadic connectives (as in $.\wedge.$, $:\leftrightarrow:$, $.:\rightarrow:.$) separate these from their arguments. The more dots, the stronger the separation. But, for two different kinds of connectives, both with the same number of dots, the above *convention of adhesion* prevails.

The following example of the application of the *convention of separation* demonstrates that the dots need not occur symmetrically.

$$\big(A \to (B \to C)\big) \to \quad (A \wedge B \to C) \Leftrightarrow$$
$$A \to (B \to C)\ .\!\!\to\!\!.\ A \wedge B \to C \Leftrightarrow$$
$$A\ .\!\!\to\!\!.\ B \to C\ :\!\!\to\!\!.\ A \wedge B \to C$$

3.2 Basic Gates

A gate is a specific graphic symbol for a logic function. As such, it expresses an idealised behaviour of a switching device. Viewing it the other way around, the behaviour of a real-world switching device is an approximation of a gate's behaviour. Gates were preliminarily introduced in Sect. 1.5 especially stating how to connect them to express a given logic formula. Not every logic function has a gate assigned to it. Those gates that do exist are called **basic**, hereby emphasising that all other logic functions are represented by graphic symbols comprised of basic gates. A basic gate is given the same name as the function it represents. If a gate has one input it is called **monadic**, if it is has two inputs it is called **dyadic**.

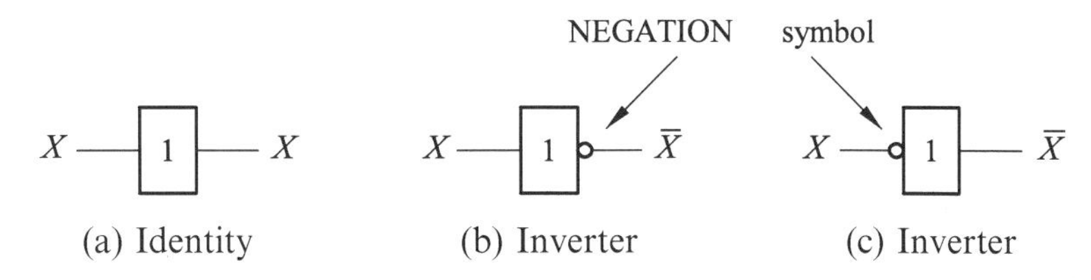

Fig. 3.4 The monadic gates. (**a**) Identity (**b**) Inverter (**c**) Inverter

There are two logic functions with a single input, the IDENTITY and the NEGATION. The IDENTITY gate, shown in Fig. 3.4a, is a rectangle with a 1 written into it. A gate representing a NEGATION is an IDENTITY gate with either the input or output signal marked by a small circle representing inversion and drawn where a lead enters or leaves the rectangle, see Fig. 3.4b, c.

Only *four* basic dyadic gates exist, called AND, OR, XOR, and EQUIVALENCE, (Fig. 3.5), each representing the equally named logic functions. But, why

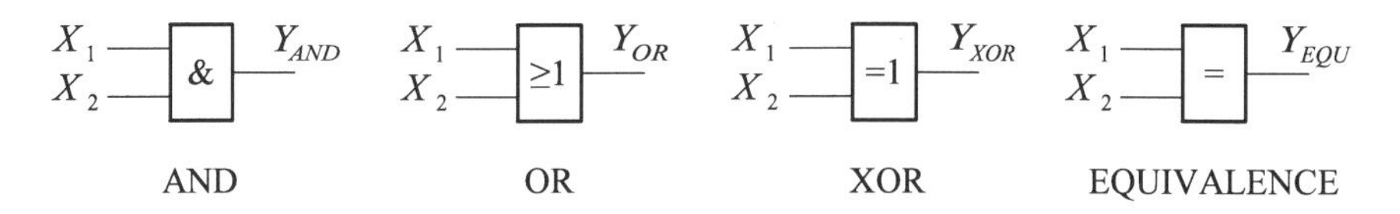

Fig. 3.5 The dyadic gates

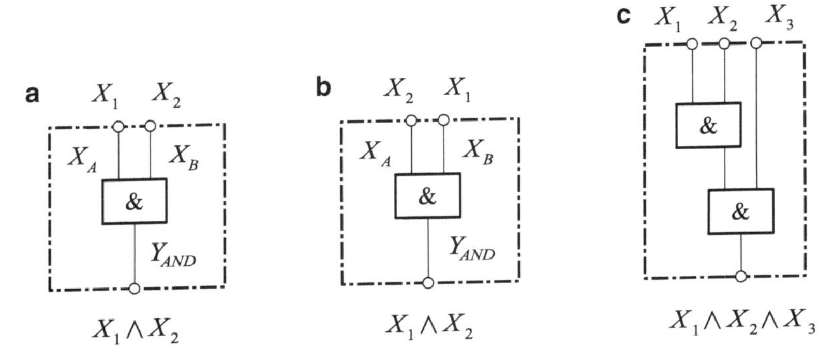

Fig. 3.6 On commutativity and associativity

are gates only defined for these four functions? For a gate to be of practical use, it must represent a function that is **associative** and **commutative**, and only the above mentioned functions have both these properties as discussed further on. As for the rather curious choice of inscriptions (&, ≥ 1, $=1$, $=$) distinguishing the gates, I'm afraid you'll just have to accept them—they are international standard as declared by the International Electrotechnical Commission (IEC 117-15).

Commutativity allows you to interchange the input signals to a gate as you like without being able to notice a change in the behaviour of the output value. For instance, assigning (X_1, X_2) to (X_A, X_B) in Fig. 3.6a will give the same result as assigning (X_2, X_1) to (X_A, X_B) in Fig. 3.6b.

Associativity permits you to *cascade* identical gate types thus creating a multiple-input gate of that type. For instance, cascading two dual-input AND-gates, as shown in Fig. 3.6c, leads to a triple-input AND-gate. Its inputs are again commutative and associative.

Due to the properties of commutativity and associativity, the four dyadic functions AND, OR, XOR, and EQUIVALENCE can be used to create gates with multiple inputs as shown in Fig. 3.7.

The output variables for these multiple-input gates can be stated quite strait forwardly as

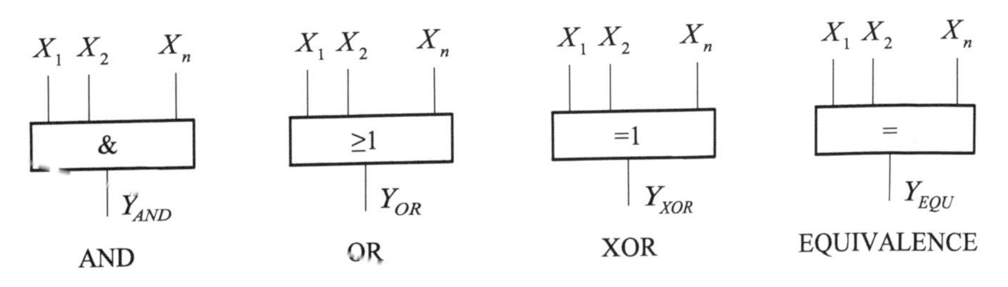

Fig. 3.7 Gates with multiple inputs

$$Y_{AND} \Leftrightarrow X_1 \wedge X_2 \wedge \cdots \wedge X_n, \qquad Y_{XOR} \Leftrightarrow X_1 \oplus X_2 \oplus \cdots \oplus X_n,$$

$$Y_{OR} \Leftrightarrow X_1 \vee X_2 \vee \cdots \vee X_n, \qquad Y_{EQU} \Leftrightarrow X_1 \leftrightarrow X_2 \leftrightarrow \cdots \leftrightarrow X_n,$$

You might like to try to formulate when the output values of these functions are **1**. This is quite easy for AND and OR, but before continuing try to state when the outputs of the multiple XOR and multiple EQUIVALENCE are **1**.

A summary of when these functions are **1** is given in the following statements.

AND: Y_{AND} is **1** *iff* all inputs are **1**.

OR: Y_{OR} is **1** *iff* at least one input is **1**.

XOR: Y_{XOR} is **1** *iff* an odd number of the inputs are **1**.

EQU: For an **even** number of inputs (greater than or equal to 2):
Y_{EQU} is **1** *iff* an even number of the inputs are **1** (0 is taken as even).
For an **odd** number of inputs:
Y_{EQU} is **1** *iff* an odd number of the inputs are **1**.

The above statements are illustrated in the tables of Fig. 3.8 for three and for four input variables. You will possibly want to compare the rows for Y_{XOR} and Y_{EQU} more carefully. There is one case which tends to cause confusion in the definition stating when the output of the EQUIVALENCE function is **1**: Given an *even* number of inputs, why is the output **1** when all inputs are **0**?

Saying that 'No inputs are **1**' is the same as saying 'Zero inputs are **1**' so that the output is **1** because an even number (albeit that this number is zero) of inputs are **1**.

From the above statements, saying when XOR and EQUIVALENCE are **1**, we deduce that their outputs are inverted for functions with an even number of input variables, and are logically equivalent for functions with an odd number of input variables:

$$Y_{EQU} \Leftrightarrow \overline{Y}_{XOR} \quad \text{for an } \textbf{even} \text{ number of inputs,}$$

$$Y_{EQU} \Leftrightarrow Y_{XOR} \quad \text{for an } \textbf{odd} \text{ number of inputs.}$$

$X_1 \leftrightarrow$	**00001111**
$X_2 \leftrightarrow$	**00110011**
$X_3 \leftrightarrow$	**01010101**
$Y_{AND} \leftrightarrow$	**00000001**
$Y_{OR} \leftrightarrow$	**01111111**
$Y_{XOR} \leftrightarrow$	**01101001**
$Y_{EQU} \leftrightarrow$	**01101001**

$X_1 \leftrightarrow$	**0000000011111111**
$X_2 \leftrightarrow$	**0000111100001111**
$X_3 \leftrightarrow$	**0011001100110011**
$X_4 \leftrightarrow$	**0101010101010101**
$Y_{AND} \leftrightarrow$	**0000000000000001**
$Y_{OR} \leftrightarrow$	**0111111111111111**
$Y_{XOR} \leftrightarrow$	**0110100110010110**
$Y_{EQU} \leftrightarrow$	**1001011001101001**

Fig. 3.8 AND, OR, XOR and EQUIVALENCE for three and four inputs

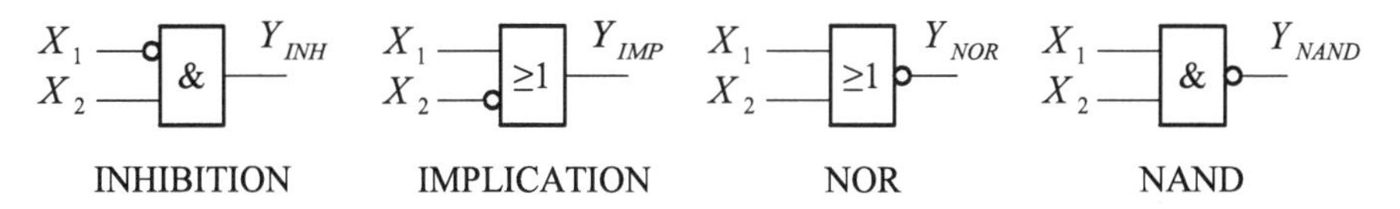

Fig. 3.9 The four derived gates

This rather inconsistent behaviour of Y_{EQU} usually lets us work with the XOR function instead of the EQUIVALENCE function.

The functions INHIBITION, IMPLICATION, NOR, and NAND can be represented by AND, OR and NOT leading to the gate representations of Fig. 3.9.

3.3 Using AND, OR and NOT

The most common functionally complete set of connectives is $\{\wedge, \vee, \neg\}$, i.e., the set of AND, OR, and NOT functions and gates. When using these functions it is useful to summarise the relevant theorems, e.g., as is done in the table of Fig. 3.10. None of these theorems, nor those following in this chapter, are proven in this text.

Basic theorems on negation	
$\overline{\mathbf{0}} \Leftrightarrow \mathbf{1}$	$\overline{\mathbf{1}} \Leftrightarrow \mathbf{0}$
$\overline{\overline{A}} \Leftrightarrow A$	
DeMorgan's theorems	
$\overline{A \wedge B} \Leftrightarrow \overline{A} \vee \overline{B}$	$\overline{A \vee B} \Leftrightarrow \overline{A} \wedge \overline{B}$
Basic theorems on a single variable	
$A \wedge \mathbf{0} \Leftrightarrow \mathbf{0}$	$A \vee \mathbf{1} \Leftrightarrow \mathbf{1}$
$A \wedge \mathbf{1} \Leftrightarrow A$	$A \vee \mathbf{0} \Leftrightarrow A$
$A \wedge A \Leftrightarrow A$	$A \vee A \Leftrightarrow A$
$A \wedge \overline{A} \Leftrightarrow \mathbf{0}$	$A \vee \overline{A} \Leftrightarrow \mathbf{1}$
Commutativity	
$A \wedge B \Leftrightarrow B \wedge A$	$A \vee B \Leftrightarrow B \vee A$
Associativity	
$A \wedge (B \wedge C) \Leftrightarrow (A \wedge B) \wedge C$	$A \vee (B \vee C) \Leftrightarrow (A \vee B) \vee C$
Distributivity	
$A \wedge (B \vee C) \Leftrightarrow (A \wedge B) \vee (A \wedge C)$	$A \vee (B \wedge C) \Leftrightarrow (A \vee B) \wedge (A \vee C)$
Some theorems on minimisation	
$(A \vee B) \wedge (A \vee \overline{B}) \Leftrightarrow A$	$(A \wedge B) \vee (A \wedge \overline{B}) \Leftrightarrow A$
$A \wedge (A \vee B) \Leftrightarrow A$	$A \vee (A \wedge B) \Leftrightarrow A$
$A \wedge (\overline{A} \vee B) \Leftrightarrow A \wedge B$	$A \vee (\overline{A} \wedge B) \Leftrightarrow A \vee B$

Fig. 3.10 Basic theorems on AND, OR and NOT

Note that the formulas in the two columns are row-wise dual pointing out that duality is inherent to logic algebra. The first formulas are on negation, DeMorgan's theorems being of prime importance. The formulas on commutativity, associativity, and distributivity are typical for the functionally complete set $\{\wedge, \vee, \neg\}$. Frequently, the theorems listed at the bottom of the table will be needed when minimising logic functions.

3.4 Basic Laws

The basic laws in switching algebra are those on commutativity, associativity, and distributivity. To be able to speak about these laws in a general manner, we use two meta-connectives, $\circ$ (called **circle**) and $\diamond$ (called **diamond**), that stand for any of the ten dyadic logic connectives. We express this as

$$\circ, \diamond \in \{\wedge, >, <, \oplus, \vee, \tilde{\vee}, \leftrightarrow, \leftarrow, \rightarrow, \tilde{\wedge}\}$$

allowing us to formulate the basic laws quite generally.

A connective $\circ$ is **commutative** if

$$\boxed{A \circ B \Leftrightarrow B \circ A.} \tag{3.1}$$

A connective $\circ$ is **associative** if

$$\boxed{A \circ (B \circ C) \Leftrightarrow (A \circ B) \circ C.} \tag{3.2}$$

The connective $\circ$ is **left-distributive** over the connective $\diamond$ if

$$\boxed{A \circ (B \diamond C) \Leftrightarrow (A \circ B) \diamond (A \circ C).} \tag{3.3}$$

The connective $\circ$ is **right-distributive** over the connective $\diamond$ if

$$\boxed{(B \diamond C) \circ A \Leftrightarrow (B \circ A) \diamond (C \circ A).} \tag{3.4}$$

A connective $\circ$ which distributes from the left *and* from the right over $\diamond$ is said to be **(completely) distributive** over $\diamond$.

The safest and quickest way to obtain all commutative, associative, and distributive logic equivalences is to write computer programs that automatically create and test all formulas of the respective kind. These programs, which were written in **APL** (and later in **J**), are not part of this text, but I strongly recommend you write such programs in a computer language of your choice.

3.5 Single-Variable Formulas

The tables in this section contain all single-variable formulas for all dyadic logic connectives anticipating the commutativity of all but IMPLICATION and INHIBITION. The commutative connectives are grouped into dual pairs: AND-OR in Fig. 3.11a, b, EQU-XOR in Fig. 3.11c, d, and NAND-NOR in Fig. 3.12a, b. INHIBITION, the only logical function whose single-variable formulas are not explicitly shown, can be developed as the duals of the formulas for IMPLICATION of Fig. 3.12c.

The AND and OR functions both have what is called a **neutral element**. To demonstrate what a neutral element is, take the operations of addition and multiplication. Writing $x + 0 = x$ shows that adding zero to any number x leaves the number unchanged—one calls zero the neutral element of addition. Similarly, 1 is the neutral element of multiplication, because multiplying any number x by 1 leaves the number unchanged: $x \times 1 = x$.

Now returning to logic functions, let $\circ$ be some dyadic logic connective. Then, ν is said to be a neutral element if $A \circ \nu \Leftrightarrow A$. As you can see from the table of Fig. 3.11a, b, the neutral element for the AND function is **1** as follows from $A \wedge \mathbf{1} \Leftrightarrow A$; due to duality, the neutral element of the OR function is **0**.

The second property to point out is *idempotents* of a function. Again, let $\circ$ be some dyadic logic connective. Then the connective $\circ$ is said to be **idempotent** if $A \circ A \Leftrightarrow A$. Both AND and OR are idempotent, meaning $A \wedge A \Leftrightarrow A$ and $A \vee A \Leftrightarrow A$.

AND (a)	OR (b)
$A \wedge \mathbf{0} \Leftrightarrow \mathbf{0}$	$A \vee \mathbf{1} \Leftrightarrow \mathbf{1}$
$A \wedge \mathbf{1} \Leftrightarrow \mathrm{A}$	$A \vee \mathbf{0} \Leftrightarrow \mathrm{A}$
$A \wedge A \Leftrightarrow \mathrm{A}$	$A \vee A \Leftrightarrow \mathrm{A}$
$A \wedge \overline{A} \Leftrightarrow \mathbf{0}$	$A \vee \overline{A} \Leftrightarrow \mathbf{1}$

EQU (c)	XOR (d)
$A \leftrightarrow \mathbf{0} \Leftrightarrow \overline{A}$	$A \oplus \mathbf{1} \Leftrightarrow \overline{A}$
$A \leftrightarrow \mathbf{1} \Leftrightarrow \mathrm{A}$	$A \oplus \mathbf{0} \Leftrightarrow \mathrm{A}$
$A \leftrightarrow A \Leftrightarrow \mathbf{1}$	$A \oplus A \Leftrightarrow \mathbf{0}$
$A \leftrightarrow \overline{A} \Leftrightarrow \mathbf{0}$	$A \oplus \overline{A} \Leftrightarrow \mathbf{1}$

Fig. 3.11 The Single-Variable Formulas for AND, OR, EQU and XOR

NAND (a)	NOR (b)
$A \tilde{\wedge} \mathbf{0} \Leftrightarrow \mathbf{1}$	$A \tilde{\vee} \mathbf{1} \Leftrightarrow \mathbf{0}$
$A \tilde{\wedge} \mathbf{1} \Leftrightarrow \overline{A}$	$A \tilde{\vee} \mathbf{0} \Leftrightarrow \overline{A}$
$A \tilde{\wedge} A \Leftrightarrow \overline{A}$	$A \tilde{\vee} A \Leftrightarrow \overline{A}$
$A \tilde{\wedge} \overline{A} \Leftrightarrow \mathbf{1}$	$A \tilde{\vee} \overline{A} \Leftrightarrow \mathbf{0}$

IMPLICATION (c)	
$A \rightarrow \mathbf{0} \Leftrightarrow \overline{A}$	$\mathbf{0} \rightarrow A \Leftrightarrow \mathbf{1}$
$A \rightarrow \mathbf{1} \Leftrightarrow \mathbf{1}$	$\mathbf{1} \rightarrow A \Leftrightarrow \mathrm{A}$
$A \rightarrow A \Leftrightarrow \mathbf{1}$	$A \rightarrow A \Leftrightarrow \mathbf{1}$
$A \rightarrow \overline{A} \Leftrightarrow \overline{A}$	$\overline{A} \rightarrow A \Leftrightarrow \mathrm{A}$

Fig. 3.12 The single-variable formulas for NAND, NOR and IMPLICATION

As you can see from Fig. 3.11c, d, EQUIVALENCE and XOR have, respectively, the neutral elements **1** and **0**, i.e., $A \leftrightarrow \mathbf{1} \Leftrightarrow A$ and $A \oplus \mathbf{0} \Leftrightarrow A$. But, note that neither EQUIVALENCE nor XOR are idempotent.

The last pair of commutative logic functions consists of NAND and NOR, their single-variable formulas being found in Fig. 3.12a, b. Neither of these functions has a neutral element, nor are they idempotent. The IMPLICATION (Fig. 3.12c), finally, is neither commutative nor idempotent, and has only a left-neutral element, this being **1** (i.e., $\mathbf{1} \rightarrow A \Leftrightarrow A$).

3.6 Commutative and Associative Laws*

The only commutative functions are AND, OR, EQU, XOR, NAND and NOR as shown in the table of Fig. 3.13. In particular, the functions INHIBITION and IMPLICATION are *not* commutative.

Associativity only holds for a subset of the commutative logic functions, i.e., for AND, OR, EQUIVALENCE, and XOR as compiled in the table of Fig. 3.14 demonstrates.

For associative connectives the placing of the parentheses is irrelevant, so that one often writes $A \circ B \circ C$ instead of either $A \circ (B \circ C)$ or $(A \circ B) \circ C$:

$$A \circ B \circ C :\Leftrightarrow A \circ (B \circ C) \Leftrightarrow (A \circ B) \circ C$$

Commutativity and associativity (taken together) are of practical importance. They influence the way we employ gates and switching devices, as touched upon in Fig. 3.6. Also, notice the correspondence in Fig. 3.15a, b between the parentheses in the logic formulas and the way the two-input gates are interconnected.

There is a difference in the real world behaviour of the circuits of Fig. 3.15a, b. Each physical realisation of a gate always causes a time lag between the input and

Commutativity	
$A \wedge B \Leftrightarrow B \wedge A$	$A \vee B \Leftrightarrow B \vee A$
$A \leftrightarrow B \Leftrightarrow B \leftrightarrow A$	$A \oplus B \Leftrightarrow B \oplus A$
$A \mathbin{\overset{\frown}{\wedge}} B \Leftrightarrow B \mathbin{\overset{\frown}{\wedge}} A$	$A \mathbin{\overset{\frown}{\vee}} B \Leftrightarrow B \mathbin{\overset{\frown}{\vee}} A$

Fig. 3.13 All the commutative laws

Associativity	
$A \wedge (B \wedge C) \Leftrightarrow (A \wedge B) \wedge C$	$A \vee (B \vee C) \Leftrightarrow (A \vee B) \vee C$
$A \leftrightarrow (B \leftrightarrow C) \Leftrightarrow (A \leftrightarrow B) \leftrightarrow C$	$A \oplus (B \oplus C) \Leftrightarrow (A \oplus B) \oplus C$

Fig. 3.14 All the associative laws

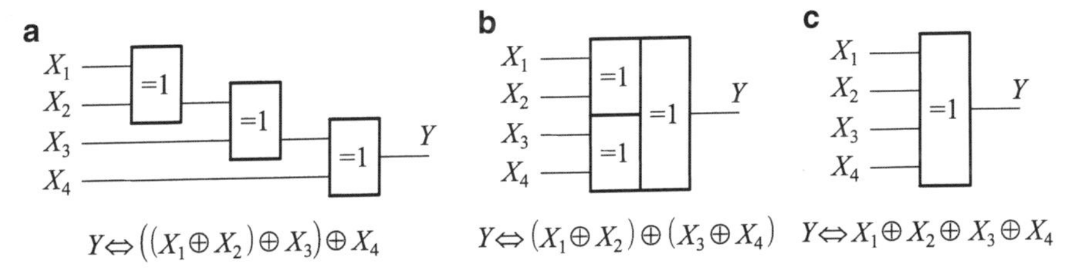

Fig. 3.15 Analogy of formula and circuit

output signals. The propagation time a signal takes from an input to the output of a circuit depends on the number of gates the signal has to pass. The propagation time can vary markedly from input to input of an asymmetric circuit, whereas it will hardly vary for the various inputs of a symmetric circuit. Symmetrical designs are obviously preferred.

3.7 Distributive Laws*

The term 'distributivity' is ambiguous. You have to state which connective distributes over which other, *and* whether the connective that distributes does so from the left or from the right. Remember, a connective that distributes over another one from the left *and* from the right is said to be **completely distributive**. Only two connectives are completely distributive: AND and OR. But they do not distribute over all other connectives. AND only distributes over AND, INHIBITIONs, XOR, and OR. OR only distributes over OR, IMPLICATIONs, EQUIVALENCE, and AND. This is documented in the table of Fig. 3.16. AND and OR are called **self-distributive** as they distribute over themselves.

Complete Distributivity	
Left Distributivity	Right Distributivity
$A \wedge (B \wedge C) \Leftrightarrow (A \wedge B) \wedge (A \wedge C)$	$(B \wedge C) \wedge A \Leftrightarrow (B \wedge A) \wedge (C \wedge A)$
$A \wedge (B > C) \Leftrightarrow (A \wedge B) > (A \wedge C)$	$(B > C) \wedge A \Leftrightarrow (B \wedge A) > (C \wedge A)$
$A \wedge (B < C) \Leftrightarrow (A \wedge B) < (A \wedge C)$	$(B < C) \wedge A \Leftrightarrow (B \wedge A) < (C \wedge A)$
$A \wedge (B \oplus C) \Leftrightarrow (A \wedge B) \oplus (A \wedge C)$	$(B \oplus C) \wedge A \Leftrightarrow (B \wedge A) \oplus (C \wedge A)$
$A \wedge (B \vee C) \Leftrightarrow (A \wedge B) \vee (A \wedge C)$	$(B \vee C) \wedge A \Leftrightarrow (B \wedge A) \vee (C \wedge A)$
$A \vee (B \vee C) \Leftrightarrow (A \vee B) \vee (A \vee C)$	$(B \vee C) \vee A \Leftrightarrow (B \vee A) \vee (C \vee A)$
$A \vee (B \leftarrow C) \Leftrightarrow (A \vee B) \leftarrow (A \vee C)$	$(B \leftarrow C) \vee A \Leftrightarrow (B \vee A) \leftarrow (C \vee A)$
$A \vee (B \rightarrow C) \Leftrightarrow (A \vee B) \rightarrow (A \vee C)$	$(B \rightarrow C) \vee A \Leftrightarrow (B \vee A) \rightarrow (C \vee A)$
$A \vee (B \leftrightarrow C) \Leftrightarrow (A \vee B) \leftrightarrow (A \vee C)$	$(B \leftrightarrow C) \vee A \Leftrightarrow (B \vee A) \leftrightarrow (C \vee A)$
$A \vee (B \wedge C) \Leftrightarrow (A \vee B) \wedge (A \vee C)$	$(B \wedge C) \vee A \Leftrightarrow (B \vee A) \wedge (C \vee A)$

Fig. 3.16 Complete distributivity

One-Sided Distributivity	
Left Distributivity	Right Distributivity
$A < (B \wedge C) \Leftrightarrow (A < B) \wedge (A < C)$	no formula
$A < (B > C) \Leftrightarrow (A < B) > (A < C)$	no formula
$A < (B < C) \Leftrightarrow (A < B) < (A < C)$	no formula
$A < (B \oplus C) \Leftrightarrow (A < B) \oplus (A < C)$	no formula
$A < (B \vee C) \Leftrightarrow (A < B) \vee (A < C)$	no formula
no formula	$(B \wedge C) > A \Leftrightarrow (B > A) \wedge (C > A)$
no formula	$(B > C) > A \Leftrightarrow (B > A) > (C > A)$
no formula	$(B < C) > A \Leftrightarrow (B > A) < (C > A)$
no formula	$(B \oplus C) > A \Leftrightarrow (B > A) \oplus (C > A)$
no formula	$(B \vee C) > A \Leftrightarrow (B > A) \vee (C > A)$
$A \rightarrow (B \vee C) \Leftrightarrow (A \rightarrow B) \vee (A \rightarrow C)$	no formula
$A \rightarrow (B \leftarrow C) \Leftrightarrow (A \rightarrow B) \leftarrow (A \rightarrow C)$	no formula
$A \rightarrow (B \rightarrow C) \Leftrightarrow (A \rightarrow B) \rightarrow (A \rightarrow C)$	no formula
$A \rightarrow (B \leftrightarrow C) \Leftrightarrow (A \rightarrow B) \leftrightarrow (A \rightarrow C)$	no formula
$A \rightarrow (B \wedge C) \Leftrightarrow (A \rightarrow B) \wedge (A \rightarrow C)$	no formula
no formula	$(B \vee C) \leftarrow A \Leftrightarrow (B \leftarrow A) \vee (C \leftarrow A)$
no formula	$(B \leftarrow C) \leftarrow A \Leftrightarrow (B \leftarrow A) \leftarrow (C \leftarrow A)$
no formula	$(B \rightarrow C) \leftarrow A \Leftrightarrow (B \leftarrow A) \rightarrow (C \leftarrow A)$
no formula	$(B \leftrightarrow C) \leftarrow A \Leftrightarrow (B \leftarrow A) \leftrightarrow (C \leftarrow A)$
no formula	$(B \wedge C) \leftarrow A \Leftrightarrow (B \leftarrow A) \wedge (C \leftarrow A)$

Fig. 3.17 One-sided distributivity

Distributivity is a scarce property among logic connectives. The only functions, other than AND and OR, that are distributive are the INHIBITIONs and the IMPLICATIONs—see the table of Fig. 3.17. The INHIBITIONs distribute over the same functions as the AND function, i.e., over AND, INHIBITIONs, XOR, and OR. But note: While the *transposed* INHIBITION distributes only from the left, the normal INHIBITION distributes only from the right. In a like manner, the IMPLICATIONs distribute over the same functions as the OR function, i.e., over OR, IMPLICATIONs, EQUIVALENCE, and AND. Here note that the normal INHIBITION distributes only from the left, while the *transposed* IMPLICATION distributes only from the right.

Although it is probably obvious, allow me to comment on those squares in the table of Fig. 3.17 containing the phrase 'no formula'. Consider the first formula, $A < (B \wedge C) \Leftrightarrow (A < B) \wedge (A < C)$, in which $<$ is left-distributive over $\wedge$. If $<$ were right-distributive over $\wedge$, which it is not, we would be able to write the formula

$(B \wedge C) < A \Leftrightarrow (B < A) \wedge (C < A)$ next to the previous one. But, please do note that the formula just written is wrong, does not exist. Each phrase 'no formula', therefore, reminds us that the alternatively distributive formula does not exist.

3.8 Generalised DeMorgan Theorems

DeMorgan's original theorems—stated in the table of Fig. 3.10—refer to the negation of AND and OR. Similar formulas—compiled in the table of Fig. 3.18—hold for all dyadic connectives.

To complement or invert a general logic expression you can quite conveniently employ the following method, called the **rule of negation**: *Replace every variable and constant by its complement and each connective by its dual connective taking care to maintain all parentheses*. For a given logic function F this is often written symbolically in the following way:

$$\neg F\left(X_1, X_2, \cdots, X_n; \mathbf{0}, \mathbf{1}; (,); \wedge, >, <, \oplus, \vee, \tilde{\vee}, \leftrightarrow, \leftarrow, \rightarrow, \tilde{\wedge}\right) \Leftrightarrow$$
$$F\left(\overline{X}_1, \overline{X}_2, \cdots, \overline{X}_n; \mathbf{1}, \mathbf{0}; (,); \vee, \leftarrow, \rightarrow, \leftrightarrow, \wedge, \tilde{\wedge}, \oplus, >, <, \tilde{\vee}\right)$$

For the four associative connectives, DeMorgan's theorems can be extended to multiple variables as shown in Fig. 3.19.

DeMorgan's Theorems	
$\overline{A \wedge B} \Leftrightarrow \overline{A} \vee \overline{B}$	$\overline{A \vee B} \Leftrightarrow \overline{A} \wedge \overline{B}$
$\overline{A > B} \Leftrightarrow \overline{A} \leftarrow \overline{B}$	$\overline{A \leftarrow B} \Leftrightarrow \overline{A} > \overline{B}$
$\overline{A < B} \Leftrightarrow \overline{A} \rightarrow \overline{B}$	$\overline{A \rightarrow B} \Leftrightarrow \overline{A} < \overline{B}$
$\overline{A \oplus B} \Leftrightarrow \overline{A} \leftrightarrow \overline{B}$	$\overline{A \leftrightarrow B} \Leftrightarrow \overline{A} \oplus \overline{B}$
$\overline{A \tilde{\vee} B} \Leftrightarrow \overline{A}\, \tilde{\wedge}\, \overline{B}$	$\overline{A \tilde{\wedge} B} \Leftrightarrow \overline{A}\, \tilde{\vee}\, \overline{B}$

Fig. 3.18 The set of DeMorgan's theorems

Generalised DeMorgan Theorems
$\overline{X_1 \wedge X_2 \wedge \cdots \wedge X_n} \Leftrightarrow \overline{X}_1 \vee \overline{X}_2 \vee \cdots \vee \overline{X}_n$
$\overline{X_1 \vee X_2 \vee \cdots \vee X_n} \Leftrightarrow \overline{X}_1 \wedge \overline{X}_2 \wedge \cdots \wedge \overline{X}_n$
$\overline{X_1 \oplus X_2 \oplus \cdots \oplus X_n} \Leftrightarrow \overline{X}_1 \leftrightarrow \overline{X}_2 \leftrightarrow \cdots \leftrightarrow \overline{X}_n$
$\overline{X_1 \leftrightarrow X_2 \leftrightarrow \cdots \leftrightarrow X_n} \Leftrightarrow \overline{X}_1 \oplus \overline{X}_2 \oplus \cdots \oplus \overline{X}_n$

Fig. 3.19 DeMorgan's theorems for n variables

The following example demonstrates how to apply the rule of negation to a logic expression and how important it is to maintain the correct sequence of operation. Given

$$\overline{\left((A \vee \overline{C}D)AB \vee C\right)\overline{D}} \Leftrightarrow$$
$$\left(\overline{A}(C \vee \overline{D}) \vee \overline{A} \vee \overline{B}\right)\overline{C} \vee D.$$

Chapter 4
Normal Forms

Normal forms are the classical logic formulas by which to calculate logic circuits. The output connective of a normal form must be associative restricting the number of normal forms to four, i.e., to formulas whose (multi-input) output connective is either AND, OR, XOR or EQU.

The normal forms split into two groups—canonical (using AND or OR output connectives) and non-canonical (using XOR or EQU output connectives). The canonical normal forms (as opposed to the non-canonical ones) are completely ordered, or well-ordered, in all their input variables. This well-ordering of all input variables is expressed with the help of logic formulas called minterms and maxterms.

As we start our excursion into the realm of normal forms with the most primitive, the canonical normal forms, we must first introduce minterms and maxterm.

4.1 Minterms and Maxterms

We consider the general events-table for n input variables $x_1, x_2, \ldots, x_n$ of Fig. 4.1 concentrating on column e. In this column we assume, without knowing whether our claim is true or not, that x_1 has the value e_1 (i.e., $x_1 = e_1$), that x_2 has the value e_2 (i.e., $x_2 = e_2$), etc., for all input variables. Using the **multiple conjunction symbol** $\bigwedge$, our assumption for column e may be written as

$$\bigwedge_{i=1}^{n}(x_i = e_i) \;\Leftrightarrow\; (x_1 = e_1) \wedge (x_2 = e_2) \wedge \cdots \wedge (x_n = e_n).$$

This conjunction, which describes exactly one column of the events table, in our case column e, is called **minterm** and we denote it as

$$\boxed{C_e(x) \Leftrightarrow \bigwedge_{i=1}^{n}(x_i = e_i) \quad \text{with} \quad e = \sum_{i=1}^{n} e_i.2^{n-i} \quad \text{and} \quad e_i \in \{0,1\}.} \tag{4.1}$$

S.P. Vingron, *Logic Circuit Design*, DOI 10.1007/978-3-642-27657-6_4,

	0	1	$\cdots$	e	$\cdots$	2^n-2	2^n-1
$x_1 =$	0	0	$\cdots$	e_1	$\cdots$	1	1
$x_2 =$	0	0	$\cdots$	e_2	$\cdots$	1	1
$\vdots$	$\vdots$	$\vdots$	$\vdots$	$\vdots$	$\vdots$	$\vdots$	$\vdots$
$x_{n-1} =$	0	0	$\cdots$	e_{n-1}	$\cdots$	1	1
$x_n =$	0	1	$\cdots$	e_n	$\cdots$	0	1
$y(x) =$	$y(0)$	$y(1)$	$\cdots$	$y(e)$	$\cdots$	$y(2^n-2)$	$y(2^n-1)$

Fig. 4.1 General events table—brief form

The minterms for two inputs, for example, are calculated as follows using the formulas $\boxed{X_i \Leftrightarrow x_i = 1 \text{ and } \overline{X}_i \Leftrightarrow x_i = 0}$, (1.2), for logic variables:

$$C_0(x_1, x_2) \Leftrightarrow (x_1 = 0) \wedge (x_2 = 0) \Leftrightarrow \overline{X}_1\overline{X}_2,$$
$$C_1(x_1, x_2) \Leftrightarrow (x_1 = 0) \wedge (x_2 = 1) \Leftrightarrow \overline{X}_1 X_2,$$
$$C_2(x_1, x_2) \Leftrightarrow (x_1 = 1) \wedge (x_2 = 0) \Leftrightarrow X_1\overline{X}_2,$$
$$C_3(x_1, x_2) \Leftrightarrow (x_1 = 1) \wedge (x_2 = 1) \Leftrightarrow X_1 X_2.$$

To calculate, say, $C_{26}(x_1, \ldots, x_6)$ you must determine the binary equivalent of 26 for 6 binary variables, $(26)_{10} = (011010)_2$, which you can do with a pocket calculator. The binary digits are the values of $(e_1, \ldots, e_6)$ allowing us to write the minterm as

$$\begin{aligned} C_{26}(x_1, \ldots, x_6) &\Leftrightarrow (x_1 = 0) \wedge (x_2 = 1) \wedge (x_3 = 1) \wedge (x_4 = 0) \\ &\qquad \wedge (x_5 = 1) \wedge (x_6 = 0) \\ &\Leftrightarrow \overline{X}_1 X_2 X_3 \overline{X}_4 X_5 \overline{X}_6. \end{aligned}$$

The **mnemonic rule for writing minterms** is: For a given column e of the *events table* the minterm $C_e(x)$ is the conjunction of all logic variables, inverted or not-inverted; they are inverted (i.e., $\overline{X}_i$) when their input variables (x_i) are assumed to be *zero* ($x_i = 0$), and not-inverted (X_i) when their input variables are assumed to be *one* ($x_i = 1$). *This mnemonic rule provides the standard procedure for writing minterms.*

The **interpretation of the minterm** $C_e(x)$ is that it is **1** (i.e., true) *iff* the measured values, the instances, of the individual input variables $x_1, x_2, \ldots, x_n$ are identical with the equally indexed, assumed values of the input event, i.e., with $e_1, e_2, \ldots, e_n$.

The **maxterm**, $D_e(x)$, is defined as the negation of the minterm, this leading to the following formula for its calculation when using DeMorgan:

$$D_e(x) \Leftrightarrow \neg C_e(x) \Leftrightarrow \neg \bigwedge_{i=1}^{n}(x_i = e_i) \Leftrightarrow \bigvee_{i=1}^{n}(x_i = 1 - e_i) \tag{4.2}$$

The maxterms for two input variables are calculated thus:

$$D_0(x_1, x_2) \Leftrightarrow (x_1 = 1 - 0) \vee (x_2 = 1 - 0) \Leftrightarrow (x_1 = 1) \vee (x_2 = 1) \Leftrightarrow X_1 \vee X_2,$$

$$D_1(x_1, x_2) \Leftrightarrow (x_1 = 1 - 0) \vee (x_2 = 1 - 1) \Leftrightarrow (x_1 = 1) \vee (x_2 = 0) \Leftrightarrow X_1 \vee \overline{X}_2,$$

$$D_2(x_1, x_2) \Leftrightarrow (x_1 = 1 - 1) \vee (x_2 = 1 - 0) \Leftrightarrow (x_1 = 0) \vee (x_2 = 1) \Leftrightarrow \overline{X}_1 \vee X_2,$$

$$D_3(x_1, x_2) \Leftrightarrow (x_1 = 1 - 1) \vee (x_2 = 1 - 1) \Leftrightarrow (x_1 = 0) \vee (x_2 = 0) \Leftrightarrow \overline{X}_1 \vee \overline{X}_2.$$

The **interpretation of the maxterm** $D_e(x)$ is that it is **0** (i.e., false) *iff* the measured values, the instances, of the individual input variables $x_1, x_2, \ldots, x_n$ are identical with the equally indexed, assumed values of the input event, i.e., with $e_1, e_2, \ldots, e_n$.

4.2 Canonical Normal Forms

The **canonical normal forms** are stated as follows, $Y(x)$ being the circuit's output, and $Y(e)$ the logic output value of column e of the events table.

$$Y(x) \Leftrightarrow \bigvee_{e=0}^{2^n-1} C_e(x)Y(e) \Leftrightarrow \bigwedge_{e=0}^{2^n-1} D_e(x) \vee Y(e). \tag{4.3}$$

To calculate the inverted values of the circuit's output you use

$$\overline{Y}(x) \Leftrightarrow \bigvee_{e=0}^{2^n-1} C_e(x)\overline{Y}(e) \Leftrightarrow \bigwedge_{e=0}^{2^n-1} D_e(x) \vee \overline{Y}(e). \tag{4.4}$$

The formulas starting with a multiple OR ($\bigvee$) have the unwieldy name **disjunctive canonical normal form** which, following Dietmeyer (1971), we refer to as (canonical) **AND-to-OR** formulas. In a like manner the so-called **conjunctive canonical normal form**, i.e., those which start with a multiple AND ($\bigwedge$), are called (canonical) **OR-to-AND** formulas.[1]

When applying the disjunctive normal form for the non-inverted output $Y(x)$, effectively only those minterms $C_e(x)$ are used (and their outputs connected disjunctively) which imply an output value $Y(e)$ of **1**; when employing the conjunctive

[1] The derivation of these formulas is discussed in-depth in Vingron (2004).

normal form, only the maxterms of those columns are conjuncted whose output value is **0**. As a first application of the canonical normal forms look back at (1.3) and (1.5), and all the formulas for logic outputs in Chap. 2.

4.3 Using Canonical Normal Forms

In general, canonical normal forms, (4.3) and (4.4), tend to yield unwieldy, excessively long, logical expressions and correspondingly complicated circuits as is shown when calculating a circuit for the events table of Fig. 4.2, here using the AND-to-OR formula of (4.3). The numbers under-bracing the minterms refer to the decimal values x of the binary input events (x_1, x_2, x_3, x_4).

$x=$	0	1	2	3	4	5	6	7	8	9	10	11	12	13	14	15
$x_1=$	0	0	0	0	0	0	0	0	1	1	1	1	1	1	1	1
$x_2=$	0	0	0	0	1	1	1	1	0	0	0	0	1	1	1	1
$x_3=$	0	0	1	1	0	0	1	1	0	0	1	1	0	0	1	1
$x_4=$	0	1	0	1	0	1	0	1	0	1	0	1	0	1	0	1
$y(x)=$	1	1	1	0	0	1	0	1	1	1	1	0	1	1	1	0

Fig. 4.2 An example events-table used to demonstrate a circuit's calculation and minimisation

$$
\begin{aligned}
Y \Leftrightarrow\ & \underbrace{\overline{X}_1\overline{X}_2\overline{X}_3\overline{X}_4}_{x=0} \vee \underbrace{\overline{X}_1\overline{X}_2\overline{X}_3X_4}_{1} \vee \underbrace{\overline{X}_1\overline{X}_2X_3\overline{X}_4}_{2} \vee \underbrace{\overline{X}_1X_2\overline{X}_3X_4}_{5} \vee \\
& \underbrace{\overline{X}_1X_2X_3X_4}_{7} \vee \underbrace{X_1\overline{X}_2\overline{X}_3\overline{X}_4}_{8} \vee \underbrace{X_1\overline{X}_2\overline{X}_3X_4}_{9} \vee \underbrace{X_1\overline{X}_2X_3\overline{X}_4}_{10} \vee \\
& \underbrace{X_1X_2\overline{X}_3\overline{X}_4}_{12} \vee \underbrace{X_1X_2\overline{X}_3X_4}_{13} \vee \underbrace{X_1X_2X_3\overline{X}_4}_{14}.
\end{aligned} \tag{4.5}
$$

Most frequently these formulas, directly obtained from a canonical normal form, can be written in a significantly simpler form. Quine (1952) proposed doing this by repeatedly using the formula $X_iA \vee \overline{X}_iA \Leftrightarrow A$. For instance, this allows us to combine the two minterm $\overline{X}_1\overline{X}_2\overline{X}_3\overline{X}_4$ and $\overline{X}_1\overline{X}_2\overline{X}_3X_4$ to yield the single conjunction $\overline{X}_1\overline{X}_2\overline{X}_3$:

$$
\underbrace{\overline{X}_1\overline{X}_2\overline{X}_3}_{A}\,\overline{X}_4 \vee \underbrace{\overline{X}_1\overline{X}_2\overline{X}_3}_{A}\,X_4 \Leftrightarrow \underbrace{\overline{X}_1\overline{X}_2\overline{X}_3}_{A}
$$

The minterms are reused multiple times in conjunction with $X_iA \vee \overline{X}_iA \Leftrightarrow A$ by duplicating them as often as necessary—employing $X \Leftrightarrow X \vee X \vee \cdots \vee X$:

$$Y \Leftrightarrow \underbrace{\overline{X}_1\overline{X}_2\overline{X}_3}_{0-1} \vee \underbrace{\overline{X}_1\overline{X}_2\overline{X}_4}_{0-2} \vee \underbrace{\overline{X}_2\overline{X}_3\overline{X}_4}_{0-8} \vee \underbrace{\overline{X}_1\overline{X}_3X_4}_{1-5} \vee \underbrace{\overline{X}_2\overline{X}_3X_4}_{1-9} \vee$$
$$\underbrace{\overline{X}_2X_3\overline{X}_4}_{2-10} \vee \underbrace{\overline{X}_1X_2X_4}_{5-7} \vee \underbrace{X_2\overline{X}_3X_4}_{5-13} \vee \underbrace{X_1\overline{X}_2\overline{X}_3}_{8-9} \vee \underbrace{X_1\overline{X}_2\overline{X}_4}_{8-10} \vee$$
$$\underbrace{X_1\overline{X}_3\overline{X}_4}_{8-12} \vee \underbrace{X_1\overline{X}_3X_4}_{9-13} \vee \underbrace{X_1X_3\overline{X}_4}_{10-14} \vee \underbrace{X_1X_2\overline{X}_3}_{12-13} \vee \underbrace{X_1X_2\overline{X}_4}_{12-14}.$$

Repeated application of this principle to the (non-canonical) conjunctions and removing duplicates leads to the final result:

$$Y \Leftrightarrow \overline{X}_2\overline{X}_3 \vee \overline{X}_2\overline{X}_4 \vee \overline{X}_2\overline{X}_3 \vee \overline{X}_2\overline{X}_4 \vee \overline{X}_3X_4 \vee$$
$$\overline{X}_3X_4 \vee \overline{X}_1X_2X_4 \vee X_1\overline{X}_3 \vee X_1\overline{X}_4 \vee X_1\overline{X}_4$$
$$Y \Leftrightarrow \overline{X}_2\overline{X}_3 \vee \overline{X}_2\overline{X}_4 \vee \overline{X}_3X_4 \vee \overline{X}_1X_2X_4 \vee X_1\overline{X}_3 \vee X_1\overline{X}_4. \tag{4.6}$$

As $XA \vee \overline{X}A \Leftrightarrow A$ cannot be applied again, the procedure terminates. Quine's minimisation method must be started from a *canonical* normal form. To illustrate this necessity, consider starting from an arbitrary disjunctive form of the above function:

$$Y \Leftrightarrow \overline{X}_1\overline{X}_2\overline{X}_3 \vee \overline{X}_1\overline{X}_2X_3\overline{X}_4 \vee \overline{X}_1X_2X_4 \vee X_1\overline{X}_3 \vee X_1X_3\overline{X}_4. \tag{4.7}$$

In this example, $XA \vee \overline{X}A \Leftrightarrow A$ cannot be applied at all. We first need to transform our example into a canonical normal form which we do by ANDing the non-canonical conjunctions by $(X_i \vee \overline{X}_i)$ for each missing variable:

$$Y \Leftrightarrow \overline{X}_1\overline{X}_2\overline{X}_3(X_4 \vee \overline{X}_4) \vee \overline{X}_1\overline{X}_2X_3\overline{X}_4 \vee \overline{X}_1X_2(X_3 \vee \overline{X}_3)X_4 \vee$$
$$X_1(X_2 \vee \overline{X}_2)\overline{X}_3(X_4 \vee \overline{X}_4) \vee X_1(X_2 \vee \overline{X}_2)X_3\overline{X}_4,$$

We then employ distributivity to obtain canonical normal forms:

$$Y \Leftrightarrow \underbrace{\overline{X}_1\overline{X}_2\overline{X}_3X_4}_{1} \vee \underbrace{\overline{X}_1\overline{X}_2\overline{X}_3\overline{X}_4}_{0} \vee \underbrace{\overline{X}_1\overline{X}_2X_3\overline{X}_4}_{2} \vee \underbrace{\overline{X}_1X_2X_3X_4}_{7} \vee$$
$$\underbrace{\overline{X}_1X_2\overline{X}_3X_4}_{5} \vee \underbrace{X_1X_2\overline{X}_3X_4}_{13} \vee \underbrace{X_1X_2\overline{X}_3\overline{X}_4}_{12} \vee \underbrace{X_1\overline{X}_2\overline{X}_3X_4}_{9} \vee$$
$$\underbrace{X_1\overline{X}_2\overline{X}_3\overline{X}_4}_{8} \vee \underbrace{X_1X_2X_3\overline{X}_4}_{14} \vee \underbrace{X_1\overline{X}_2X_3\overline{X}_4}_{10}.$$

This **expansion to normal forms** can itself become arduous. Minimisation has for years on end been a topic of intensive research and is covered in nearly all books on logic circuit design and switching theory.

4.4 Zhegalkin Normal Form

The output connective of a Zhegalkin normal form is a multiple XOR connective. Each inputs to the XOR is a conjunction of various input variables, these, importantly, always being non-inverted. This latter property is an asset making it possible to work with single-rail instead of double-rail logic. Double rail logic provides leads for the non-inverted as well as for the inverted input signals, whereas single-rail logic provides leads only for one of these, usually only for the non-inverted signals. Natural applications of Zhegalkin normal forms are the realisation of adders and error-correcting codes.

Instead of writing the Zhegalkin polynomial (as it is also called) in its general form, we detail it here for two, three and four input variables, shown in (4.8)–(4.10), respectively. After we discuss them, you should find it easy to expand these formulas for five and more input variables.

$$\begin{aligned} F(X_1, X_2) \Leftrightarrow \\ &\Phi_{00} \oplus \\ &X_1\Phi_{10} \oplus X_2\Phi_{01} \oplus \\ &X_1X_2\Phi_{11} \end{aligned} \tag{4.8}$$

$$\begin{aligned} F(X_1, X_2, X_3) \Leftrightarrow \\ &\Phi_{000} \oplus \\ &X_1\Phi_{100} \oplus X_2\Phi_{010} \oplus X_3\Phi_{001} \oplus \\ &X_1X_2\Phi_{110} \oplus X_1X_3\Phi_{101} \oplus X_2X_3\Phi_{011} \oplus \\ &X_1X_2X_3\Phi_{111} \end{aligned} \tag{4.9}$$

$$\begin{aligned} F(X_1, X_2, X_3, X_4) \Leftrightarrow \\ &\Phi_{0000} \oplus \\ &X_1\Phi_{1000} \oplus X_2\Phi_{0100} \oplus X_3\Phi_{0010} \oplus X_4\Phi_{0001} \oplus \\ &X_1X_2\Phi_{1100} \oplus X_1X_3\Phi_{1010} \oplus X_1X_4\Phi_{1001} \oplus \\ &\quad X_2X_3\Phi_{0110} \oplus X_2X_4\Phi_{0101} \oplus X_3X_4\Phi_{0011} \oplus \\ &X_1X_2X_3\Phi_{1110} \oplus X_1X_2X_4\Phi_{1101} \oplus \\ &\quad X_1X_3X_4\Phi_{1011} \oplus X_2X_3X_4\Phi_{0111} \oplus \\ &X_1X_2X_3X_4\Phi_{1111} \end{aligned} \tag{4.10}$$

A Zhegalkin normal form is characterised by logical constants $\Phi \in \{\mathbf{0}, \mathbf{1}\}$ called Zhegalkin coefficients. For a function of n input variables, $X_1, \ldots, X_n$, there are 2^n Zhegalkin coefficients. These we denote as $\Phi_{x_1 x_2 \ldots x_n}$ where each subscript x_i is either 0 or 1. Each Zhegalkin coefficient is ANDed with conjunctions of input variables—*iff* a subscript x_i of a Zhegalkin coefficient is 1, then the input variable X_i is contained in the conjunction of input variables.

By these rules, a coefficient $\Phi_{0,\ldots,0}$ is never ANDed with any input variables. If k of n subscripts of $\Phi_{x_1 x_2 \ldots x_n}$ are 1 there are $\binom{n}{k}$ conjunctions consisting of k input variables each.

The prime question is, of course, how to determine the logic value, either **0** or **1**, of each and every Zhegalkin coefficient $\Phi_{x_1 x_2 \ldots x_n}$ for a given problem. We use the example of Fig. 4.3 to demonstrate how this is done. The top part of sub-figure (a), consisting of rows for x, X_1, X_2, X_3, and the emphasized row for the output variable $Y(x)$, comprises the events graph of the circuit for which we want to design the Zhegalkin circuit. First, transpose the *rows* for the input variables X_1, X_2, X_3, to obtain their *column* arrangement as shown in the sub-figure. The next step is to develop the XOR-**logic-triangle** by following the rule $A_{i+1,j} \Leftrightarrow A_{i,j} \oplus A_{i,j+1}$ pictured in sub-figure (b). The leftmost column of the logic triangle contains the values of the Zhegalkin coefficients whose indices are specified by the logic values of the input variables of the associated row. Using the Zhegalkin coefficients of the logic triangle from top to bottom allows us to write the Zhegalkin normal form as follows:

$$Y \Leftrightarrow \mathbf{1} \oplus X_2 X_3 \oplus X_1 \oplus X_1 X_2 \oplus X_1 X_2 X_3. \tag{4.11}$$

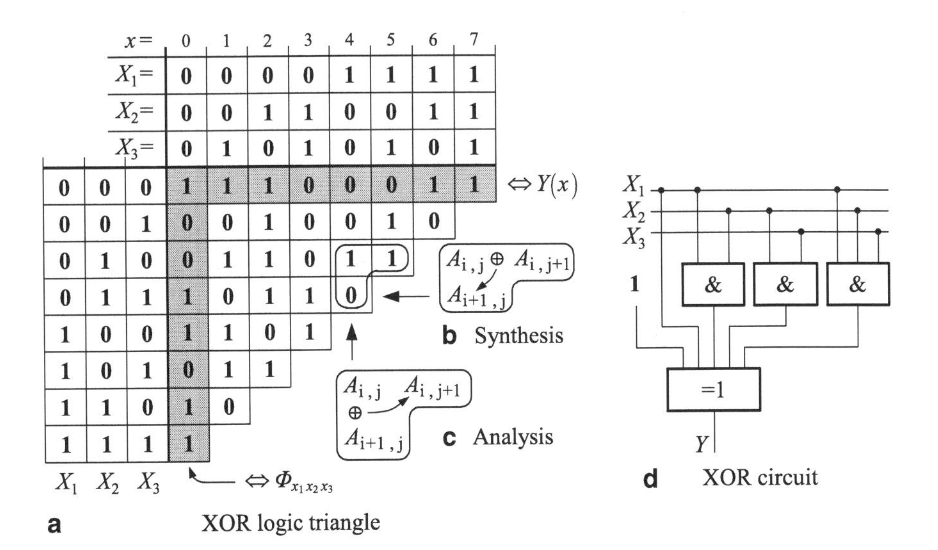

Fig. 4.3 An example showing how to obtain the Zhegalkin coefficients $\Phi_{x_1 x_2 x_3}$ for a three-input circuit whose events table is given

This formula leads to the logic circuit of sub-figure (d). The process of developing a logic circuit from, say, an events graph is called **synthesis**. **Analysis**, on the other hand, refers to finding the events graph which describes the behaviour of a given logic circuit.

Suppose we are presented with the logic circuit of Fig. 4.3d, and are required to develop the events graph that describes the circuit's input–output behaviour. You start by writing the *columns* for the input variables X_1, X_2, X_3. Into the leftmost column for the logic triagle (emphasized in Fig. 4.3a) you enter **1** in the appropriate row for the conjunctions of input variables of the logic circuit. Then, as cyclic permutation holds for an XOR-equivalence (meaning if $A \oplus B \Leftrightarrow C$, then $B \oplus C \Leftrightarrow A$, and $C \oplus A \Leftrightarrow B$) we use, instead of the synthesis formula, $A_{i+1,j} \Leftrightarrow A_{i,j} \oplus A_{i,j+1}$, the analysis formula, $A_{i,j+1} \Leftrightarrow A_{i,j} \oplus A_{i+1,j}$, as shown in Fig. 4.3c. Working upwards in the logic triangle, we come to its top row which contains the sought values for the output variable Y.

4.5 Dual Zhegalkin Normal Form

Duality allows us to replace *all* connectives and truth values by their duals in a given tautology thereby obtaining a neu tautology called dual to the original one. The mutually dual connectives and truth values are listed in Fig. 3.3. By this duality principle, we can write the Zhegalkin polynomial in its dual form, which is shown here for three input variables.

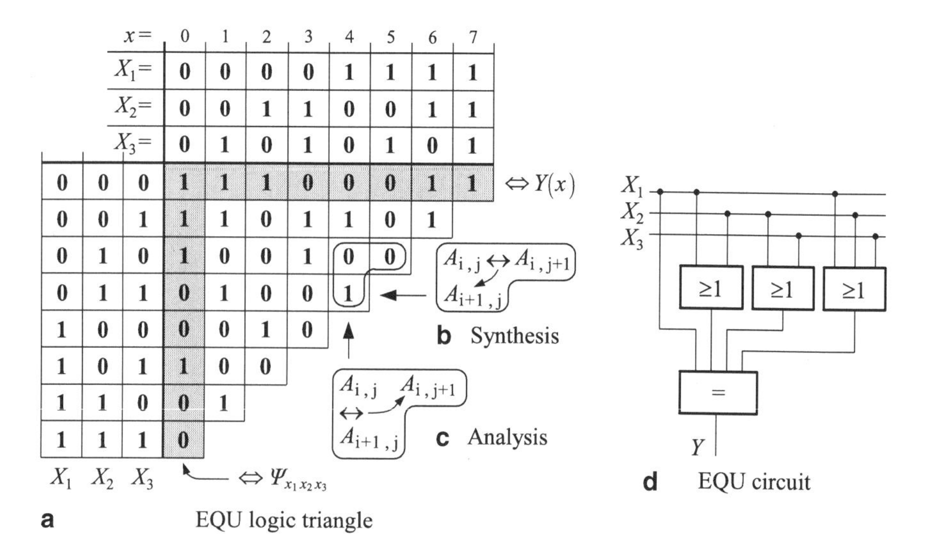

Fig. 4.4 Developing the dual Zhegalkin coefficients $\Psi_{x_1x_2x_3}$ for the given events table of a three-input circuit

$$\boxed{\begin{aligned} &F(X_1, X_2, X_3) \Leftrightarrow \\ &\quad \Psi_{000} \leftrightarrow \\ &\quad X_1 \vee \Psi_{100} \leftrightarrow X_2 \vee \Psi_{010} \leftrightarrow X_3 \vee \Psi_{001} \leftrightarrow \\ &\quad X_1 \vee X_2 \vee \Psi_{110} \leftrightarrow X_1 \vee X_3 \vee \Psi_{101} \leftrightarrow X_2 \vee X_3 \vee \Psi_{011} \leftrightarrow \\ &\quad X_1 \vee X_2 \vee X_3 \Psi_{111} \end{aligned}} \tag{4.12}$$

The dual Zhegalkin coefficients, $\Psi_{x_1x_2x_3}$, are obtained from a logic triangle of EQUIVALENCES similar to the way in which the original Zhegalkin coefficients are calculated, and detailed in Fig. 4.4 for the synthesis and analysis process. Here too using the dual Zhegalkin coefficients of the logic triangle from top to bottom allows us to write the dual Zhegalkin normal form as follows:

$$Y \Leftrightarrow X_2 \vee X_3 \leftrightarrow X_1 \leftrightarrow X_1 \vee X_2 \leftrightarrow X_1 \vee X_2 \vee X_3. \tag{4.13}$$

Chapter 5
Karnaugh Maps

In the previous chapter we used an events graph to *specify* a combinational circuit, and normal forms to *calculate* its logical design, here touching on the practical necessity to minimise (or simplify) the formulas obtained when employing *canonical* normal forms. In contrast to this somewhat formal approach, Karnaugh (1953), building on Veitch (1952), presented a graphic specification method with a marked theoretical potential by which to stimulate deeper insight and the development of new concepts. Among the more important aspects of the **K-map**, as the Karnaugh map is frequently called, are the following points discussed in this chapter:

- Introduce and visualise Karnaugh sets (K-sets);
- Develop new theorems, and prove given ones;
- Directly calculate the AND-to-OR and OR-to-AND formulas in their *simplified* form.

There are a number of further applications and modifications of the K-map, some discussed as we go along.

5.1 How to Draw a Karnaugh Map

A Karnaugh map is drawn as a rectangle subdivided into rows and columns by input variables. The look of a Karnaugh map depends primarily on the number n of input variables being depicted, the way they are arranged around its perimeter, and whether they are expressed as numeric variables—$x_1, \ldots, x_n \in \{0, 1\}$—or logic variables—$X_1, \ldots, X_n \in \{\mathbf{0}, \mathbf{1}\}$. First, consider the K-maps of Fig. 5.1 for three input variables. The fields are partitioned into cells, each representing an input event. The small numbers x, taken from $\mathcal{E} \in \{0, 1, 2, \ldots, 7\}$, in the top left corner of each cell are the decimal equivalents of the input events (x_1, x_2, x_3). The more prominent numbers, placed in the centre of a cell, are the numeric or logic output values invoked by the input event the cell represents.

S.P. Vingron, *Logic Circuit Design*, DOI 10.1007/978-3-642-27657-6_5,

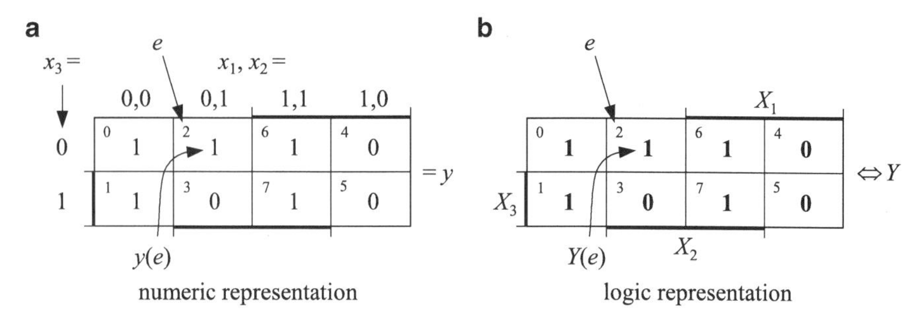

Fig. 5.1 K-maps for three input variables specifying the events graph of in the *top* part of Fig. 4.3a

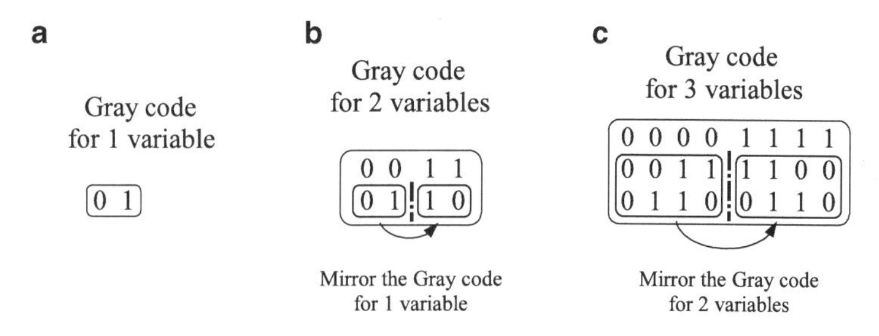

Fig. 5.2 Showing how a *Gray* code is developed for one, two and three variables

In this text, the input variables of a K-map are arranged as follows: The first half (or the first half plus 1, in the case of an uneven number of input variables) are assigned to the columns; the remaining input variables are assigned to the rows. Each column is encoded by a combination of zero- or one-values of input variables, the succession of these combinations following a Gray code. A **Gray code** is developed as shown in Fig. 5.1, the property of interest being that when switching from one code tuple to another, only a single variable changes its value. The rows of a K-map are, of course, organised in a like manner. Two cells of a K-map are said to be **adjacent** or **neighbouring** if they differ in the value of exactly one input variable of their binary input events.

The K-maps for one, two, three, four and five variables are shown in Fig. 5.2 in their logic representation. The bars on the perimeter mark those rows and columns for which the values of the associated numeric variables are 1, as follows from $X_i \Leftrightarrow x_i = 1$. Note that for five input variables the areas marked by the X_3-bar are disconnected. This property of disconnectedness is typical for K-maps of five and more input variables, and makes their visual application less intuitive. You might like to take the trouble of drawing the *numeric* representation of the K-maps of Fig. 5.2 verifying the fact that the rows and columns are encoded according to the Gray code of Fig. 5.3. Furthermore, I encourage you to draw the numeric and logic K-maps for six input variables.

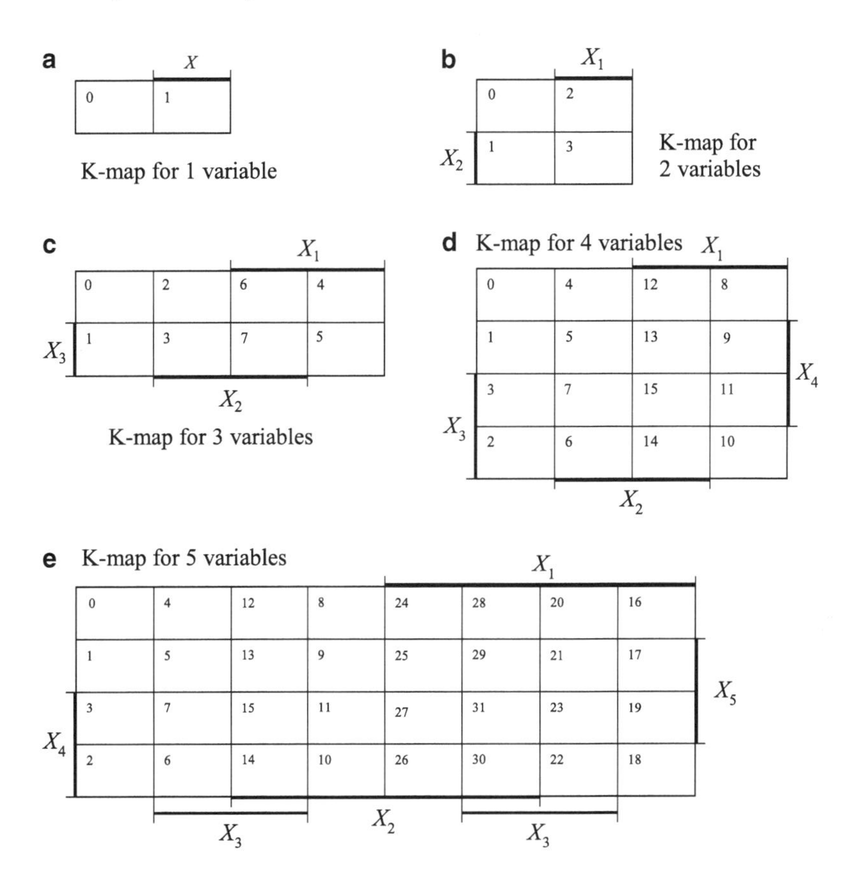

Fig. 5.3 Karnaugh maps for one, two, three, four and five variables

In the following sections we take a closer look at various applications of the K-map.

5.2 Karnaugh Set and Conjunctive Term

The Karnaugh map is a Venn diagram depicting a finite number of elements—namely, the input events x for a given number n of input variables. An input event x is an n-tuple of binary, numeric values of the ordered input variables, $(x_1, x_2, \ldots, x_n)$. An input event can be written in its binary form, $(x_1, x_2, \cdots, x_n)$, or in its decimal form, x, where $x = \sum_{i=1}^{n} x_i \cdot 2^{n-i}$. The input events tile the area of a K-map and are, as we have seen in the previous section, arranged in a very special order.

When calculating circuits specified by a K-map you will usually be grouping the input events whose cells contain an output value of 1 into a number of so-called **Karnaugh sets** (or **K-sets**) (Fig. 5.4). As you will see, a specific K-set is

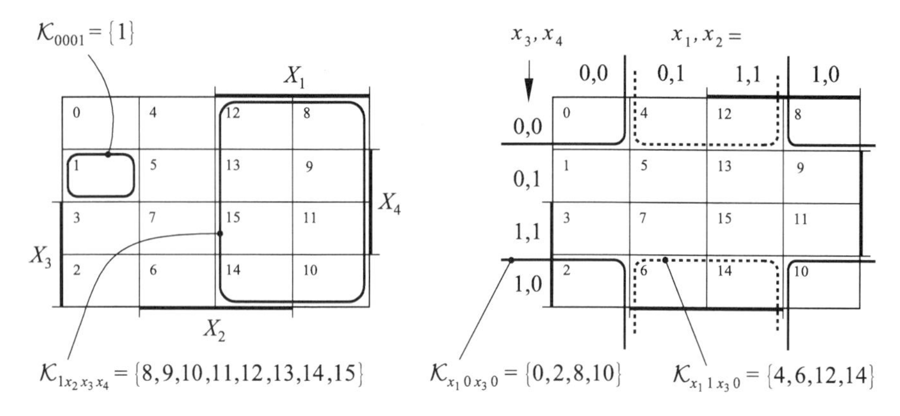

Fig. 5.4 Picturing example K-sets

easily described by the conjunction of certain non-inverted or inverted logic input variables. First let us define and interpret a *general conjunctive term* we call a **generalised minterm**

$$C_{\varphi_1 \ldots \varphi_n}(x_1, \ldots, x_n) \Leftrightarrow \bigwedge_{i=1}^{n}(x_i = \varphi_i), \quad \varphi_i \in \{0, 1, x_i\}. \tag{5.1}$$

Taking n to be 4, the following conjunctions $\overline{X}_1\overline{X}_2\overline{X}_3X_4$, or X_1, or $\overline{X}_2\overline{X}_4$, for example, can be developed from the above definition for $C_{\varphi_1 \ldots \varphi_n}$ by choosing appropriate values for φ_i:

$$C_{0001} \Leftrightarrow (x_1 = 0) \wedge (x_2 = 0) \wedge (x_3 = 0) \wedge (x_4 = 1) \Leftrightarrow \overline{X}_1\overline{X}_2\overline{X}_3X_4,$$

$$C_{1,x_2,x_3,x_4} \Leftrightarrow (x_1 = 1) \wedge (x_2 = x_2) \wedge (x_3 = x_3) \wedge (x_4 = x_4)$$
$$\Leftrightarrow X_1 \wedge \mathbf{1} \wedge \mathbf{1} \wedge \mathbf{1} \Leftrightarrow X_1,$$

$$C_{x_1,0,x_3,0} \Leftrightarrow (x_1 = x_1) \wedge (x_2 = 0) \wedge (x_3 = x_3) \wedge (x_4 = 0)$$
$$\Leftrightarrow \overline{X}_2\overline{X}_4.$$

A generalised minterm has the important and characteristic property of being true *iff* the instances[1] of the *describing* input variables equal their specified values, 0 or 1. For example, the describing variables of $C_{x_1,0,x_3,0} \Leftrightarrow \overline{X}_2\overline{X}_4 \Leftrightarrow (x_2 = 0) \wedge (x_4 = 0)$ are x_2 and x_4. Only if the measured values of x_2 and x_4 are zero, and these values are substituted for x_2 and x_4 in $(x_2 = 0) \wedge (x_4 = 0)$ does $C_{x_1,0,x_3,0}$ become true. The values of the non-describing input variables, x_1 and x_3, in our example, are irrelevant for the truth value of the generalised minterm $C_{x_1,0,x_3,0}$.

[1]On the concept of an **instance**, please refer to Sect. 1.4.

The set of instances for which the generalised minterm $C_{\varphi_1 \ldots \varphi_n}(x_1, \ldots, x_n)$ is true is called a **Karnaugh set** $\mathcal{K}_{\varphi_1 \ldots \varphi_n}$. Either of the following formulas, equivalent in their meaning, can be used to define the K-set $\mathcal{K}_{\varphi_1 \ldots \varphi_n}$.

$$\begin{aligned} &(x_1, \ldots, x_n) \in \mathcal{K}_{\varphi_1 \ldots \varphi_n} \Leftrightarrow C_{\varphi_1 \ldots \varphi_n}(x_1, \ldots, x_n), \quad \varphi_i \in \{0, 1, x_i\}, \\ &\mathcal{K}_{\varphi_1 \ldots \varphi_n} = \big\{(x_1, \ldots, x_n) \big| C_{\varphi_1 \ldots \varphi_n}(x_1, \ldots, x_n)\big\}, \quad \varphi_i \in \{0, 1, x_i\}. \end{aligned} \tag{5.2}$$

Using φ as an abbreviation for $(\varphi_1 \ldots \varphi_n)$ in which each φ_i stands for one of the symbols 0,1, or x_i, we can write the above equations, (5.2), as

$$\begin{aligned} &x \in \mathcal{K}_\varphi \Leftrightarrow C_\varphi(x) \quad \text{or} \\ &\mathcal{K}_\varphi = \{x | C_\varphi(x)\} \quad \text{with} \\ &\qquad \varphi = (\varphi_1 \ldots \varphi_n) \quad \text{and} \quad \varphi \in \{0, 1, x_i\}. \end{aligned} \tag{5.3}$$

We call $(\varphi_1 \ldots \varphi_n)$, as well as its abbreviation φ, a **multiple input event**. To get a feeling for the concept of *multiple input event* and its specification $(\varphi_1, \ldots, \varphi_n)$, with $\varphi_i \in \{0, 1, x_i\}$, take a look at the following examples in which n is 4. If each φ_i is a given constant, either 0 or 1, then $(\varphi_1, \varphi_2, \varphi_3, \varphi_4)$ is some individual input event, say, $(0, 1, 1, 1)$. Now assume that only three variables φ_i are assigned specific values, e.g., φ_1 is 0, φ_2 is 1, φ_3 is 1, while φ_4 remains unspecified (φ_4 is x_4), then $(\varphi_1, \varphi_2, \varphi_3, \varphi_4)$ becomes $(0, 1, 1, x_4)$. As x_4 can be either 0 or 1 ($x_4 \in \{0, 1\}$), the multiple event $(0, 1, 1, x_4)$ represents any of the individual input events $(0, 1, 1, 0)$, or $(0, 1, 1, 1)$. Similarly, the multiple event $(1, x_2, 0, x_4)$, containing two variable symbols, stands for any of the four individual input events $(1, 0, 0, 0)$, $(1, 0, 0, 1)$, $(1, 1, 0, 0)$, $(1, 1, 0, 1)$. Generalising, if the multiple event contains k variable symbols $x_{i_1}, x_{i_2}, \ldots, x_{i_k}$, then $(\varphi_1, \ldots, \varphi_n)$ stands for 2^k individual input events $(e_1, \ldots, e_n)$.

Of general interest are the smallest and largest K-sets. A smallest K-set, $\mathcal{K}_e$, called an **elementary K-set**, contains only a single input event, e, and is thus defined by one of the equivalent notations in which each element of the n-tuple, $(e_1, \ldots, e_n)$ stands for either 0 or 1,

$$\begin{aligned} &\mathcal{K}_e = \big\{(x_1, \ldots, x_n) \big| (x_1 = e_1) \wedge \cdots \wedge (x_n = e_n)\big\} \quad \text{or} \\ &\mathcal{K}_e = \{x | x = e\} \quad \text{with} \\ &\qquad x = \sum_{i=1}^{n} x_i \cdot 2^{n-i} \quad \text{and} \quad x_1, \ldots, x_n \in \{0, 1\}, \\ &\qquad e = \sum_{i=1}^{n} e_i \cdot 2^{n-i} \quad \text{and} \quad e_1, \ldots, e_n \in \{0, 1\}. \end{aligned} \tag{5.4}$$

Of prime importance when working with Karnaugh maps are the largest K-sets, the **maximum Karnaugh sets**. A K-map for n input variables has n maximum K-sets. Their importance and their respective association with an input variable warrants using special symbols to characterise them—$\mathcal{X}_1, \cdots, \mathcal{X}_n$. A **maximum K-set** $\mathcal{X}_i$ has those input events in common whose i-th input variable x_i is 1. This can be expressed equivalently in a number of ways:

$$\boxed{\begin{aligned} \mathcal{X}_i &= \{(x_1, \cdots, x_n) | x_i = 1\}, \\ (x_1, \cdots, x_n) \in \mathcal{X}_i &\Leftrightarrow x_i = 1, \\ x \in \mathcal{X}_i &\Leftrightarrow X_i. \end{aligned}} \tag{5.5}$$

Figure 5.4 shows, as an example, the elementary K-set $\mathcal{K}_1 = \mathcal{K}_{0001}$, and the maximum K-set $\mathcal{X}_1$, this being the shorter and alternative notation for $\mathcal{K}_{1x_2x_3x_4}$ as follows from $x \in \mathcal{K}_{1x_2x_3x_4} \Leftrightarrow (x_1 = 1) \wedge (x_2 = x_2) \wedge (x_3 = x_3) \wedge (x_4 = x_4) \Leftrightarrow X_1 \wedge \mathbf{1} \wedge \mathbf{1} \wedge \mathbf{1} \Leftrightarrow X_1 \Leftrightarrow x \in \mathcal{X}_i$. Note: In a K-map the input events of the columns or rows covered by the bars of an input variable being one picture and comprise a maximum K-set.

The Venn diagram is optimally adapted to visualise the **intersection**, $\mathcal{S} \cap \mathcal{T}$, and **union**, $\mathcal{S} \cup \mathcal{T}$, of two sets $\mathcal{S}$ and $\mathcal{T}$. These set operations are defined by AND and OR operations, respectively, according to:

$$x \in (\mathcal{S} \cap \mathcal{T}) \Leftrightarrow (x \in \mathcal{S}) \wedge (x \in \mathcal{T}), \tag{5.6}$$

$$x \in (\mathcal{S} \cup \mathcal{T}) \Leftrightarrow (x \in \mathcal{S}) \vee (x \in \mathcal{T}), \tag{5.7}$$

The usefulness of maximum K-sets lies in the fact that **any K-set can be expressed as the intersection of maximum K-sets or their complements, as the case may be**. For instance, $\mathcal{K}_{x_1 1 x_3 0}$ can be expressed as the intersection of $\mathcal{X}_2$ and $\overline{\mathcal{X}_4}$ as can easily be seen:

$$\begin{aligned} x \in \mathcal{K}_{x_1,1,x_3,0} &\overset{(5.2)}{\Leftrightarrow} (x_1 = x_1) \wedge (x_2 = 1) \wedge (x_3 = x_3) \wedge (x_4 = 0) \\ &\overset{(1.2)}{\Leftrightarrow} \mathbf{1} \wedge X_2 \wedge \mathbf{1} \wedge \overline{X}_4 \\ &\Leftrightarrow X_2 \wedge \overline{X}_4 \\ &\overset{(5.5)}{\Leftrightarrow} (x \in \mathcal{X}_2) \wedge (x \in \overline{\mathcal{X}_4}). \\ &\overset{(5.6)}{\Leftrightarrow} x \in (\mathcal{X}_2 \cap \overline{\mathcal{X}_4}) \\ \mathcal{K}_{x_1,1,x_3,0} &= \mathcal{X}_2 \cap \overline{\mathcal{X}_4}. \end{aligned}$$

The complement, $\overline{\mathcal{K}}_{x_1,1,x_3,0}$, of the K-set $\mathcal{K}_{x_1,1,x_3,0}$ is easily calculated, but its elements are best read from the K-map of Fig. 5.4:

$$\overline{\mathcal{K}}_{x_1 1 x_3 0} = \overline{\mathcal{X}_2 \cap \overline{\mathcal{X}}_4} = \overline{\mathcal{X}}_2 \cup \mathcal{X}_4 = \{1, 3, 4, 5, 6, 7, 9, 11, 12, 13, 14, 15\}.$$

The general notation for the complement of K-set $\mathcal{K}_{\varphi_1 \ldots \varphi_n}$ is one of the equivalent notations:

$$\boxed{\begin{aligned} &(x_1, \ldots, x_n) \in \overline{\mathcal{K}}_{\varphi_1 \ldots \varphi_n} \Leftrightarrow \overline{C}_{\varphi_1 \ldots \varphi_n}(x_1, \ldots, x_n), \quad \varphi_i \in \{0, 1, x_i\}, \\ &\overline{\mathcal{K}}_{\varphi_1 \ldots \varphi_n} = \{(x_1, \ldots, x_n) | \overline{C}_{\varphi_1 \ldots \varphi_n}(x_1, \ldots, x_n)\}, \quad \varphi_i \in \{0, 1, x_i\}. \end{aligned}} \tag{5.8}$$

The negation of a conjunctive term is always a disjunctive term, so that we define the negation of the generalised minterm $C_{\varphi_1 \ldots \varphi_n}(x_1, \ldots, x_n)$ as a **generalised maxterm** $D_{\varphi_1 \ldots \varphi_n}(x_1, \ldots, x_n)$,

$$\boxed{D_{\varphi_1 \ldots \varphi_n}(x_1, \ldots, x_n) \Leftrightarrow \overline{C}_{\varphi_1 \ldots \varphi_n}(x_1, \ldots, x_n).} \tag{5.9}$$

To obtain a **simpler notation** for a K-set—say, for $\mathcal{K}_{x_1,1,x_3,0}$—the indices, i, of the variables, x_i, are omitted letting the *position* in which the variable is written represent the value of the omitted index i. By this convention we may write $\mathcal{K}_{x1x0}$ instead of $\mathcal{K}_{x_1,1,x_3,0}$.

5.3 Proving and Developing Theorems

There are three ways to prove logic theorems. Firstly, algebraically, by using postulates and already proved theorems (a method we shan't look into in this text[2]). Secondly, by truth tables, touched on below. And, thirdly, with the help of K-maps, the main topic pointed out in this section. With the help of K-maps we can not only prove theorems, but also develop new ones.

The Karnaugh map is an excellent tool for developing logically equivalent expressions, so-called theorems. The basic idea to the method is this: *All logical descriptions of a certain area of a Karnaugh map are logically equivalent.* We accept logic formulas thus obtained as proven. Let us work through a few examples.

We start by choosing some area in a K-map, for instance, the area of loop X_1 in Fig. 5.5. Trivially, one logical description of this area is X_1 itself. A second way to describe this area, also shown in Fig. 5.5a, is to argue that the area X_1 can also be obtained by *uniting* (which means using the $\vee$-operation) the loops $X_1\overline{X}_2$ and X_1X_2. So, we can also describe the area of X_1 as $X_1\overline{X}_2 \vee X_1X_2$; these logic expressions, being descriptions of one and the same area, are logically equivalent:

$$X_1 \Leftrightarrow X_1\overline{X}_2 \vee X_1X_2.$$

[2]But considered quite extensively in Vingron (2004).

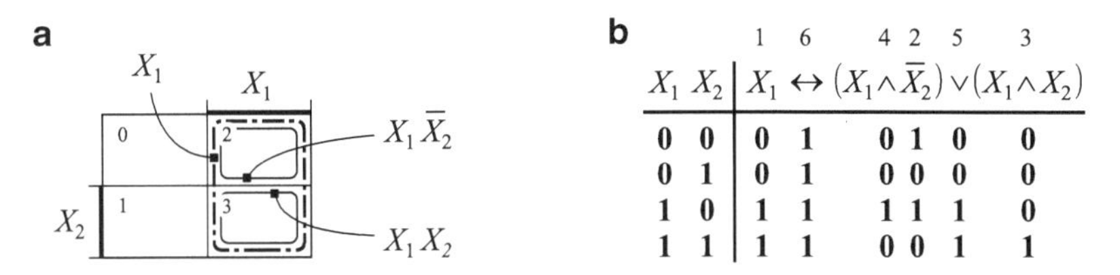

X_1	X_2	X_1 [1]	$\leftrightarrow$ [6]	$(X_1 \wedge$ [4]	$\overline{X}_2)$ [2]	$\vee$ [5]	$(X_1 \wedge X_2)$ [3]
0	**0**	**0**	**1**	**0**	**1**	**0**	**0**
0	**1**	**0**	**1**	**0**	**0**	**0**	**0**
1	**0**	**1**	**1**	**1**	**1**	**1**	**0**
1	**1**	**1**	**1**	**0**	**0**	**1**	**1**

Fig. 5.5 Proving the theorem $X_1 \Leftrightarrow X_1\overline{X}_2 \vee X_1X_2$, (**a**) using a K-map, (**b**) using a truth table

To prove the theorem by **truth table**, see Fig. 5.5b, we calculate its logic expression for all combinations of its variables, writing these combinations to the left of the logic expression. Below each logic connective we write the result of its calculation, necessitating that the AND-connectives ($\wedge$) is always written explicitly. As there is often no *unique* sequence by which the logic connectives must be evaluated, it is good practice to number the logic connectives according to the sequence used. The logic connective with the highest number is the connective under which the resulting truth value of the logic expression is written. In our example, the sequence of evaluation was chosen in the following way. The logic expression is read from left to right, evaluating any connective possible, a process we call a *pass* of the logic expression. This lets us write the values for $\overset{1}{X_1}$ and $\overset{2}{\overline{X}_2}$, and to evaluate $\overset{3}{\wedge}$. In further passes we can evaluate $\overset{4}{\wedge}$, then $\overset{5}{\vee}$, and finally $\overset{6}{\leftrightarrow}$. As the EQUIVALENCE ($\leftrightarrow$) is true for all combinations of the input variables, it is proved correct, allowing us to replace $\leftrightarrow$ with $\Leftrightarrow$.

For good measure we take a brief look at formula

$$X_1 \Leftrightarrow (X_1 \vee \overline{X}_2) \wedge (X_1 \vee X_2)$$

which is dual to the one just proved above, and needs no further proof due to the principle of duality. Nevertheless it is instructive to note, as shown in Fig. 5.6a, that the disjunction $X_1 \vee \overline{X}_2$ is pictured as the area of the union $\mathcal{X}_1 \cup \overline{\mathcal{X}}_2$ of the maximum K-set $\mathcal{X}_1$ and the complement, $\overline{\mathcal{X}}_2$, of the maximum K-set $\mathcal{X}_2$. In a like manner, the disjunction $X_1 \vee X_2$ is expressed as the union $\mathcal{X}_1 \cup \mathcal{X}_2$. The intersection of these

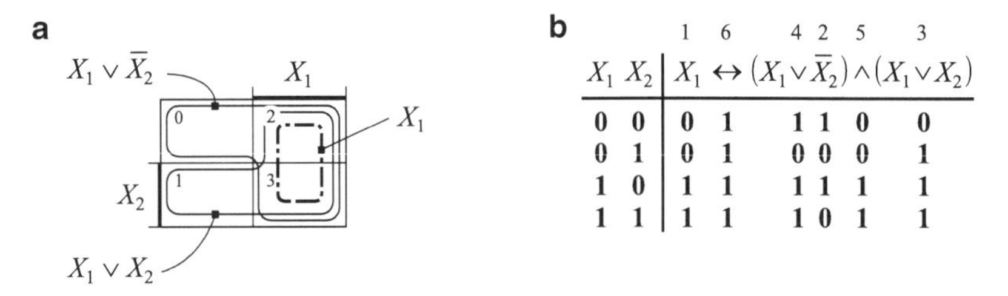

X_1	X_2	X_1 [1]	$\leftrightarrow$ [6]	$(X_1 \vee$ [4]	$\overline{X}_2)$ [2]	$\wedge$ [5]	$(X_1 \vee X_2)$ [3]
0	**0**	**0**	**1**	**1**	**1**	**0**	**0**
0	**1**	**0**	**1**	**0**	**0**	**0**	**1**
1	**0**	**1**	**1**	**1**	**1**	**1**	**1**
1	**1**	**1**	**1**	**1**	**0**	**1**	**1**

Fig. 5.6 Proving the theorem $X_1 \Leftrightarrow (X_1 \vee \overline{X}_2)(X_1 \vee X_2)$, (**a**) using a K-map, (**b**) using a truth table

two unions, $(\mathcal{X}_1 \cup \overline{\mathcal{X}}_2) \cap (\mathcal{X}_1 \cup \mathcal{X}_2)$, equals $\mathcal{X}_1$ which in logical notation is the above theorem.

5.4 Evaluating Karnaugh Maps

The output $Y(x)$ of a circuit defined by a K-map is evaluated by collecting those cells that contain an output value of 1 into K-sets. The larger and fewer the K-sets, the simpler the formula describing the circuit's output. On the other hand, if you want to calculate the inverted value of a circuit's output, $\overline{Y}(x)$, you can do this by collecting all cells that contain an output value of 0 into K-sets. Those cells x whose output values $y(x)$ are 1 comprise the **setting input events** $\mathcal{E}_1$, whereas those cells whose output $y(x)$ are 0 comprise the **resetting input events** $\mathcal{E}_0$,

$$\mathcal{E}_1 = \{x | y(x) = 1\}, \qquad \text{or, equivalently,} \qquad x \in \mathcal{E}_1 \Leftrightarrow y(x) = 1, \tag{5.10}$$

$$\mathcal{E}_0 = \{x | y(x) = 0\}, \qquad \text{or, equivalently,} \qquad x \in \mathcal{E}_0 \Leftrightarrow y(x) = 0. \tag{5.11}$$

These equations are used to calculate the logic variables $Y(x) \overset{(1.2)}{\Leftrightarrow} y(x) = 1$ and $\overline{Y}(x) \overset{(1.2)}{\Leftrightarrow} y(x) = 0$,

$$Y(x) \Leftrightarrow x \in \mathcal{E}_1, \qquad \text{and} \qquad \overline{Y}(x) \Leftrightarrow x \in \mathcal{E}_0, \tag{5.12}$$

assuming we know what to do with the expressions $x \in \mathcal{E}_1$ and $x \in \mathcal{E}_0$. For the sake of simplicity, we next concentrate only on *setting* input events $\mathcal{E}_1$. The principle is to *cover* $\mathcal{E}_1$ with any number of K-sets whose union equals $\mathcal{E}_1$. A **cover**

$$\boxed{\begin{aligned} \mathcal{K}(\mathcal{E}_1) &= \{\mathcal{K}_{\varphi(1)}, \cdots, \mathcal{K}_{\varphi(L)}\} \qquad \text{with} \\ \mathcal{E}_1 &= \mathcal{K}_{\varphi(1)} \cup \cdots \cup \mathcal{K}_{\varphi(L)}. \end{aligned}} \tag{5.13}$$

Two sets, in our case $\mathcal{E}_1$ and the union $\mathcal{K}_{\varphi(1)} \cup \cdots \cup \mathcal{K}_{\varphi(L)}$, are equal if they have the same elements, i.e.,

$$x \in \mathcal{E}_1 \Leftrightarrow x \in (\mathcal{K}_{\varphi(1)} \cup \cdots \cup \mathcal{K}_{\varphi(L)}). \tag{5.14}$$

The right side, by the definition of union, (5.7), leads to

$$x \in \mathcal{E}_1 \Leftrightarrow (x \in \mathcal{K}_{\varphi(1)}) \vee \cdots \vee (x \in \mathcal{K}_{\varphi(L)}). \tag{5.15}$$

Now using the conjunctive terms, $C_{\varphi(1)}, \ldots, C_{\varphi(L)}$, which define the K-sets $\mathcal{K}_{\varphi(1)}, \ldots, \mathcal{K}_{\varphi(L)}$ according to (5.3), we can write

$$x \in \mathcal{E}_1 \Leftrightarrow C_{\varphi(1)} \vee \cdots \vee C_{\varphi(L)} \tag{5.16}$$

and with (5.12) we get the preliminary result

$$Y(x) \Leftrightarrow C_{\varphi(1)} \vee \cdots \vee C_{\varphi(L)}. \tag{5.17}$$

The only drawback of this formula is that it does not state the restriction, (5.13), imposed on the choice of the K-sets, $\mathcal{K}_{\varphi(1)}, \ldots, \mathcal{K}_{\varphi(L)}$, whose conjunctive terms, $C_{\varphi(1)}, \ldots, C_{\varphi(L)}$, are used. To incorporate this information it is better to write (5.17) as

$$Y(x) \Leftrightarrow \bigvee_{\varphi:\mathcal{K}_\varphi \in \mathcal{K}(\mathcal{E}_1)} C_\varphi(x), \tag{5.18}$$

noting that $\varphi : \mathcal{K}_\varphi \in \mathcal{K}(\mathcal{E}_1)$ is read: *those* φ *for which* $\mathcal{K}_\varphi \in \mathcal{K}(\mathcal{E}_1)$. With no further ado we now generalise (5.18) to all so-called **evaluation formulas**, these being **the core results for evaluating K-sets**:

$$\boxed{\begin{aligned} Y(x) &\Leftrightarrow \bigvee_{\varphi:\mathcal{K}_\varphi \in \mathcal{K}(\mathcal{E}_1)} C_\varphi(x) \Leftrightarrow \bigwedge_{\varphi:\mathcal{K}_\varphi \in \mathcal{K}(\mathcal{E}_0)} D_\varphi(x) \\ \overline{Y}(x) &\Leftrightarrow \bigvee_{\varphi:\mathcal{K}_\varphi \in \mathcal{K}(\mathcal{E}_0)} C_\varphi(x) \Leftrightarrow \bigwedge_{\varphi:\mathcal{K}_\varphi \in \mathcal{K}(\mathcal{E}_1)} D_\varphi(x) \end{aligned}} \tag{5.19}$$

These evaluation formulas parallel the canonical normal forms (4.3) and (4.4) but are far more versatile because they allow us to work with a cover of arbitrarily chosen K-sets while the canonical normal forms only use a cover of elementary K-sets. To obtain minimised logic expressions for the output Y, the evaluation formulas suggest, as the examples show, choosing covers consisting of the largest possible K-sets and using the fewest of these K-sets necessary. The largest possible K-sets are called **prime sets**, and due to their importance we denote them as $\mathcal{P}_\varphi$ instead if $\mathcal{K}_\varphi$. More formally, a prime-set is defined as a K-set that is not a subset of another K-set.

Let us now look at some examples starting with the K-map of Fig. 5.7. Without considering any consequences, we choose the K-sets $\mathcal{K}_{1x0}$ and $\mathcal{K}_{0x1}$, shown in sub-figure (a), to start creating a cover of the setting input events $\mathcal{E}_1$. As cells 1 and 7 are not yet covered, we further choose the K-sets $\mathcal{K}_{000}$ and $\mathcal{K}_{11x}$, shown in sub-figure (b), to cover these cells. By (5.19) Y is written as the disjunction of the conjunctive terms C_φ defining the K-sets $\mathcal{K}_\varphi$ of the chosen cover $\mathcal{K}_1(\mathcal{E}_1) = \{\mathcal{K}_{000}, \mathcal{K}_{1x0}, \mathcal{K}_{0x1}, \mathcal{K}_{11x}\}$,

$$Y \Leftrightarrow \overline{X}_1\overline{X}_2\overline{X}_3 \vee X_1\overline{X}_3 \vee \overline{X}_1 X_3 \vee X_1 X_2. \tag{5.20}$$

This result is already a considerable achievement compared with the formula we obtain when using the AND-to-Or canonical normal form (4.3) which, in effect, uses the cover of the *elementary* K-sets

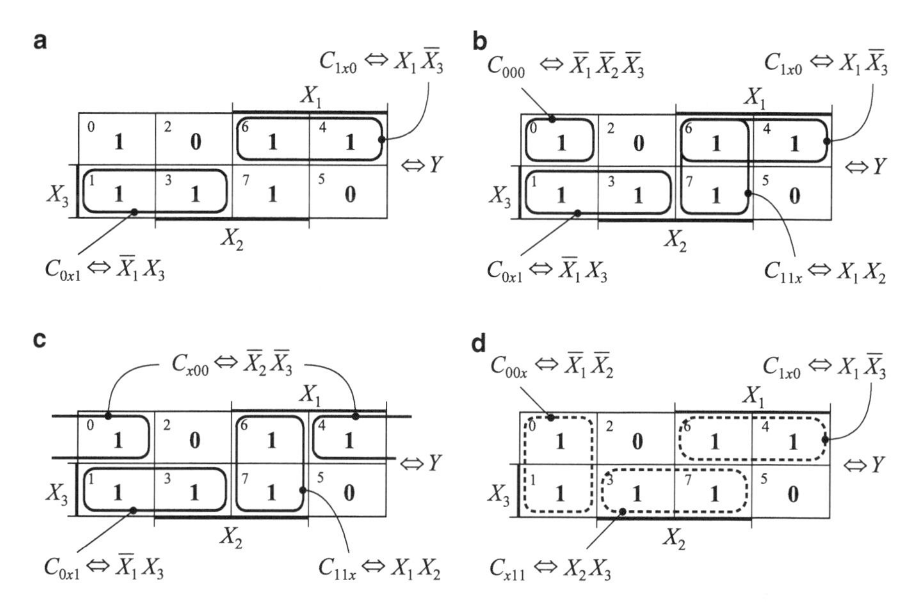

Fig. 5.7 Choosing covers on an example K-map

$$\mathcal{K}_2(\mathcal{E}_1) = \{\mathcal{K}_0, \mathcal{K}_1, \mathcal{K}_3, \mathcal{K}_4, \mathcal{K}_6, \mathcal{K}_7\}$$

and thus provides the formula

$$Y \Leftrightarrow \overline{X}_1\overline{X}_2\overline{X}_3 \vee \overline{X}_1\overline{X}_2 X_3 \vee \overline{X}_1 X_2 X_3 \vee X_1\overline{X}_2\overline{X}_3 \vee X_1 X_2\overline{X}_3 \vee X_1 X_2 X_3.$$

However, (5.20) is not a minimal result. To obtain one, we rely on the pronounced ability of the human brain to recognise patterns, in our case, patterns of covers consisting of a minimal number of prime set. The two existing such covers, $\mathcal{K}_3(\mathcal{E}_1) = \{\mathcal{K}_{x00}, \mathcal{K}_{0x3}, \mathcal{K}_{11x}\}$ and $\mathcal{K}_4(\mathcal{E}_1) = \{\mathcal{K}_{00x}, \mathcal{K}_{x11}, \mathcal{K}_{1x0}\}$, for our example are shown in sub-figures (c) and (d), the respective equations for the output Y being:

$$Y \Leftrightarrow \overline{X}_2\overline{X}_3 \vee \overline{X}_1 X_3 \vee X_1 X_2, \tag{5.21}$$

$$Y \Leftrightarrow \overline{X}_1\overline{X}_2 \vee X_2 X_3 \vee X_1\overline{X}_3. \tag{5.22}$$

The covers $\mathcal{K}_1(\mathcal{E}_1)$, $\mathcal{K}_3(\mathcal{E}_1)$ and $\mathcal{K}_4(\mathcal{E}_1)$, used above, are so-called **irredundant covers** meaning that each of their K-sets are necessary to cover $\mathcal{E}_1$. The latter two covers, $\mathcal{K}_3(\mathcal{E}_1)$ and $\mathcal{K}_4(\mathcal{E}_1)$, are **minimal irredundant covers**: They contain the smallest number of K-sets.

Alternatively, to the above evaluations, Y can also be calculated by (5.19) as an OR-to-AND formula using a cover of $\mathcal{E}_0$. The only such cover in our example is $\mathcal{K}(\mathcal{E}_0) = \{\mathcal{K}_{010}, \mathcal{K}_{101}\}$ allowing us to write the output as

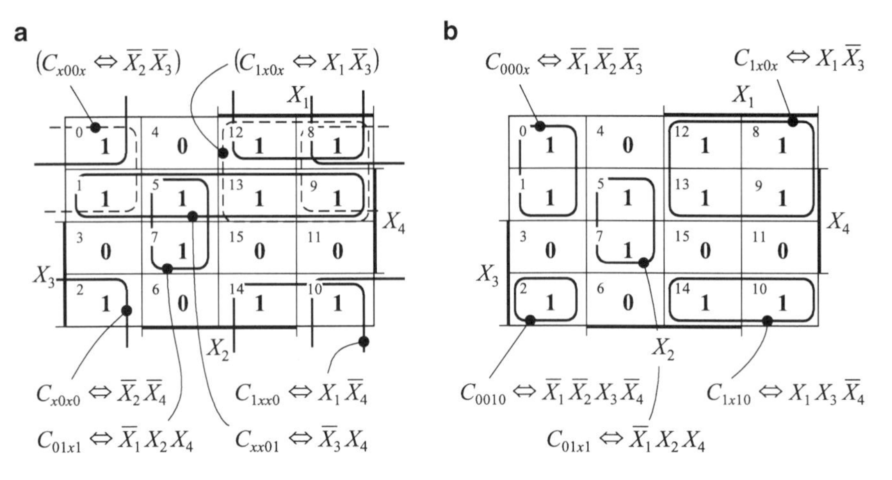

Fig. 5.8 Reviewing the example discussed in Sect. 4.3

$$Y \Leftrightarrow (X_1 \vee \overline{X}_2 \vee X_3)(\overline{X}_1 \vee X_2 \vee \overline{X}_3). \tag{5.23}$$

Whether you use one of the AND-to-OR formulas, (5.21) or (5.22), or the OR-to-AND formula, 5.23, to realise the solution is primarily a question of the technology employed. In Vingron (2004) (Chap. 7, Sect. 5.3 ‘Obtaining Conjunctive Formulas’) the procedure is discussed more generally.

We next review the example of Sect. 4.3—it is depicted in the K-maps of Fig. 5.8. Sub-figure (a) contains all *prime* sets of $\mathcal{E}_1$. These six prime sets provide us, according to (5.19) and without any trouble, with the output equation (4.6):

$$Y \Leftrightarrow \overline{X}_2\overline{X}_3 \vee \overline{X}_2\overline{X}_4 \vee \overline{X}_3 X_4 \vee \overline{X}_1 X_2 X_4 \vee X_1\overline{X}_3 \vee X_1\overline{X}_4. \tag{4.6}$$

Moreover, the prime sets $\mathcal{P}_{x00x}$ and $\mathcal{P}_{1x0x}$, defined by the conjunctive terms C_{x00x} and C_{1x0x}, are recognisably redundant (i.e., can be omitted) which, again by (5.19), leads to the minimised output formula

$$Y \Leftrightarrow \overline{X}_2\overline{X}_4 \vee \overline{X}_3 X_4 \vee \overline{X}_1 X_2 X_4 \vee X_1\overline{X}_4. \tag{5.24}$$

After this very conveniently obtained result, it might be instructive to show how the *non-canonical* AND-to-OR formula (4.7),

$$Y \Leftrightarrow \overline{X}_1\overline{X}_2\overline{X}_3 \vee \overline{X}_1\overline{X}_2 X_3\overline{X}_4 \vee \overline{X}_1 X_2 X_4 \vee X_1\overline{X}_3 \vee X_1 X_3\overline{X}_4 \tag{4.7}$$

was derived: This formula pictures the cover chosen in sub-figure (b).

5.5 Karnaugh Trees and Map-Entered Variables

Standard K-maps have *constants* entered into their cells—0 and 1, if you are working with numeric values, or **0** and **1**, if you are using logic values. But you can equally well enter a numeric or logic *variable* into a cell instead of a constant. Such variables are called **map-entered variables**. A K-map containing map-entered variables is usually referred to as a **reduced K-map**. As we will see, one can switch back and forth from the standard to a reduced K-map. But note, whereas the standard K-map is unique, there are many equivalent reduced K-maps. The next two figures are intended to give an impression of various ways one can develop reduced K-maps for any given standard K-map. For no deeper reason, we use the standard K-map of Fig. 5.9. When referring to its sub-figures we content ourselves with using their parenthesized letters, (a) to (d), without referring to these as sub-figures.

Consider Fig. 5.9. One frequently refers to the number of input variables of a K-map as its **dimension**. By this convention, (a) pictures a 3-dimensional, (b_1) and (b_2) are 2-dimensional, while (b) and (c) are 1-dimensional K-maps. How do

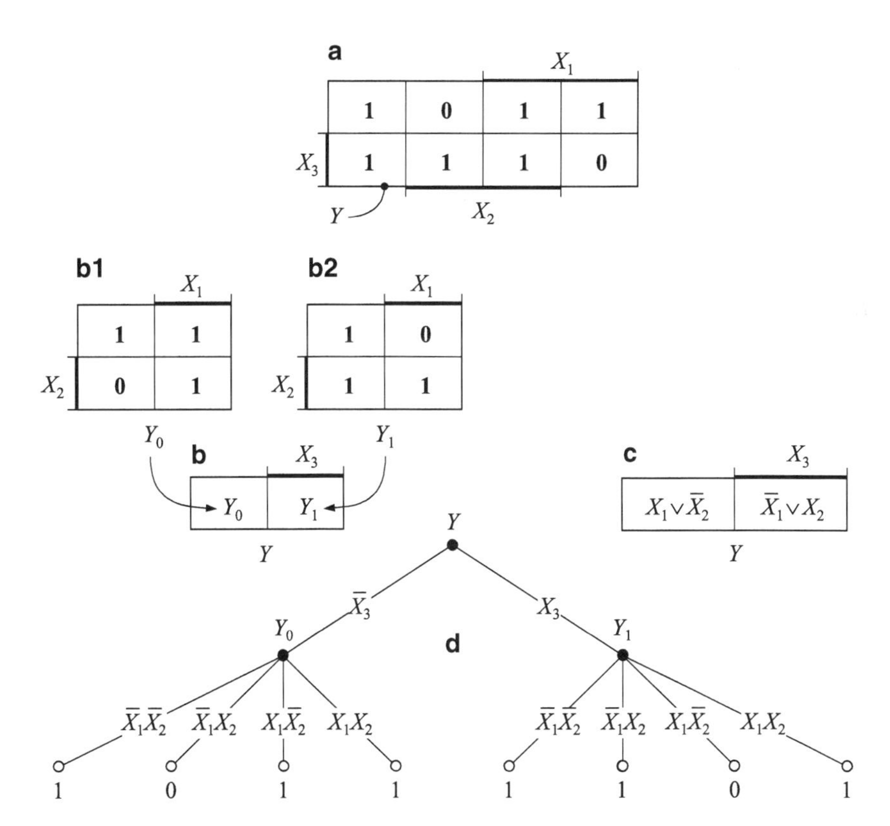

Fig. 5.9 From **standard K-map**, (**a**), to **reduced K-map**, (**b**) or (**c**), and to the **K-tree**, (**d**), of the reduced K-map

we transform the 3-dimensional K-map (a) into the 1-dimensional K-map (b)? The latter contains the two partial output variables Y_0 and Y_1 whose values are determined in the top and bottom rows of (a), respectively. The values of these variables are only dependant on the input variables X_1 and X_2, a dependence conveniently expressed in the K-maps (b_1) and (b_2), allowing us to calculate Y_0 and Y_1 as $Y_0 \Leftrightarrow X_1 \vee \overline{X}_2$ and $Y_1 \Leftrightarrow \overline{X}_1 \vee X_2$. In (c), these functions, $X_1 \vee \overline{X}_2$ and $\overline{X}_1 \vee X_2$, are entered into the cells instead of Y_0 and Y_1, respectively, X_1 and X_2, in this context, being called **map entered variables**.

The interdependence of the K-maps (b), (b_1) and (b_2) is aptly pictured by what I choose to call a **Karnaugh tree** shown in (d). As is customary in mathematics we draw trees upside down: root at the top, leaves at the bottom.

Figure 5.10 demonstrates the development of a 2-dimensional reduced K-map in which X_3 was chosen as the map-entered variable. Each cell in (b) stands for a column in (a) thus immediately explaining the **1**s in (b). In column $\overline{X}_1 X_2$ of (a), the values **0** and **1** coincide with the values of X_3 allowing us to enter X_3 into cell $\overline{X}_1 X_2$

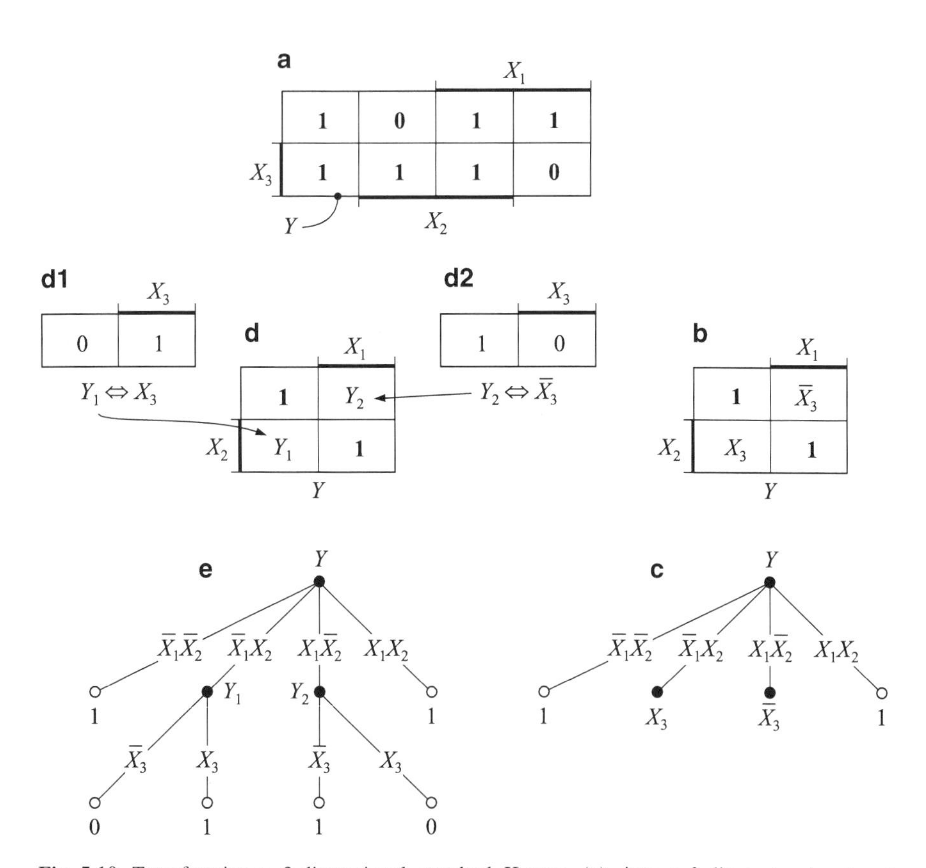

Fig. 5.10 Transforming a 3-dimensional standard K-map, (**a**), into a 2-dimensional reduced K-map, (**b**) or (**d**), and their respective K-trees, (**c**) or (**e**)

of (b), On the other hand, the values **1** and **0** in column $X_1\overline{X}_2$ of (a) coincide with the *inverted* values of X_3 so that we must enter $\overline{X}_3$ in cell $X_1\overline{X}_2$ of (b). The K-tree which coincides with (b) is shown in (c). An alternative representation to (b) and (c) is shown in (d)—together with (d_1) and (d_2)—and (e).

Reduced K-maps can be very helpful. For instance, developing a circuit whose output z is 1 if the 4-digit binary number $x = x_1 \cdot 2^3 + x_2 \cdot 2^2 + x_3 \cdot 2^1 + x_4 \cdot 2^0$ is greater than the 4-digit binary number $y = y_1 \cdot 2^3 + y_2 \cdot 2^2 + y_3 \cdot 2^1 + y_4 \cdot 2^0$ is an appropriate example. Specifying this problem would normally require an 8-dimensional K-map with 16 columns, for x_1, x_2, x_3, x_4, and 16 rows, for y_1, y_2, y_3, y_4. Using reduced K-maps for this example simplifies its representation and evaluation considerably.

Whether two numbers are distinct is decided in the highest position in which two digits are distinct. You thus start by comparing the highest weighted bits, in our case x_1 and y_1. If they are equal you compare the next lower bits, x_2 and y_2. You carry on in this manner towards lower weighted bits until you come across unequal bits, these deciding which of the numbers is the larger one.

To simplify the problem, we first assume the two highest bits of x and y to be equal ($x_1 = y_1$ and $x_2 = y_2$) so that we need only compare the two lowest bits. This is done in the 4-dimensional K-map of Fig. 5.11a using the above algorithm, and deciding for each cell individually whether to enter 0 or 1. The 1 in cell 4 states that $x > y$ because x_4, being 1, is larger than y_4, which is 0, while x_3 and y_3, both being 0, are equal. In the cells 8, 9, 12, 13 $x > y$ because $x_3 > y_3$. The 1 in cell 14 states that $x > y$ because $x_4 > y_4$ while $x_3 = y_3$. Evaluating the discussed K-map, we get:

$$Z^{(2)}_{x>y} \Leftrightarrow X_3\overline{Y}_3 \vee (X_3 \vee \overline{Y}_3)X_4\overline{Y}_4. \tag{5.25}$$

Now consider the possibility of the two higher order bits being unequal as discussed in Fig. 5.11b. If they *are* equal, which is the case in the main diagonal of the K-map, then we are referred to the interim result $Z^{(2)}_{x>y}$ of the prior K-map

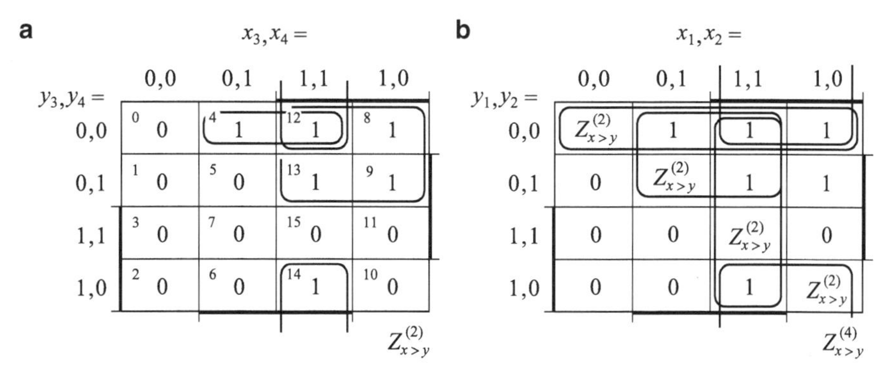

Fig. 5.11 Using reduced K-maps to specify when one 4-digit binary number, $x = x_1 \cdot 2^3 + x_2 \cdot 2^2 + x_3 \cdot 2^1 + x_4 \cdot 2^0$, is greater than another, $y = y_1 \cdot 2^3 + y_2 \cdot 2^2 + y_3 \cdot 2^1 + y_4 \cdot 2^0$

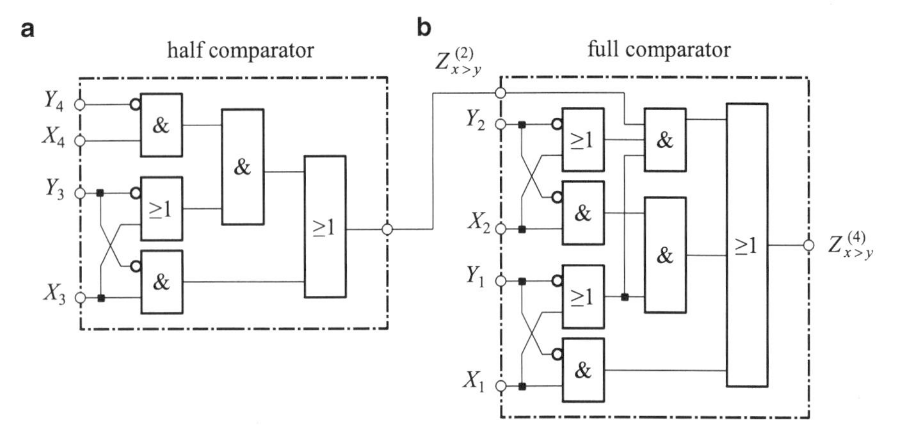

Fig. 5.12 The circuit, derived from (5.25) and (5.26), that emits a 1-signal when a 4-digit binary number, $x = x_1 \cdot 2^3 + x_2 \cdot 2^2 + x_3 \cdot 2^1 + x_4 \cdot 2^0$, is greater than another, $y = y_1 \cdot 2^3 + y_2 \cdot 2^2 + y_3 \cdot 2^1 + y_4 \cdot 2^0$. (**a**) half comparator (**b**) full comparator

by entering $Z^{(2)}_{x>y}$ into the cells of the main diagonal. If the higher order bits are unequal, the K-map has 0s and 1s entered as in the prior K-map. When evaluating a reduced K-map, a K-map containing map-entered variables, you first calculate the expression for 1-cells. Then every map-entered variable is ANDed with the conjunctive term of a K-set containing the map-entered variable and 1s.[3] The largest such K-sets are shown in Fig. 5.11b. The result obtained in our case is:

$$\begin{aligned} Z^{(4)}_{x>y} &\Leftrightarrow X_1\overline{Y}_1 \vee (X_1 \vee \overline{Y}_1)X_2\overline{Y}_2 \\ &\quad \vee (X_1X_2 \vee \overline{Y}_1X_2 \vee X_1\overline{Y}_2 \vee \overline{Y}_1\overline{Y}_2)Z^{(2)}_{x>y}, \\ Z^{(4)}_{x>y} &\Leftrightarrow X_1\overline{Y}_1 \vee (X_1 \vee \overline{Y}_1)X_2\overline{Y}_2 \\ &\quad \vee (X_1 \vee \overline{Y}_1)(X_2 \vee \overline{Y}_2)Z^{(2)}_{x>y}. \end{aligned} \tag{5.26}$$

The circuit obtained from (5.25) and (5.26) is shown in Fig. 5.12.

[3]This procedure is explained in quite some detail in Vingron (2004), Sect. (20.3) 'Evaluating Reduced K-maps'.

Chapter 6
Adjacency and Consensus

This chapter is a precursor to algebraic minimisation taken up in the next chapter. The operators and methods developed here are inspired by the visual methods used in evaluating K–maps. It is helpful to have a clear intuitive grasp of the concepts we want to use formally, so, in Sect. 6.1 we introduce the two prime concepts, **adjacency** and **consensus**, in a visual and informal way. Only in the next two sections do we define these concepts formally. The last section, Sect. 6.4, is devoted to the question, when one K-set is a (proper) **subset** of another.

6.1 Adjacent K-Sets and their Consensus*

Adjacency is a property associating two K-sets, this property allowing us to define a third and special K-set called a *consensus*, its elements calved out of the adjacent K-sets.

> Informally, two K-sets are said to be **adjacent** if they are disjoint (i.e., have no elements in common), and if they have, at least partially, a boarder or edge in common in a K-map. (6.1)

Adjacency was originally conceived for elementary K-sets $\mathcal{K}_e$, as those pictured in Fig. 6.1a, where $\mathcal{K}_0$ and $\mathcal{K}_2$, as well as $\mathcal{K}_5$ and $\mathcal{K}_{13}$ are adjacent. This original adjacency concept of two *elementary* K-sets is extended to the adjacency of any two K-sets as stated above. The K-sets $(\mathcal{K}_A, \mathcal{K}_B)$, $(\mathcal{K}_A, \mathcal{K}_C)$, $(\mathcal{K}_A, \mathcal{K}_{14})$, $(\mathcal{K}_B, \mathcal{K}_C)$ of Fig. 6.1b show that adjacent K-sets need not be of the same size and need not be aligned. The criterion of importance is for them to have, at least partially, a boarder in common. Just to round off the picture, no two K-sets of the three shown in Fig. 6.2a are adjacent as they are not pairwise disjoint. Neither are any of the elementary K-sets of Fig. 6.2b adjacent as no two have a boarder in common.

S.P. Vingron, *Logic Circuit Design*, DOI 10.1007/978-3-642-27657-6_6,

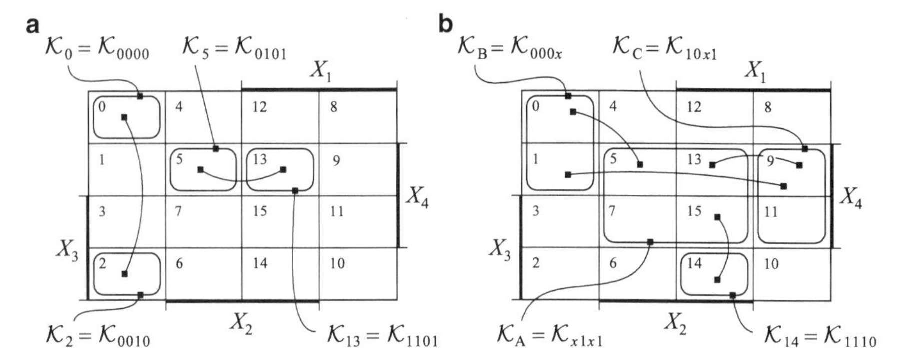

Fig. 6.1 Examples of adjacent K-sets

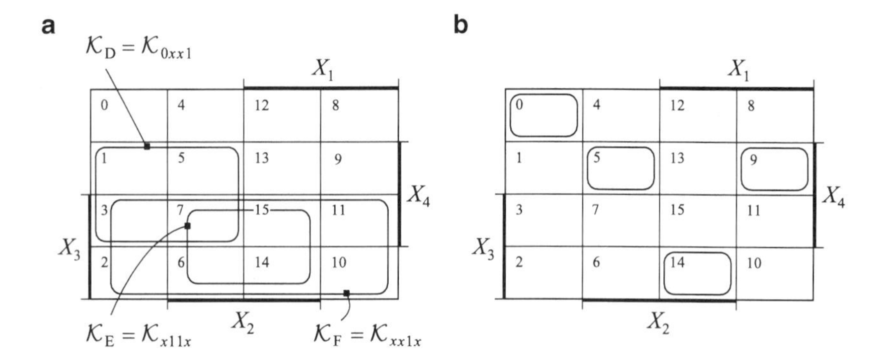

Fig. 6.2 Examples of non-adjacent K-sets

Now, to get a first impression of the consensus, take a look at Fig. 6.3 which shows all but one of the consensuses for the sets of Fig. 6.1. You might want to see if you can find the general rule by which to find or define consensuses—if so interrupt your reading here ...

> The **consensus** of two given *adjacent* K-sets is the largest K–set that can be constructed exclusively of elements of both the given adjacent K–sets. (6.2)

By this definition, you will notice that $\mathcal{K}_{0.2}$ is the consensus of $(\mathcal{K}_0, \mathcal{K}_2)$ while $\mathcal{K}_{5.13}$ is the consensus of $(\mathcal{K}_5, \mathcal{K}_{13})$ as shown in Fig. 6.3a. This also gives a first impression of how adjacent K–sets are incorporated into a larger K-set, one called consensus. The incorporated K-sets are proper subsets of the consensus, and of no further relevance in a minimisation process. For this reason we shall later search for and eliminate K-sets that are proper subsets of other K-set.

In Fig. 6.3b you can see how the K-set $\mathcal{K}_{14}$ is incorporated into the consensus $\mathcal{K}_{A.14}$, and $\mathcal{K}_C$ is incorporated into the consensus $\mathcal{K}_{AC}$. Not always is a consensus

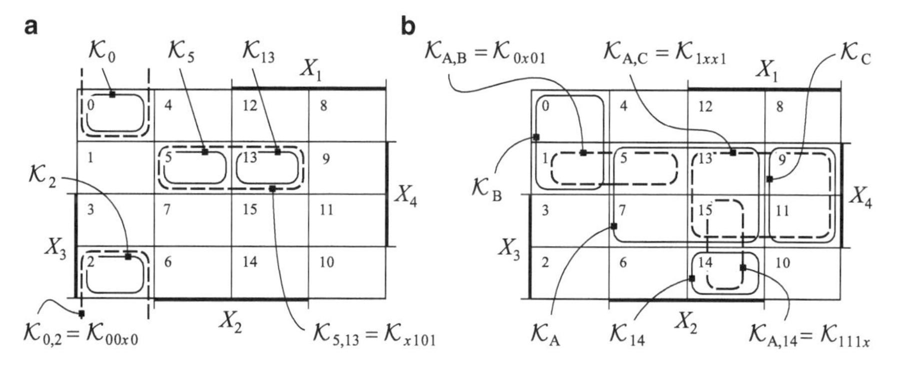

Fig. 6.3 Examples of consensuses

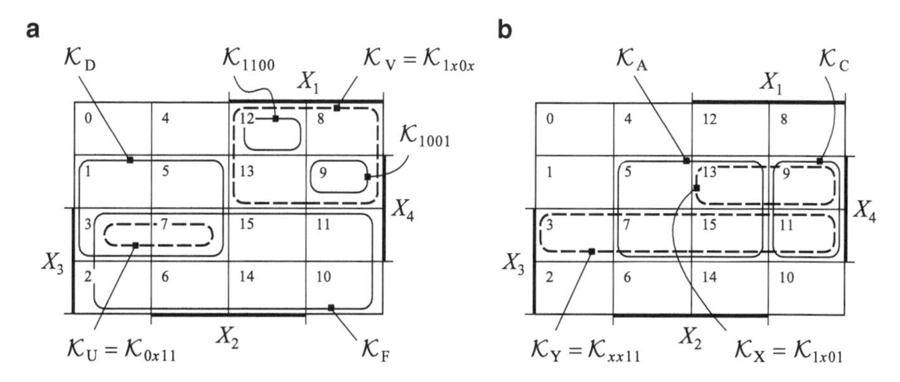

Fig. 6.4 K-sets that are not consensuses

successful in incorporating one or both of the K-sets of which it is constructed as you can see by viewing $\mathcal{K}_{AB}$. The one K-set missing in Fig. 6.3b, the K-set $\mathcal{K}_{BC} = \{1, 9\}$ of $(\mathcal{K}_B, \mathcal{K}_C)$, was only left away so as not to overload the picture.

In contrast to the consensus examples of Fig. 6.3, the examples of Fig. 6.4 present counter examples. $\mathcal{K}_U$ cannot be the consensus of $\mathcal{K}_D$ and $\mathcal{K}_F$, for these K-sets—not being disjoint—do not have a consensus. $\mathcal{K}_9$ and $\mathcal{K}_{12}$, on the other hand, *are* disjoint. But, not being adjacent, they also have no consensus, so that $\mathcal{K}_V$ cannot be their consensus. Another argument can also be put forward: $\mathcal{K}_V$ cannot be the consensus of $\mathcal{K}_9$ and $\mathcal{K}_{12}$ as $\mathcal{K}_V$ does not consist *solely* of elements of $\mathcal{K}_9$ and $\mathcal{K}_{12}$.

But, why is neither $\mathcal{K}_X$ nor $\mathcal{K}_Y$ of Fig. 6.4b a consensus? $\mathcal{K}_X$ is not the *largest* K-set whose elements are taken exclusively from $\mathcal{K}_A$ and $\mathcal{K}_C$—whereas $\mathcal{K}_{AC}$ (as shown in Fig. 6.3b) is. We can state this more simply by saying: As $\mathcal{K}_X$ is a subset of $\mathcal{K}_{AC}$, $\mathcal{K}_X$ cannot be the consensus. $\mathcal{K}_Y$, on the other hand, is too big: It contains an element that belongs neither to $\mathcal{K}_A$ nor to $\mathcal{K}_C$.

6.2 Formalising Adjacency

All the information on a generalised minterm $C_{\varphi_1 \ldots \varphi_n}$ and on a K-set $\mathcal{K}_{\varphi_1 \ldots \varphi_n}$ is contained in their index n-tuple, the multiple input event $(\varphi_1, \ldots, \varphi_n)$ in which each φ_i stands for one of the symbols 0,1,x_i. Therefore, it suffices to work with the index n-tuples $(\varphi_1, \ldots, \varphi_n)$ of the generalised minterms or the K-sets instead of with the generalised minterms or K-sets themselves. And this we shall subsequently do when formalising the concepts of adjacency and consensus.

The formal definition of the adjacency of elementary K-sets $\mathcal{K}_e$ and $\mathcal{K}_\varepsilon$ is quite simple: They are adjacent if their n-tuples $(e_1, \ldots, e_n)$ and $(\varepsilon_1, \ldots, \varepsilon_n)$ differ in exactly one position. But this definition does not point in the right direction when trying to generalise the concept of adjacency to fit all K-sets: We need to stress the *difference* in a certain position of the n-tuples, not the *equality* of all other positions. We therefore reformulate the above definition as follows:

Two **elementary K-sets** $\mathcal{K}_e$ and $\mathcal{K}_\varepsilon$ are said to be **adjacent** if there exists one and only one mutual position in their n-tuples $e := (e_1, \ldots, e_n)$ and $\varepsilon := (\varepsilon_1, \ldots, \varepsilon_n)$, with each $e_i, \varepsilon_i \in \{0, 1\}$, in which one n-tuple has the value 0, the other the value 1. (6.3)

We now extend this definition to apply to general K-sets.

Two **arbitrary K-sets** $\mathcal{K}_\varphi$ and $\mathcal{K}_\psi$ are said to be **adjacent** if there exists one and only one mutual position in their n-tuples $\varphi := (\varphi_1, \ldots, \varphi_n)$ and $\psi := (\psi_1, \ldots, \psi_n)$, with each $\varphi_i, \psi_i \in \{0, 1, x_i\}$, in which one n-tuple has the value 0, the other the value 1. (6.4)

The only difference between these definitions, (6.3) and (6.4), are the ranges $\{0, 1\}$ and $\{0, 1, x_i\}$ of the n-tuples' variables (while their names—e, ε, φ, or ψ—are irrelevant). When using the range $\{0, 1\}$, equally positioned elements in the n-tuples $(e_1, \ldots, e_n)$ and $(\varepsilon_1, \ldots, \varepsilon_n)$ must be equal (both 0, or both 1) if they are not complementary. On the other hand, when using the range $\{0, 1, x_i\}$, and considering those positions of the n-tuples $(\varphi_1, \ldots, \varphi_n)$ and $(\psi_1, \ldots, \psi_n)$ that are not complementary, they can contain equal symbols (both 0, both 1, or both x_i), or the symbol pairing 0 and x_i, or 1 and x_i. The definition (6.4) subsumes (6.3).

To express (6.4) more concisely, we introduce the **adjacency operator**, ⓐ, as defined in Fig. 6.5. Its arguments vary over the range $\{0, 1, x_i\}$, producing 0 or 1 as shown in the body of the table. As the table mirrors across its main diagonal, the operator is commutative. Applying the adjacency operator ⓐ to the n-tuples (or vectors) φ and ψ is explained as applying it to equally positioned elements of the n-tuples (operators with this property are called **scalar operators**). The result is referred to as the **adjacency vector** α:

$$\boxed{\alpha := \varphi ⓐ \psi := (\varphi_1 ⓐ \psi_1, \ldots, \varphi_n ⓐ \psi_n)} \tag{6.5}$$

	$\circledcirc$	ψ_i: 0	1	x_i
	0	0	1	0
φ_i	1	1	0	0
	x_i	0	0	0

Fig. 6.5 Defining the **adjacency operator** $\circledcirc$. The body of the table defines the value of each component α_i of the adjacency vector which is expressed as $\alpha_i := \varphi_i \circledcirc \psi_i$

a

$\mathcal{K}_0$:	0	0	0	0
$\mathcal{K}_2$:	0	0	1	0
α:	0	0	1	0

b

$\mathcal{K}_{14}$:	1	1	1	0
$\mathcal{K}_C$:	1	0	x_3	1
α:	0	1	0	1

c

$\mathcal{K}_D$:	0	x_2	x_3	1
$\mathcal{K}_F$:	x_1	1	1	x_4
α:	0	0	0	0

Fig. 6.6 Examples of (**a**) adjacent, (**b**) disjoint, and (**c**) non-disjoint K-sets

Presumably, no further explanation is necessary to see that adjacency, as defined in (6.4), can now be expressed with the help of the adjacency *vector* as follows:

Two **arbitrary K-sets** $\mathcal{K}_\varphi$ and $\mathcal{K}_\psi$ are **adjacent** if the adjacency vector $\alpha = \varphi \circledcirc \psi$ contains a single 1). (6.6)

It is of course reassuring to affirm that the formal description of adjacency, (6.6), coincides with its intuitive concept as put forth in the previous section. To this end consider the pairs of K-sets in Fig. 6.6.

Do note that the number of 1s in adjacency vector α in the table of Fig. 6.6 is 1, and that this successfully reflects the fact that the elementary K-sets $\mathcal{K}_0$ and $\mathcal{K}_2$, pictured in Fig. 6.1a, are adjacent. The adjacency vector in the table of Fig. 6.6b has two 1s so that, according to (6.6), we cannot expect the K-sets $\mathcal{K}_C$ and $\mathcal{K}_{14}$ to be adjacent although these K-sets are disjoint, as you can see in Fig. 6.1b. The last example in Fig. 6.6, the table in Fig. 6.6c, has an adjacency vector α containing no 1s. The associated K-sets, $\mathcal{K}_D$ and $\mathcal{K}_F$ are *not disjoint*, as a glance at Fig. 6.2a shows.

The examples of Fig. 6.6 are put into perspective by the following

Theorem on disjoint K-sets: Two K-sets $\mathcal{K}_{\varphi_1 \ldots \varphi_n}$ and $\mathcal{K}_{\psi_1 \ldots \psi_n}$ are disjoint *iff* the adjacency vector $(\alpha_1, \ldots, \alpha_n)$ of the index n-tuples $(\varphi_1, \ldots, \varphi_n)$ and $(\psi_1, \ldots, \psi_n)$ contains one or more 1s. (6.7)

As the proof of this theorem is quite simple, I leave it to you—but I do encourage you to take the necessary time out for the proof.

6.3 Formalising Consensus

Formalising the intuitive concept of consensus, as put forth in (6.2) and Fig. 6.3, is easier than one might think when first confronted with the problem. We start by expressing some adjacent K-sets and their consensuses of Fig. 6.3 by their defining n-tuples, see Fig. 6.7. To obtain a simpler notation, the indices of the variables x_i are omitted.

a

$\mathcal{K}_0$:	0	0	0	0
$\mathcal{K}_2$:	0	0	1	0
$\mathcal{K}_{0,2}$:	0	0	x	0

b

$\mathcal{K}_A$:	x	1	x	1
$\mathcal{K}_B$:	0	0	0	x
$\mathcal{K}_{A,B}$:	0	x	0	1

c

$\mathcal{K}_A$:	x	1	x	1
$\mathcal{K}_C$:	1	0	x	1
$\mathcal{K}_{A,C}$:	1	x	x	1

Fig. 6.7 Expressing adjacent K-sets and their consensus of Fig. 6.3 by their defining quadruples

Think of the quadruples of the first two rows of the tables of Fig. 6.7 as being instances of $(\varphi_1, \varphi_2, \varphi_3, \varphi_4)$ and $(\psi_1, \psi_2, \psi_3, \psi_4)$. Which of these general quadruples of adjacent K-sets represents the first row, and which the second, is irrelevant as any consensus operator to be developed will most certainly be commutative. Each quadruple of the bottom row represents the consensus, the general form of which we write as $(\kappa_1, \kappa_2, \kappa_3, \kappa_4)$.

Each of the tables of Fig. 6.7 contains at least some of the information needed to evaluate a component, κ_i, of a consensus vector from the equally positioned components, φ_i and ψ_i, of adjacent K-sets. For instance, consider the table of Fig. 6.7a: When φ_i and ψ_i are 0 (as in columns 1, 2, and 4), the associated component κ_i of the consensus is 0. On the way to defining a consensus operator, we draw a table such as that of Fig. 6.8a, the rows of which stand for $\varphi_i \in \{0, 1, x\}$, and the columns for $\psi_i \in \{0, 1, x\}$; the body of the table contains the associated values of κ_i. According to Fig. 6.7a, we thus enter 0 for κ_i into cell $(\varphi_i, \psi_i) = (0, 0)$ of Fig. 6.8a. In column 3 of Fig. 6.7a the symbols (0, 1) for (φ_i, ψ_i) of the adjacent K-sets are assigned the symbol x for component κ_i of the consensus. Accordingly, in row $\varphi_i = 0$ and column $\psi_i = 1$ of Fig. 6.8a we enter the symbol x. As φ_i and ψ_i

a

	κ	ψ_i: 0	1	x
φ_i	0	0	x	
	1	x		
	x			

b

	κ	ψ_i: 0	1	x_i
φ_i	0	0	x_i	0
	1	x_i	1	1
	x_i	0	1	x_i

Fig. 6.8 Defining the **consensus operator** κ. The body of the table (**b**) defines the value of each component κ_i of the adjacency vector which is expressed as $\kappa_i := \varphi_i \,\kappa\, \psi_i$. (**a**) κ_i of Fig. 6.7a (**b**) $\kappa_i := \varphi_i \,\kappa\, \psi_i$

$\mathcal{K}_D$:	0	x	x	0	$\mathcal{K}_9$:	1	0	0	1
$\mathcal{K}_F$:	x	x	1	x	$\mathcal{K}_{12}$:	1	1	0	0
$\mathcal{K}_U$:	0	x	1	1	$\mathcal{K}_V$:	1	x	0	x

Fig. 6.9 Applying the consensus operator to two non-adjacent K-sets produces a K-set that we do not accept as a consensus—see Fig. 6.4a

can be interchanged (the consensus operation is of necessity commutative), x must also be entered into cell $(1,0)$ of (φ_i, ψ_i). Considering the columns of all the tables of Fig. 6.7 allows us to complete the consensus table as shown in Fig. 6.8b, and it is this table that we use to define the **consensus operator** ⓚ.

Applying the consensus operator ⓚ to the n-tuples (or multiple events) φ and ψ of adjacent K-sets $\mathcal{K}_\varphi$ and $\mathcal{K}_\psi$ is explained as applying it to equally positioned elements of the n-tuples, the result being the **consensus vector** κ,

$$\boxed{\kappa := \varphi ⓚ \psi := (\varphi_1 ⓚ \psi_1, \ldots, \varphi_n ⓚ \psi_n)} \tag{6.8}$$

The consensus operator ⓚ allows us to define the concept of consensus formally:

The K-set $\mathcal{K}_\kappa$ is the **consensus** of the *adjacent* K-sets $\mathcal{K}_\varphi$ and $\mathcal{K}_\psi$ *iff* $\kappa = \varphi ⓚ \psi$ and ⓚ is defined by the table of Fig. 6.8b. (6.9)

Note that a consensus is only calculated from *adjacent* K-sets. If you apply the consensus operator ⓚ to two K-sets that are *not* adjacent, the K-set you get will *not* be a consensus. For example, see Fig. 6.9.

6.4 When Is One K-Set a Subset of Another?

In the process of finding all prime sets, we wish to eliminate those K-sets incorporated into (i.e., which are subsets of) a consensus. It is thus necessary to have an algorithm, a formal procedure, by which to decide if a given K-set is a subset of another. To formulate such a procedure, we introduce the **subset operator** ⓢ, as defined in Fig. 6.10.

	ⓢ	ψ_i: 0	1	x_i
	0	1	0	1
φ_i	1	0	1	1
	x_i	0	0	1

Fig. 6.10 Defining the **subset operator** ⓢ. The body of the table defines the value of each component σ_i of the adjacency vector which is expressed as $\sigma_i := \varphi_i ⓢ \psi_i$

$\mathcal{K}_5$:	0	1	0	1
$\mathcal{K}_{5,13}$:	x_1	1	0	1
$\mathcal{K}_5 \subseteq \mathcal{K}_{5,13}$:	1	1	1	1

$\mathcal{K}_C$:	1	0	x_3	1
$\mathcal{K}_{A,C}$:	1	x_2	x_3	1
$\mathcal{K}_C \subseteq \mathcal{K}_{A,C}$:	1	1	1	1

$\mathcal{K}_{A,B}$:	0	x_2	0	1
$\mathcal{K}_A$:	x_1	1	x_3	1
$\mathcal{K}_{A,B} \nsubseteq \mathcal{K}_A$:	1	0	1	1

$\mathcal{K}_B$:	0	0	0	x_4
$\mathcal{K}_A$:	x_1	1	x_3	1
$\mathcal{K}_B \nsubseteq \mathcal{K}_A$:	1	0	1	0

Fig. 6.11 Illustrating (6.11)

As with the other operators—the adjacency operator ⓐ and consensus operator ⓚ—its arguments vary over the range $\{0, 1, x_i\}$. But, ⓢ is not commutative. Applying the subset operator ⓢ to the n-tuples (or vectors) φ and ψ is explained as applying it to equally positioned elements of the n-tuples (so it is a **scalar operator**). The result is referred to as the **subset vector** σ:

$$\boxed{\sigma := \varphi \circledS \psi := (\varphi_1 \circledS \psi_1, \ldots, \varphi_n \circledS \psi_n)} \tag{6.10}$$

Employing the subset operator, the claim is that

> A K-set $\mathcal{K}_\varphi$ is a **subset** of a K-set $\mathcal{K}_\psi$—symbolically, $\mathcal{K}_\varphi \subseteq \mathcal{K}_\psi$—*iff* the subset vector $\sigma = \varphi \circledS \psi$ consists exclusively of 1s. (6.11)

We shan't prove (6.11) in this text because the predicate logic required has not been covered. But do take a look at the illustrating examples of Fig. 6.11—$\mathcal{K}_5 \subseteq \mathcal{K}_{3,13}$ and $\mathcal{K}_C \subseteq \mathcal{K}_{A,C}$ on the one hand, and $\mathcal{K}_{A,B} \nsubseteq \mathcal{K}_A$ and $\mathcal{K}_B \nsubseteq \mathcal{K}_A$ on the other.

Chapter 7
Algebraic Minimisation

Minimising relatively small switching functions (of no more than, say, six input variables) can be done quite efficiently by K-map. Larger problems should be solved by computer. To be able to write computer programs for minimisation, we need to (a) develop an algorithm by which to find a *full cover*, and (b) a further algorithm by which to select a *minimal cover* from the full cover.

There are a large number of algorithms for finding a full cover. The one put forth in Section 7.1 was developed by Vingron (2004). It is based on the concept of consensus, so that we need not start with a canonical normal form. Section 7.2 presents Vingron's (2004) procedure for finding a *minimal* cover. Both algorithms—the one for finding the full cover, and the one for finding the minimal cover—are well suited for developing computer programs; but such programs are not contained in this book as it is not intended to give preference to any computer language.

7.1 Finding the Full Cover*

The first step to take in minimising the logic expression of any given set of setting input events $\mathcal{E}_1$ (or $\mathcal{E}_0$, if you so wish to work with the resetting input events) is to find the *full cover* of prime sets of $\mathcal{E}_1$ which we denote as $\mathcal{P}_{max}(\mathcal{E}_1)$. You can specify $\mathcal{E}_1$ in any way you find appropriate. For the example of this section we want to minimise the function depicted in Fig. 7.1 in which we have pre-collected the setting input events of $\mathcal{E}_1$ into the K-sets $\mathcal{K}_{011010}$, $\mathcal{K}_{0x1000}$, $\mathcal{K}_{1x1100}$, $\mathcal{K}_{111x10}$, $\mathcal{K}_{11x10x}$, $\mathcal{K}_{x1x00x}$, $\mathcal{K}_{1xx0xx}$.

Now, in formulating the algorithm, we associate it closely to our example and its further development in the following figures.

S.P. Vingron, *Logic Circuit Design*, DOI 10.1007/978-3-642-27657-6_7,

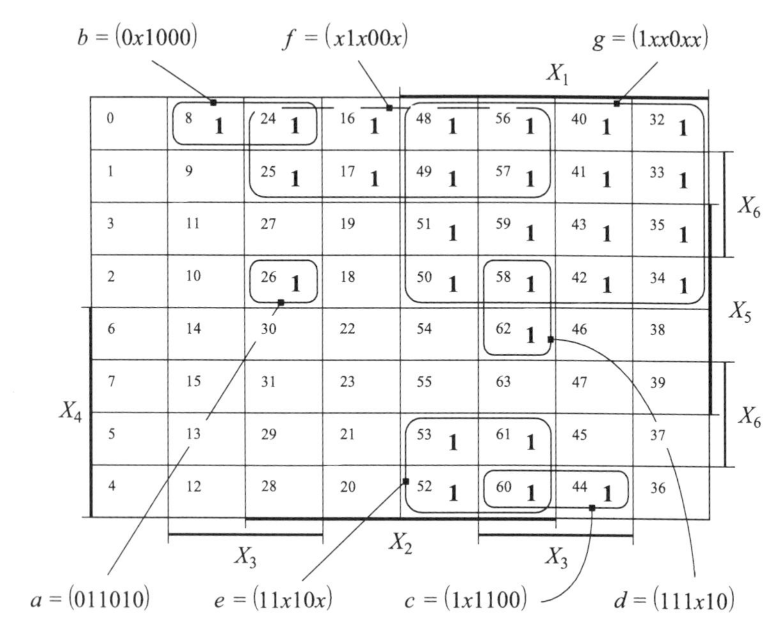

Fig. 7.1 Visualising a function to be minimised

Iterative consensus algorithm for finding the full cover $\mathcal{P}_{\max}(\mathcal{E}_1)$ of the setting input events $\mathcal{E}_1$.

(a) Start by listing the multiple input events $(\varphi_1, \cdots, \varphi_n)$ of K-sets that cover $\mathcal{E}_1$ (e.g., list 1 of Fig. 7.2). The duplicates of multiple input events in this list are dropped, as are all the multiple input events which, according to (6.11), specify K-sets that are subsets of other K-sets of the list. Call the resulting list a *Basic List* (again, list 1 of Fig. 7.2).

(b) Each of the multiple input events of the *Basic List* is checked against the others for adjacency according to (6.6). For each pair of multiple input events that are adjacent you establish their consensus according to (6.9) appending it to the *Basic List* (see list 2 of Fig. 7.2).

(c) The duplicates of multiple input events in the last list are dropped, as are all the multiple input events which, according to (6.11), specify K-sets that are subsets of other K-sets of the list. This provides us with a new *Basic List* (e.g., list 3 of Fig. 7.2).

(d) Repeat steps (b) and (c) until two successive *Basic Lists* are identical, in which case the algorithm terminates.

(7.1)

	List 1		List 2		List 3
a	011010	a	011010	d	111x10
b	0x1000	b	0x1000	f	x1x00x
c	1x1100	c	1x1100	g	1xx0xx
d	111x10	d	111x10	a ⓚ b	0110x0
e	11x10x	e	11x10x	a ⓚ g	x11010
f	x1x00x	f	x1x00x	b ⓚ g	xx1000
g	1xx0xx	g	1xx0xx	c ⓚ d	1111x0
		a ⓚ b	0110x0	c ⓚ g	1x1x00
		a ⓚ d	x11010	e ⓚ g	11xx0x
		a ⓚ f	0110x0		
		a ⓚ g	x11010		
		b ⓚ g	xx1000		
		c ⓚ d	1111x0		
		c ⓚ f	111x00		
		c ⓚ g	1x1x00		
		d ⓚ e	1111x0		
		d ⓚ f	1110x0		
		e ⓚ f	11xx0x		
		e ⓚ g	11xx0x		

Fig. 7.2 Illustrating step (**a**) and the iterative steps (**b**) and (**c**) of (7.1)

List 1	List 2	List 3	List 4	$C_{\varphi_1\ldots\varphi_6}$
011010	111x10	x1x00x	x1x00x	$X_2\overline{X}_4\overline{X}_5$
0x1000	x1x00x	1xx0xx	1xx0xx	$X_1\overline{X}_4$
1x1100	1xx0xx	xx1000	xx1000	$X_3\overline{X}_4\overline{X}_5\overline{X}_6$
111x10	0110x0	1x1x00	1x1x00	$X_1X_3\overline{X}_5\overline{X}_6$
11x10x	x11010	11xx0x	11xx0x	$X_1X_2\overline{X}_5$
x1x00x	xx1000	111xx0	111xx0	$X_1X_2X_3\overline{X}_6$
1xx0xx	1111x0	x110x0	x110x0	$X_2X_3\overline{X}_4\overline{X}_6$
	1x1x00			
	11xx0x			

Fig. 7.3 Results of the iteration

List 1 of the table of Fig. 7.3 is the original basic list of our example, obtained by step (a), while lists 2–4 are the successive basic lists each obtained by the iterative steps (b) and (c). This example represents one of those relatively rare cases where the result is a single irredundant cover, meaning that $\mathcal{P}_{\max}(\mathcal{E}_1) = \mathcal{P}_{\min}(\mathcal{E}_1)$. I encourage you to visualise this result in a K-map as that of Fig. 7.1. However, in general, the resulting list contains an abundance of prime sets allowing us to choose from them a number of different covers. Of these the *minimal covers* are of prime interest with respect to minimisation.

7.2 Finding Minimal Covers*

The procedure presented in this section allows us to directly find a minimal cover $\mathcal{P}_{\text{min}}(\mathcal{E}_1) \subseteq \mathcal{P}_{\text{max}}(\mathcal{E}_1)$ among all the prime sets. In fact, the first cover found is a minimal cover, allowing us to terminate the search if we wish. This method assumes we have determined the full cover $\mathcal{P}_{\text{max}}(\mathcal{E}_1)$, for instance, by the *iterative consensus algorithm* (7.1) of the previous section.

Let us use a simple example of a switching function of four input variables for which $\mathcal{E}_1 = \{2, 3, 5, 6, 7, 9, 11, 13\}$, and further assume we have determined all its prime implicants—in this simple case, possibly with the help of a K-map (Fig. 7.4):

$$\mathcal{P}_A = \{5, 7\} \qquad \mathcal{P}_B = \{9, 11\} \qquad \mathcal{P}_C = \{3, 11\}$$

$$\mathcal{P}_D = \{5, 13\} \qquad \mathcal{P}_E = \{9, 13\} \qquad \mathcal{P}_F = \{2, 3, 6, 7\}$$

We start the procedure by writing the **table of covered input events**, as shown in Fig. 7.5a (the equivalent to McCluskey's (1956) *prime implicant table*). Each row represents a prime set $\mathcal{P}_\varphi$, while each column stands for an input event e of the setting input events $\mathcal{E}_1$. The intersection of a column e and a row $\mathcal{P}_\varphi$ contains the truth value (**0** or **1**) of the proposition $e \in \mathcal{P}_\varphi$.

A table of covered input events can be simplified if it contains one or more *essential prime sets*. An **essential prime set** is a prime set that contains at least one input event e—call it a **unique input event**—not contained in any other prime set. The column of a unique input event contains only a single **1**, the row in which it occurs marking the essential prime set. An essential prime set is an element of every irredundant cover. Having found the essential prime sets, we need not search for them any further, and may thus remove them from the table of prime sets. Removing all essential prime sets from the table allows us to also remove the columns of all the input events that are elements of the essential prime sets. Our table of covered input

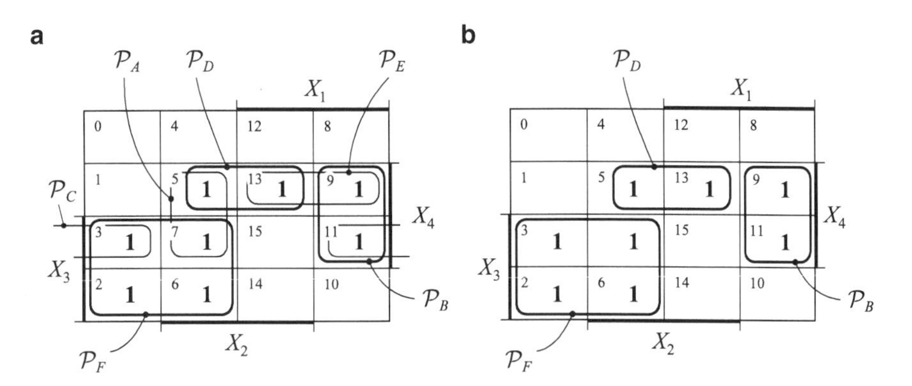

Fig. 7.4 The maximum and minimum covers of a given function

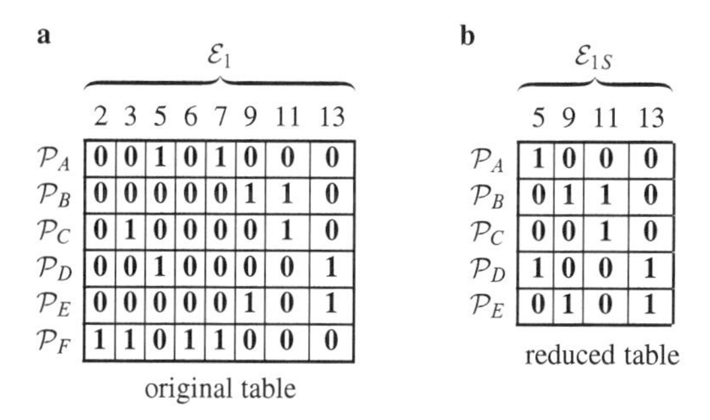

a original table, $\mathcal{E}_1$

	2	3	5	6	7	9	11	13
$\mathcal{P}_A$	0	0	1	0	1	0	0	0
$\mathcal{P}_B$	0	0	0	0	0	1	1	0
$\mathcal{P}_C$	0	1	0	0	0	0	1	0
$\mathcal{P}_D$	0	0	1	0	0	0	0	1
$\mathcal{P}_E$	0	0	0	0	0	1	0	1
$\mathcal{P}_F$	1	1	0	1	1	0	0	0

b reduced table, $\mathcal{E}_{1S}$

	5	9	11	13
$\mathcal{P}_A$	1	0	0	0
$\mathcal{P}_B$	0	1	1	0
$\mathcal{P}_C$	0	0	1	0
$\mathcal{P}_D$	1	0	0	1
$\mathcal{P}_E$	0	1	0	1

Fig. 7.5 Table of covered input events

events contains $\mathcal{P}_F$ as the only essential prime set. Removing it and the columns of all input events it contains leads to Fig. 7.5b.

The last step is to develop the **table of possible covers**—see **Fig. 7.6**—for the prime sets of the reduced table of covered input events. The basic idea is to test for each **union** ($\cup$) **of combinations of prime sets**, which such union contains all setting input events e of $\mathcal{E}_{1S}$ of Fig. 7.5b. Figure 7.6 details how such a test can be carried out. The combinations of prime sets are organised into groups as shown in the column 'Possible covers of $\mathcal{E}_{1S}$' . Group 1 contains per row a single prime set of the five prime sets $\mathcal{P}_A, \mathcal{P}_B, \mathcal{P}_C, \mathcal{P}_D, \mathcal{P}_E$—there are thus $\binom{5}{1} = 5$ rows in group 1 actually making it a repetition of Fig. 7.5b. Group 2 contains, row-wise, all combinations (without repetition) of two of the prime sets from $\mathcal{P}_A, \mathcal{P}_B, \mathcal{P}_C, \mathcal{P}_D, \mathcal{P}_E$ so that there are $\binom{5}{2} = 10$ rows in group 2. Similarly, group 3 consists of the $\binom{5}{3} = 10$ rows of the combination (without repetition) of three of the prime sets of $\mathcal{P}_A, \mathcal{P}_B, \mathcal{P}_C, \mathcal{P}_D, \mathcal{P}_E$, and so on for all remaining groups.

Now concentrate on the columns marked by the input events e—5, 9, 11, 13—of $\mathcal{E}_{1S}$. The **1**s state which input events e are elements of the union of combinations of prime sets. For group 1 of Fig. 7.6 the **1**s for which $e \in \mathcal{E}_{1S}$ are copied from Fig. 7.5b. The logic vectors in the rows of the $\mathcal{E}_{1S}$-column are obtained by *scalar disjunction* of those vectors indicated in the combinations. **Scalar disjunction** of logic vectors refers to the disjunction of equally indexed elements. E.g., the logic vector of row 6, $(\mathbf{1}, \mathbf{1}, \mathbf{1}, \mathbf{0})$, is the scalar disjunction of the vectors of rows 1 (for $\mathcal{P}_A$) and 2 (for $\mathcal{P}_B$)—$(\mathbf{1}, \mathbf{0}, \mathbf{0}, \mathbf{0}) \vee (\mathbf{0}, \mathbf{1}, \mathbf{1}, \mathbf{0}) \Leftrightarrow (\mathbf{1}, \mathbf{1}, \mathbf{1}, \mathbf{0})$. Results, once calculated, may be used again. For instance, interpreting $\mathcal{P}_A \cup \mathcal{P}_B \cup \mathcal{P}_E$ of row 18 as $(\mathcal{P}_A \cup \mathcal{P}_B) \cup \mathcal{P}_E$ allows us to calculate its logic vector from rows 6 (for $\mathcal{P}_A \cup \mathcal{P}_B$) and 5 (for $\mathcal{P}_E$)—$(\mathbf{1}, \mathbf{1}, \mathbf{1}, \mathbf{0}) \vee (\mathbf{0}, \mathbf{1}, \mathbf{0}, \mathbf{1}) \Leftrightarrow (\mathbf{1}, \mathbf{1}, \mathbf{1}, \mathbf{1})$.

The table is developed row by row, until the first logic vector containing only **1**s is encountered. The prime sets of the union are a partial[1] minimal cover $\mathcal{P}_{\min}(\mathcal{E}_S)$—see row 11. If there are more minimal covers than one, they are all to be found in the same group. So, if you want to find all minimal covers, you let the algorithm continue till the end of the group. The complete result is obtained by augmenting

[1]The essential covers having not yet been considered.

Group	Row	Possible covers of $\mathcal{E}_{1S}$	5	9	11	13	Comments
			$\mathcal{E}_{1S}$				
1	1	$\mathcal{P}_A$	**1**	**0**	**0**	**0**	
	2	$\mathcal{P}_B$	**0**	**1**	**1**	**0**	
	3	$\mathcal{P}_C$	**0**	**0**	**1**	**0**	
	4	$\mathcal{P}_D$	**1**	**0**	**0**	**1**	
	5	$\mathcal{P}_E$	**0**	**1**	**0**	**1**	
2	6	$\mathcal{P}_A \cup \mathcal{P}_B$	**1**	**1**	**1**	**0**	
	7	$\mathcal{P}_A \cup \mathcal{P}_C$	**1**	**0**	**1**	**0**	
	8	$\mathcal{P}_A \cup \mathcal{P}_D$	**1**	**0**	**0**	**1**	
	9	$\mathcal{P}_A \cup \mathcal{P}_E$	**1**	**1**	**0**	**1**	
	10	$\mathcal{P}_B \cup \mathcal{P}_C$	**0**	**1**	**1**	**0**	
	11	$\mathcal{P}_B \cup \mathcal{P}_D$	**1**	**1**	**1**	**1**	$\mathcal{P}_{\min}(\mathcal{E}_{1S})$
	12	$\mathcal{P}_B \cup \mathcal{P}_E$	**0**	**1**	**1**	**1**	
	13	$\mathcal{P}_C \cup \mathcal{P}_D$	**1**	**0**	**1**	**1**	
	14	$\mathcal{P}_C \cup \mathcal{P}_E$	**0**	**1**	**1**	**1**	
	15	$\mathcal{P}_D \cup \mathcal{P}_E$	**1**	**1**	**0**	**1**	
3	16	$\mathcal{P}_A \cup \mathcal{P}_B \cup \mathcal{P}_C$	**1**	**1**	**1**	**0**	
	(17)	$\mathcal{P}_A \cup \mathcal{P}_B \cup \mathcal{P}_D$	**1**	**1**	**1**	**1**	$11 \subset 17$
	18	$\mathcal{P}_A \cup \mathcal{P}_B \cup \mathcal{P}_E$	**1**	**1**	**1**	**1**	$\mathcal{P}_{\text{irr}}^{(1)}(\mathcal{E}_{1S})$
	19	$\mathcal{P}_A \cup \mathcal{P}_C \cup \mathcal{P}_D$	**1**	**0**	**1**	**1**	
	20	$\mathcal{P}_A \cup \mathcal{P}_C \cup \mathcal{P}_E$	**1**	**1**	**1**	**1**	$\mathcal{P}_{\text{irr}}^{(2)}(\mathcal{E}_{1S})$
	21	$\mathcal{P}_A \cup \mathcal{P}_D \cup \mathcal{P}_E$	**1**	**1**	**0**	**1**	
	(22)	$\mathcal{P}_B \cup \mathcal{P}_C \cup \mathcal{P}_D$	**1**	**1**	**1**	**1**	$11 \subset 22$
	23	$\mathcal{P}_B \cup \mathcal{P}_C \cup \mathcal{P}_E$	**0**	**1**	**1**	**1**	
	(24)	$\mathcal{P}_B \cup \mathcal{P}_D \cup \mathcal{P}_E$	**1**	**1**	**1**	**1**	$11 \subset 24$
	25	$\mathcal{P}_C \cup \mathcal{P}_D \cup \mathcal{P}_E$	**1**	**1**	**1**	**1**	$\mathcal{P}_{\text{irr}}^{(3)}(\mathcal{E}_{1S})$
4	(26)	$\mathcal{P}_A \cup \mathcal{P}_B \cup \mathcal{P}_C \cup \mathcal{P}_D$	**1**	**1**	**1**	**1**	$11 \subset 26$
	(27)	$\mathcal{P}_A \cup \mathcal{P}_B \cup \mathcal{P}_C \cup \mathcal{P}_E$	**1**	**1**	**1**	**1**	$18, 20 \subset 27$
	(28)	$\mathcal{P}_A \cup \mathcal{P}_B \cup \mathcal{P}_D \cup \mathcal{P}_E$	**1**	**1**	**1**	**1**	$11 \subset 28$
	(29)	$\mathcal{P}_A \cup \mathcal{P}_C \cup \mathcal{P}_D \cup \mathcal{P}_E$	**1**	**1**	**1**	**1**	$20, 25 \subset 29$
	(30)	$\mathcal{P}_B \cup \mathcal{P}_C \cup \mathcal{P}_D \cup \mathcal{P}_E$	**1**	**1**	**1**	**1**	$11 \subset 30$
5	(31)	$\mathcal{P}_A \cup \mathcal{P}_B \cup \mathcal{P}_C \cup \mathcal{P}_D \cup \mathcal{P}_E$	**1**	**1**	**1**	**1**	$\mathcal{P}_{\max}(\mathcal{E}_{1S})$

Fig. 7.6 Table of all possible covers

the covers found here by the previously omitted essential covers—in our example by $\mathcal{P}_F$—giving us $\mathcal{P}_{\min}(\mathcal{E}_1) = \{\mathcal{P}_B, \mathcal{P}_D, \mathcal{P}_F\}$, the result visualised in Fig. 7.4b.

If you feel a desire to find *all irredundant covers*, continue with the algorithm as shown in Fig. 7.6, discarding those solutions whose subsets have already proven to be irredundant covers. Thus, row 17 is not accepted as a solution as $\{\mathcal{P}_B, \mathcal{P}_D\}$ of row 11 is a subset of $\{\mathcal{P}_A, \mathcal{P}_B, \mathcal{P}_D\}$ of row 17—$\{\mathcal{P}_B, \mathcal{P}_D\} \subset \{\mathcal{P}_A, \mathcal{P}_B, \mathcal{P}_D\}$.

For the example $\mathcal{E}_1 = \{2, 3, 5, 6, 7, 9, 11, 13\}$ of this section, we found the minimal cover to be $\mathcal{P}_{\min}(\mathcal{E}_1) = \{\mathcal{P}_B, \mathcal{P}_D, \mathcal{P}_F\}$ which is more precisely expressed as $\mathcal{P}_{\min}(\mathcal{E}_1) = \{\mathcal{P}_{10x1}, \mathcal{P}_{x101}, \mathcal{P}_{0x1x}\}$. Then, by the evaluation formulas of (5.19) we may write the disjunctive form as

$$Y(x) \Leftrightarrow \bigvee_{\varphi:\mathcal{P}_\varphi \in \mathcal{P}_{\min}(\mathcal{E}_1)} C_\varphi(x) \Leftrightarrow C_{10x1} \vee C_{x101} \vee C_{0x1x}$$
$$\Leftrightarrow X_1\overline{X}_2X_4 \vee X_2\overline{X}_3X_4 \vee \overline{X}_1X_3. \tag{7.2}$$

For the same example, knowing $\mathcal{E}_1$, we can write its complement $\mathcal{E}_0$ as $\mathcal{E}_0 = \{0, 1, 4, 8, 10, 12, 14, 15\}$. From this, we can calculate the minimal cover $\mathcal{P}_{\min}(\mathcal{E}_0) = \{\mathcal{P}_{000x}, \mathcal{P}_{111x}, \mathcal{P}_{xx00}\}$, which I encourage you to do. Then, again (5.19) enables us to write the conjunctive form of the given function:

$$Y(x) \Leftrightarrow \bigwedge_{\varphi:\mathcal{P}_\varphi \in \mathcal{P}_{\min}(\mathcal{E}_0)} D_\varphi(x) \Leftrightarrow D_{000x} D_{111x} D_{xxoo}$$
$$\Leftrightarrow (X_1 \vee X_2 \vee X_3)(\overline{X}_1 \vee \overline{X}_2 \vee \overline{X}_3)(X_3 \vee X_4). \tag{7.3}$$

7.3 Minimisation Considering Don't Cares*

Some switching problems don't allow us to define output values for all input events. We call such problems **incompletely specified**, referring to the unspecified output values as '**don't cares**' and denoting them as δ_x, the index x marking the input event whose output value is unspecified. The values—0 or 1—of each δ_x can be chosen independently of one another. But do note that to calculate a circuit for an incompletely specified problem each δ_x must be given a specific value. Most frequently the index x is omitted.

To see how to cope with incompletely specified problems let us discuss Fig. 7.7 for which we want to **develop a minimal cover**, $\mathcal{P}_{\min}(\mathcal{E}_1)$, of the setting input events, $\mathcal{E}_1$. As before, $\mathcal{E}_1$ is the set of all input events x whose cells in a K-map

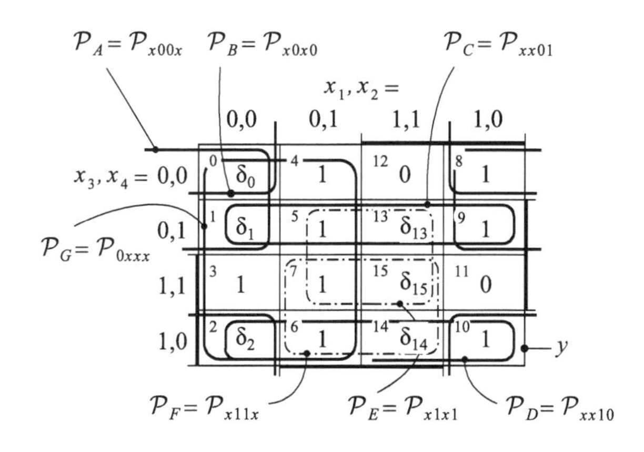

Fig. 7.7 An incompletely specified function we wish to minimise. The unspecified output values are marked as δ for 'don't care'

$\mathcal{E}_{1\delta}$	List 1	List 2	List 3	List 4	$\mathcal{P}_{max}(\mathcal{E}_{1\delta})$
0	0000	000x	00xx	x00x	$\mathcal{P}_{x00x} = \mathcal{P}_A$
1	0001	00x0	0x0x	x0x0	$\mathcal{P}_{x0x0} = \mathcal{P}_B$
3	0011	0x00	x00x	xx01	$\mathcal{P}_{xx01} = \mathcal{P}_C$
2	0010	x000	0xx0	xx10	$\mathcal{P}_{xx10} = \mathcal{P}_D$
4	0100	00x1	x0x0	x1x1	$\mathcal{P}_{x1x1} = \mathcal{P}_E$
5	0101	0x01	0xx1	x11x	$\mathcal{P}_{x11x} = \mathcal{P}_F$
7	0111	x001	xx01	0xxx	$\mathcal{P}_{0xxx} = \mathcal{P}_G$
6	0110	001x	0x1x		
13	1101	0x11	xx10		
15	1111	0x10	01xx		
14	1110	x010	x1x1		
8	1000	010x	x11x		
9	1001	01x0			
10	1010	01x1			
		x101			
		011x			
		x111			
		x110			
		11x1			
		1x01			
		111x			
		1x10			
		100x			
		10x0			

Fig. 7.8 Table of the iterative consensus method for developing the full cover $\mathcal{P}_{max}(\mathcal{E}_{1\delta}) = \{\mathcal{P}_A, \mathcal{P}_B, \mathcal{P}_C, \mathcal{P}_D, \mathcal{P}_E, \mathcal{P}_F, \mathcal{P}_G\}$ of $\mathcal{E}_{1\delta}$

contain an output value of 1—in our example $\mathcal{E}_1 = \{3, 4, 5, 6, 7, 8, 9, 10\}$. Take $\mathcal{E}_\delta$ to be the set of the don't care input events, in our case $\mathcal{E}_\delta = \{0, 1, 2, 13, 14, 15\}$. We solve the problem in two steps. **First**, we assign all don't care outputs—δ_0, δ_1, δ_2, δ_{13}, δ_{14}, δ_{15}—the value of 1, effectively working with an new set, call it $\mathcal{E}_{1\delta}$, of setting input events $\mathcal{E}_{1\delta} = \{3, 4, 5, 6, 7, 8, 9, 10; 0, 1, 2, 13, 14, 15\}$. For this set we develop the maximum cover of prime sets, $\mathcal{P}_{\max}(\mathcal{E}_{1\delta})$, preferably, by using the iterative consensus algorithm (7.1) as shown in Fig. 7.8.

In the **second step** we search for and drop those prime sets of $\mathcal{P}_{\max}(\mathcal{E}_{1\delta})$ that are redundant in the sense that dropping them leaves all *setting* input events $x \in \mathcal{E}_1$ still covered by at least one (remaining) prime set of $\mathcal{P}_{\max}(\mathcal{E}_{1\delta})$. The search starts by developing the table of covered input events shown in Fig. 7.9.

In this table we look for *essential* prime sets, dropping them and all columns containing their input events. The only essential prime set is $\mathcal{P}_G$, its input events 3 and 4 not being elements of any of the other prime sets. By dropping the row for $\mathcal{P}_G$, and the columns for the input events 3, 4, 5, 6, 7 we obtain the reduced table of covered input events of Fig. 7.10a. There are two conspicuous rows in Fig. 7.10a—those for $\mathcal{P}_E$ and $\mathcal{P}_F$, both containing only **0**s. These rows represent *redundant* prime sets. Eliminating them leads to the reduced table of Fig. 7.10b.

	$\mathcal{E}_1$								
	3	4	5	6	7	8	9	10	
$\mathcal{P}_{x00x}$	**0**	**0**	**0**	**0**	**0**	**1**	**1**	**0**	$\mathcal{P}_A$
$\mathcal{P}_{x0x0}$	**0**	**0**	**0**	**0**	**0**	**1**	**0**	**1**	$\mathcal{P}_B$
$\mathcal{P}_{xx01}$	**0**	**0**	**1**	**0**	**0**	**0**	**1**	**0**	$\mathcal{P}_C$
$\mathcal{P}_{xx10}$	**0**	**0**	**0**	**1**	**0**	**0**	**0**	**1**	$\mathcal{P}_D$
$\mathcal{P}_{x1x1}$	**0**	**0**	**1**	**0**	**1**	**0**	**0**	**0**	$\mathcal{P}_E$
$\mathcal{P}_{x11x}$	**0**	**0**	**0**	**1**	**1**	**0**	**0**	**0**	$\mathcal{P}_F$
$\mathcal{P}_{0xxx}$	**1**	**1**	**1**	**1**	**1**	**0**	**0**	**0**	$\mathcal{P}_G$

Fig. 7.9 Table of covered input events showing which setting input events $x \in \mathcal{P}_1$ are covered by prime sets of $\mathcal{P}_{\max}(\mathcal{E}_{1\delta})$

(a)

	$\mathcal{E}_{1S}$			
	8	9	10	
$\mathcal{P}_A$	**1**	**1**	**0**	
$\mathcal{P}_B$	**1**	**0**	**1**	
$\mathcal{P}_C$	**0**	**1**	**0**	
$\mathcal{P}_D$	**0**	**0**	**1**	
$\mathcal{P}_E$	**0**	**0**	**0**	*redundant*
$\mathcal{P}_F$	**0**	**0**	**0**	*redundant*

(b)

	$\mathcal{E}_{1S}$		
	8	9	10
$\mathcal{P}_A$	**1**	**1**	**0**
$\mathcal{P}_B$	**1**	**0**	**1**
$\mathcal{P}_C$	**0**	**1**	**0**
$\mathcal{P}_D$	**0**	**0**	**1**

Fig. 7.10 Reduced table of covered input events

Group	Row	Possible covers of $\mathcal{E}_{1S}$	5	9	11	Comments
1	1	$\mathcal{P}_A$	**1**	**1**	**0**	
	2	$\mathcal{P}_B$	**1**	**0**	**1**	
	3	$\mathcal{P}_C$	**0**	**1**	**0**	
	4	$\mathcal{P}_D$	**0**	**0**	**1**	
2	**5**	$\mathcal{P}_A \cup \mathcal{P}_B$	**1**	**1**	**1**	$\mathcal{P}_{\min}^{(a)}(\mathcal{E}_{1S})$
	6	$\mathcal{P}_A \cup \mathcal{P}_C$	**1**	**1**	**0**	
	7	$\mathcal{P}_A \cup \mathcal{P}_D$	**1**	**1**	**1**	$\mathcal{P}_{\min}^{(b)}(\mathcal{E}_{1S})$
	8	$\mathcal{P}_B \cup \mathcal{P}_C$	**1**	**1**	**1**	$\mathcal{P}_{\min}^{(c)}(\mathcal{E}_{1S})$
	9	$\mathcal{P}_B \cup \mathcal{P}_D$	**1**	**0**	**1**	
	10	$\mathcal{P}_C \cup \mathcal{P}_D$	**0**	**1**	**1**	

(The columns 5, 9, 11 are grouped under $\mathcal{E}_{1S}$.)

Fig. 7.11 Table of all possible covers

The table of all possible covers, Fig. 7.11, is developed as explained in the previous section. As we are only interested in minimal covers, the first already occurring in group 2, we need develop the table for no further group. After augmenting the thus found results by the essential prime set $\mathcal{P}_G$ we, as expected, obtain the same complete result as from the K-map of Fig. 7.7.

$$\mathcal{P}_{\min}^{(a)}(\mathcal{E}_1) = \{\mathcal{P}_A, \mathcal{P}_B, \mathcal{P}_G\} = \{\mathcal{P}_{x00x}, \mathcal{P}_{x0x0}, \mathcal{P}_{0xxx}\},$$

$$\mathcal{P}_{\min}^{(b)}(\mathcal{E}_1) = \{\mathcal{P}_A, \mathcal{P}_D, \mathcal{P}_G\} = \{\mathcal{P}_{x00x}, \mathcal{P}_{xx10}, \mathcal{P}_{0xxx}\},$$

$$\mathcal{P}_{\min}^{(c)}(\mathcal{E}_1) = \{\mathcal{P}_B, \mathcal{P}_C, \mathcal{P}_G\} = \{\mathcal{P}_{x0x0}, \mathcal{P}_{xx01}, \mathcal{P}_{0xxx}\}.$$

For these minimal covers you can then use the AND-to-OR evaluation formula (5.19) to obtain the results:

$$Y^{(a)} \Leftrightarrow \Bigg(\bigvee_{\varphi:\mathcal{P}_\varphi \in \mathcal{P}_{\min}^{(a)}(\mathcal{E}_1)} C_\varphi \Bigg) \Leftrightarrow \overline{X}_2\,\overline{X}_3 \vee \overline{X}_2\,\overline{X}_4 \vee \overline{X}_1,$$

$$Y^{(b)} \Leftrightarrow \Bigg(\bigvee_{\varphi:\mathcal{P}_\varphi \in \mathcal{P}_{\min}^{(b)}(\mathcal{E}_1)} C_\varphi \Bigg) \Leftrightarrow \overline{X}_2\,\overline{X}_3 \vee X_3\,\overline{X}_4 \vee \overline{X}_1,$$

$$Y^{(c)} \Leftrightarrow \Bigg(\bigvee_{\varphi:\mathcal{P}_\varphi \in \mathcal{P}_{\min}^{(c)}(\mathcal{E}_1)} C_\varphi \Bigg) \Leftrightarrow \overline{X}_2\,\overline{X}_4 \vee \overline{X}_2\,X_4 \vee \overline{X}_1.$$

The parentheses in the above formulas are only employed to increase readability. Although the formulas for $Y^{(a)}$, $Y^{(b)}$, and $Y^{(c)}$ differ, their values are always identical which warrants using the same variable Y in all three cases.

Chapter 8
Design by Composition*

The normal forms discussed in Chap. 4 allow us to design combinational circuits using only *basic elementary connectives*: The *canonical* normal forms employ AND, OR, and NOT, while the *Zhegalkin* normal forms realise circuits with XOR and AND or, respectively, with EQU and OR. Taking circuit design to a higher level, we would want to use the normal forms to design small circuit **modules** if only we had a procedure that would allow us to *put these modules together* so as to realise a specified problem. Such a process is referred to as **composing a circuit** (using previously developed and usually well-tested modules). In the simplest case, these modules can themselves be elementary connectives, e.g., NAND or NOR. But the power of a composition procedure lies in its ability to cope with more complicated modules. The composition procedure presented here is taken from Vingron (2004). The converse to composition is **decomposition**, a procedure developed by Ashenhurst's (1959) in which he showed how to split a switching function into two smaller functions, each, hopefully, being easier to realise.

8.1 The Basic Concept

To simplify explanations, we shall always use one and the same function as a module in each composition step of an example. As our main considerations turn on this function, I choose to call it a **hinge function**. For example, say we want to develop a circuit that realises the EQUIVALENCE specified in Fig. 8.1a. Using NOR intuitively as a hinge function, we could (with luck) develop the circuit pictured in Fig. 8.1b. On the other hand, taking the *normally closed* pneumatic valve as a hinge function, we could realise the EQUIVALENCE as shown on Fig. 8.1c. Of course, using a *single* hinge function is only possible if it is functionally complete.

The basic idea is to develop the circuit *from the output to the inputs*. This is emphasised in Fig. 8.2a where the **specified function** f is decomposed into the **hinge function** h at the output of the circuit, and the so-called **generic function** g the outputs of which are used as inputs to the hinge function. The name *generic*

S.P. Vingron, *Logic Circuit Design*, DOI 10.1007/978-3-642-27657-6_8,

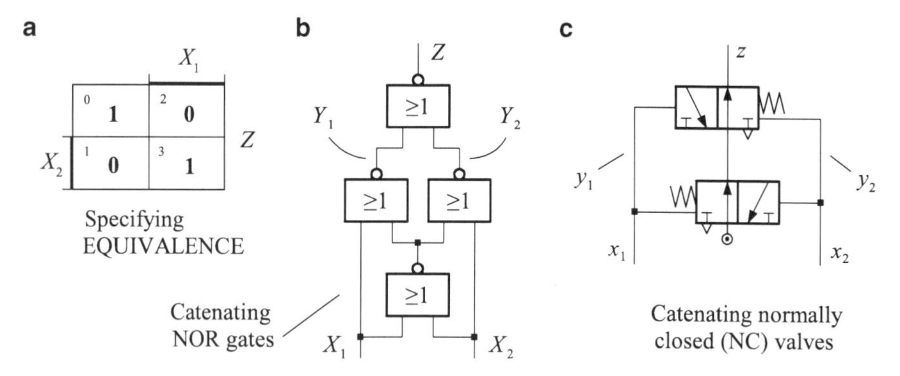

Fig. 8.1 Realisations by single hinge functions

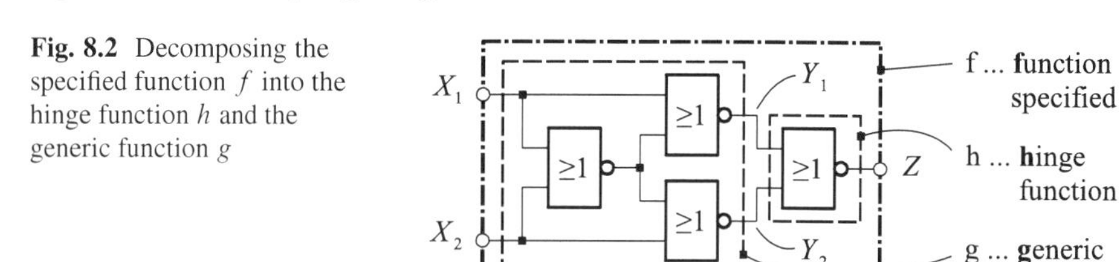

Fig. 8.2 Decomposing the specified function f into the hinge function h and the generic function g

function is chosen to remind us that there is not a singe, uniquely defined function g that we need only calculate or determine, but rather, g is a function we can choose from a whole class of functions that together with the hinge function h are a realisation of the specified function f. The generic function g is a multiple-output function with as many outputs as the hinge function h has inputs.

The mathematician refers to the functions h and g of Fig. 8.2 as being *catenated*, or chained together, to create the function f. Catenation can be pictured very descriptively using mappings so that we can do all our reasoning graphically.

8.2 Catenation

The word *catenation* originates from *'catena'*, the Latin word for *chain*. As a technical term, *catenation* refers to the chaining together of two functions. To get a feeling for this chaining together of functions consider the example shown in Fig. 8.3. The functions we wish to catenate (not identical with the equally named functions of the previous section), $g : \mathcal{X} \mapsto \mathcal{Y}$ and $h : \mathcal{Y} \mapsto \mathcal{Z}$, are defined in Fig. 8.3a, this representation being possible because $\mathcal{Y}$ is simultaneously the co-domain of $g : \mathcal{X} \mapsto \mathcal{Y}$ and the domain of $h : \mathcal{Y} \mapsto \mathcal{Z}$. While Fig. 8.3a visualises the process of catenating, Fig. 8.3b shows the result of the catenation, the composite function $hg : \mathcal{X} \mapsto \mathcal{Z}$. The composite function hg is constructed by *following the arrows* from $\mathcal{X}$ via $\mathcal{Y}$ to $\mathcal{Z}$. For instance, if we follow the arrow in Fig. 8.3a from x_1

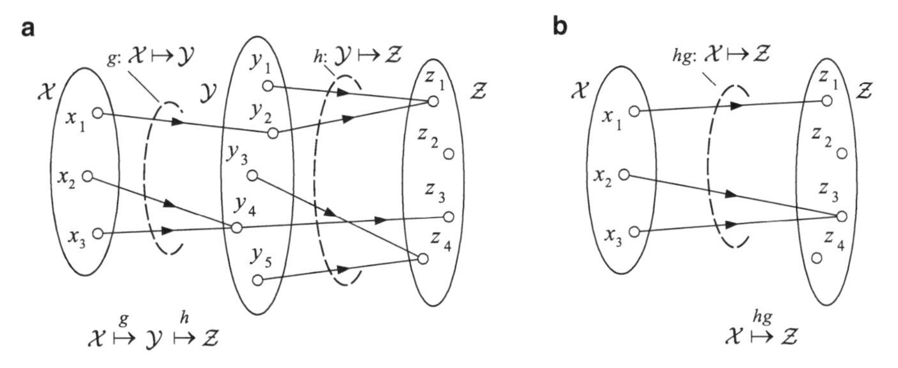

Fig. 8.3 An example demonstrating the catenation of functions

we come to y_2, and from here via the next arrow to z_1. The composite function hg summarises these successive mappings, simply stating that x_1 maps to z_1.

In a formal definition of catenation, we want to make sure that arrows are correctly chained by only using those elements of $\mathcal{Y}$ that are images of g, i.e., those y for which $y = g(x)$ holds. We thus formulate

> Given arbitrary functions $g : \mathcal{X} \mapsto \mathcal{Y}$ and $h : \mathcal{Y} \mapsto \mathcal{Z}$, their **catenation** $hg : \mathcal{X} \mapsto \mathcal{Z}$ (hg being read as 'h follows g') is specified as the set hg of ordered pairs (x, z) such that
>
> $$(x, z) \in hg :\Leftrightarrow \bigvee_{y \in \mathcal{Y}} \big((x, y) \in g \wedge (y, z) \in h\big) \tag{8.1}$$
>
> for any substitution instances $x \in \mathcal{X}$ and $z \in \mathcal{Z}$.

The *whole* definition (8.1) is often abbreviated by the picturesque notation

$$\mathcal{X} \overset{hg}{\mapsto} \mathcal{Z} = \mathcal{X} \overset{g}{\mapsto} \mathcal{Y} \overset{h}{\mapsto} \mathcal{Z}, \tag{8.2}$$

which emphasises that $\mathcal{Y}$ is simultaneously the co-domain of g and the domain of h. Without further proof, let me state that the above definition allows us to deduce the following explanation for the conventional notation $h(g(x))$:

$$h\big(g(x)\big) = (hg)(x). \tag{8.3}$$

The principle property of catenation is *associativity*, without which the catenation of circuits would not work. Associativity means, given three functions such that

$$\mathcal{X} \overset{g}{\mapsto} \mathcal{Y} \overset{h}{\mapsto} \mathcal{Z} \overset{r}{\mapsto} \mathcal{W},$$

the sequence in which catenation is carried out is irrelevant:

$$rhg = r(hg) = (rh)g. \tag{8.4}$$

8.3 Visualising the Composition Problem

When composing a specified function f, we first need to decide on, and sometimes analyse, the hinge function(s) we wish to employ. If we want to use the normally closed valve as hinge function (as in Fig. 8.1), we must develop its Karnaugh map. Applying all possible values, 0 and 1, for the input variables y_1, y_2, and y_3 (Fig. 8.4a, b) allows us to obtain the output values for z entered into the K-map of Fig. 8.4c. From this K-map we read the logic formula $Z \Leftrightarrow (\overline{Y}_1 \vee Y_2)Y_3$, which in turn leads to the composite symbol of Fig. 8.4d. For future reference, and this will turn out to be the prime objective of the analysis of the hinge function, we read from the K-map the set $\mathcal{Y}_0$ of input events (y_1, y_2, y_3) that cause the output z to be 0, as well as the set $\mathcal{Y}_1$ of those input events (y_1, y_2, y_3) that cause the output z to be 1 (Fig. 8.4e).

Now let us attempt to use the concept of catenation to compose the EQUIVALENCE (see Fig. 8.1a, c) when using the normally closed valve of Fig. 8.4 as hinge function. We first draw the function f, the function we wish to compose (the EQUIVALENCE of Fig. 8.1a), as a mapping from $\mathcal{X}$ to $\mathcal{Z}$, Fig. 8.5a. The catenation process is started by drawing the mapping $h : \mathcal{Y} \mapsto \mathcal{Z}$ of the NC valve (see Fig. 8.5b) which follows effortlessly from the K-map of this hinge function (repeated in Fig. 8.5d).

The actual problem we face is how to find the generic function g such that $hg = f$ is fulfilled. Interpreting this graphically, we want to know to draw arrows from $\mathcal{X}$ to $\mathcal{Y}$. We consider this problem in Fig. 8.5. For instance, note that f maps $(0, 0) \in \mathcal{X}$ to $1 \in \mathcal{Z}$ as pictured in sub-figure (a). In the catenation diagram of

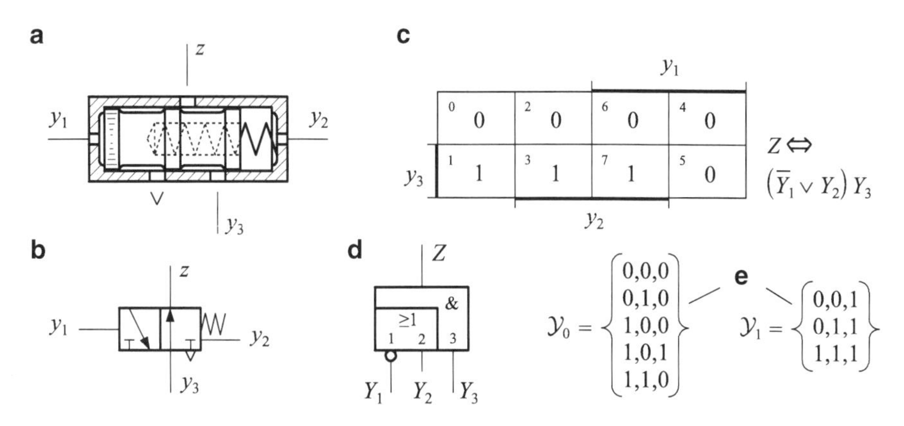

Fig. 8.4 The NC-valve as hinge function

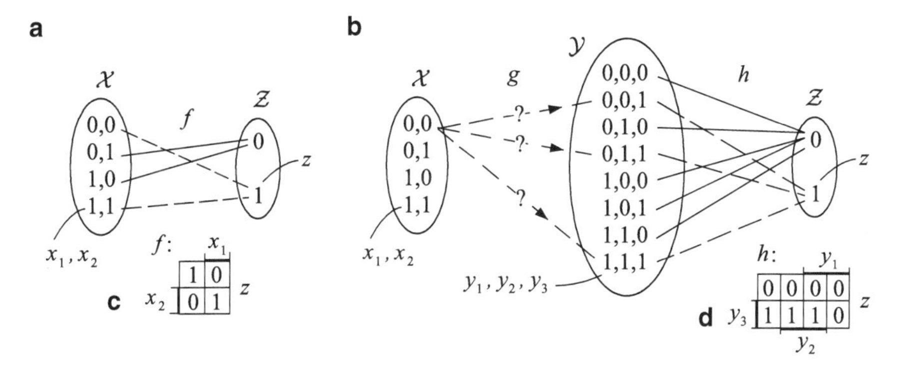

Fig. 8.5 The NC-valve as hinge function h

sub-figure (b), g must map $(0, 0) \in \mathcal{X}$ to one (but, of course, only to one) input event (y_1, y_2, y_3) that itself is mapped to $1 \in \mathcal{Z}$ by h. Each of these destinations, (y_1, y_2, y_3), is equally valid with respect to catenation. May I encourage you to extend the above considerations to each element (x_1, x_2) of $\mathcal{X}$ in sub-figure (b), using the corresponding mappings of sub-figure (a) to decide which elements $(y_1, y_2, y_3) \in \mathcal{Y}$ you may map to. In the next section we formulate the above considerations in a more formal way.

8.4 Choosing a Generic Function

Assume the function we wish to design, namely $f : \mathcal{X} \mapsto \mathcal{Z}$, to have n input variables. For a simpler notation, we write its binary input events $(x_1, \cdots, x_n) \in \mathcal{X}$ in their decimal form x, where $x = \sum_{i=1}^{n} x_i \cdot 2^{n-i}$. Similarly, we take the generic function $g : \mathcal{Y} \mapsto \mathcal{Z}$ to have m inputs $y_1, \cdots, y_m$, and abbreviate its binary input events $(y_1, \cdots, y_m)$ to their decimal form y, with $y = \sum_{i=1}^{m} x_i \cdot 2^{m-i}$.

To obtain a generic function $g : \mathcal{X} \mapsto \mathcal{Y}$, its domain $\mathcal{X}$ and co-domain $\mathcal{Y}$ are partitioned thus (noting that $z \in \mathcal{Z}$ and $\mathcal{Z} = \{0, 1\}$):

$$\mathcal{X}_z := \{x | f(x) = z\}, \qquad \mathcal{Y}_z := \{x | h(x) = z\}, \qquad \text{with} \quad z \in \{0, 1\}, \tag{8.5}$$

these partitions being used as domains of partial functions

$$f_z : \mathcal{X}_z \mapsto \{z\}, \qquad h_z : \mathcal{Y}_z \mapsto \{z\}, \qquad \text{with} \quad z \in \{0, 1\}, \tag{8.6}$$

$\mathcal{X}_z$ and $\mathcal{Y}_z$ being partitions of $\mathcal{X}$ and $\mathcal{Y}$, respectively, have the basic properties:

$$\mathcal{X}_0 \cup \mathcal{X}_1 = \mathcal{X} \qquad \text{and} \qquad \mathcal{X}_0 \cap \mathcal{X}_1 = \emptyset,$$
$$\mathcal{Y}_0 \cup \mathcal{Y}_1 = \mathcal{Y} \qquad \text{and} \qquad \mathcal{Y}_0 \cap \mathcal{Y}_1 = \emptyset.$$

The partitions $\mathcal{X}_z$ and $\mathcal{Y}_z$, (8.5), allow us to define the partial generic functions g_0 and g_1 as

$$g_z : \mathcal{X}_z \mapsto \mathcal{Y}_z \qquad \text{with} \quad z \in \{0, 1\}. \tag{8.7}$$

Due to the above definitions, (8.5)–(8.7), we may write

$$\begin{aligned} f_0 \cup f_1 &= f, & f_0 \cap f_1 &= \emptyset, \\ h_0 \cup h_1 &= h, & h_0 \cap h_1 &= \emptyset, \\ g_0 \cup g_1 &= g, & g_0 \cap g_1 &= \emptyset, \end{aligned}$$

and, most importantly,

$$\boxed{g_z : \mathcal{X}_z \mapsto \mathcal{Y}_z \Rightarrow h_z g_z = f_z, \qquad \text{with} \quad z \in \{0, 1\}} \tag{8.8}$$

which states that *any arbitrarily chosen* function g_z from $\mathcal{X}_z = \{x | f(x) = z\}$ to $\mathcal{Y}_z = \{x | h(x) = z\}$—where $z \in \{0, 1\}$—fulfils the catenation condition $h_z g_z = f_z$. The co-domains $\mathcal{Y}_0$ and $\mathcal{Y}_1$ of the partial generic functions g_0 and g_1 are determined from the hinge functions (as, e.g., shown in Fig. 8.4) and are thus called '0 **hinge-set**' and '1 **hinge-set**', respectively. Assuming the hinge function $h : \mathcal{Y} \mapsto \{0, 1\}$ to have m binary inputs $(y_1, \ldots, y_m)$ so that it is described by an m-dimensional K-map whose cells contain either 0 or 1,

- The 0 **hinge-set** $\mathcal{Y}_0 = \{(y_1, \ldots, y_m) | h(y_1, \ldots, y_m) = 0\}$ consists of the binary descriptions $(y_1, \ldots, y_m)$ of all cells containing an output value of 0;
- The 1 **hinge-set** $\mathcal{Y}_1 \{(y_1, \ldots, y_m) | h(y_1, \ldots, y_m) = 1\}$ consists of the binary descriptions $(y_1, \ldots, y_m)$ of all cells containing an output value of 1.

The above considerations on catenation are conveniently expressed when using K-maps, as the continuation of our example of the previous section shows in Fig. 8.6. The output values of 1 in the specified function f correspond to arbitrary elements (y_1, y_2, y_3) of $\mathcal{Y}_1$ in equivalent cells of g; the output values of 0 in the specified function f correspond to arbitrary elements (y_1, y_2, y_3) of $\mathcal{Y}_0$ in equivalent cells of g. Generalising on this example, we can formulate the following

Procedure for developing the K-map of a generic function g when given the K-map for the specified function f, and the hinge sets $\mathcal{Y}_0$ and $\mathcal{Y}_1$ of a hinge function h:

(a) If a cell e of the K-map for f contains an output value z of 0, then the equally positioned cell e of the K-map for g contains an arbitrary element or tuple of $\mathcal{Y}_0$.
(b) If a cell e of the K-map for f contains an output value z of 1, then the equally positioned cell e of the K-map for g contains an arbitrary element or tuple of $\mathcal{Y}_1$.

(8.9)

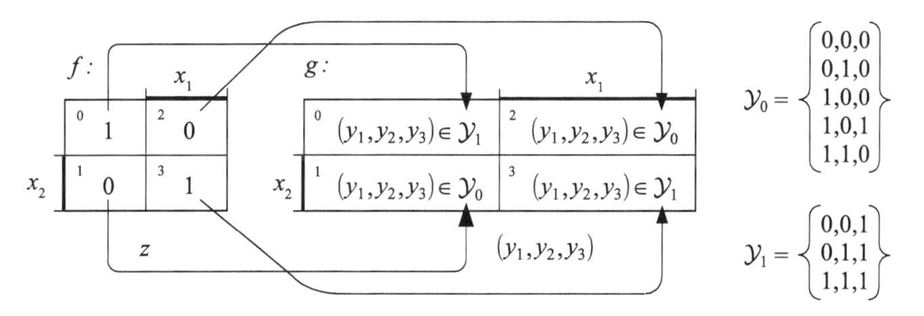

Fig. 8.6 Continuing the example of the previous section on how to define a generic function g using K-maps

8.5 Composing a Circuit: Example 1

Here we return to the introductory example of this chapter, Fig. 8.1c, to see how the EQUIVALENCE can be developed on the basis of the NC-valve. We start with the K-map of the EQUIVALENCE of Fig. 8.7a, the function f to be synthesised, and with the hinge sets $\mathcal{Y}_0$ and $\mathcal{Y}_1$ of the hinge function we wish to employ (detailed in Fig. 8.4e).

To develop a generic function g we repeat the K-map for f, except that, according to procedure (8.9), each inscribed 1 is replaced by an arbitrarily chosen triple y_1, y_2, y_3 of $\mathcal{Y}_1$, and each inscribed 0 is replaced by an arbitrarily chosen

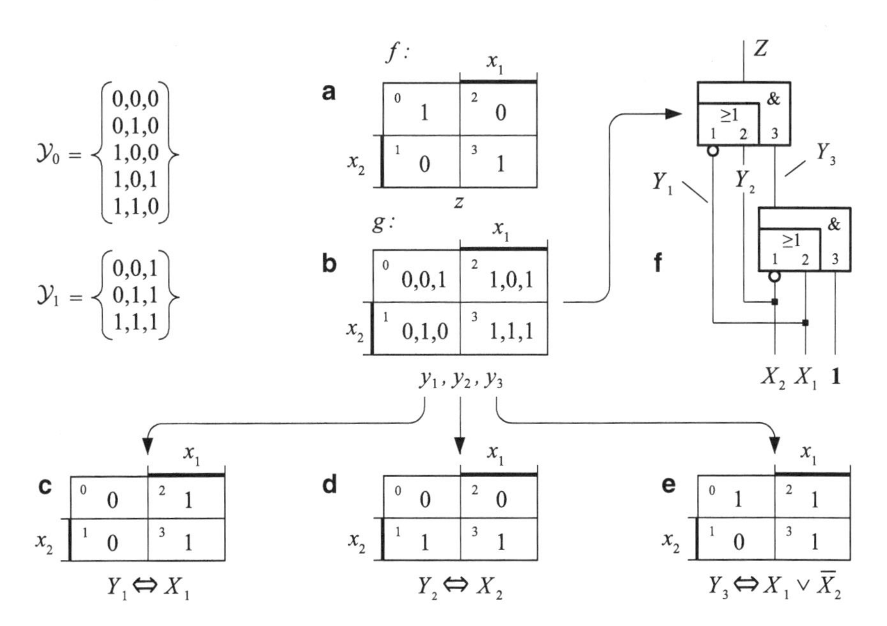

Fig. 8.7 Continuing the example of the previous section on how to define a generic function g using K-maps

triple y_1, y_2, y_3 of $\mathcal{Y}_0$. Our choice for these triples is shown in (b).[1] The triple-output K-map thus obtained is split into the three K-maps (c), (d), and (e) of the last row. Although the choice for the triples y_1, y_2, y_3 is arbitrary—within the limitations of procedure (8.9)—it is clearly not random. The triples were carefully selected from $\mathcal{Y}_0$ and $\mathcal{Y}_1$ to achieve the IDENTITIES $Y_1 \Leftrightarrow X_1$ and $Y_2 \Leftrightarrow X_2$. I strongly recommend varying the choices for y_1, y_2, y_3 to see the effect on the result obtained.

The composed circuit of Fig. 8.7f is drawn by starting at the circuit's output with the composite symbol for the hinge function. The inputs Y_1, Y_2, Y_3 are taken from the K-maps of Fig. 8.7c–e, noting that $Y_3 \Leftrightarrow X_1 \vee \overline{X}_2$ can be realised by the composite function of the NC-valve.

Actually, the above example was terminated prematurely. The processing of a K-map should continue unless the K-map represents an IDENTITY or a constant, i.e., FALSITY or TRUTH. The K-map of Fig. 8.7e represents an IMPLICATION, so, why terminate processing? Clearly, because our composite logic function can realise the IMPLICATION. Nevertheless, let us continue the process in a formal way where it was interrupted, using the opportunity to take a closer look at how one might choose n-tuples from the hinge sets $\mathcal{Y}_0$ and $\mathcal{Y}_1$.

In Fig. 8.8 we continue developing the function specified in the K-map of Fig. 8.7e. The composition procedure requires us to replace the 1s in the cells of (a)[2] by elements, i.e., ordered triples, of $\mathcal{Y}_1$. As each triple has 1 as its last element,

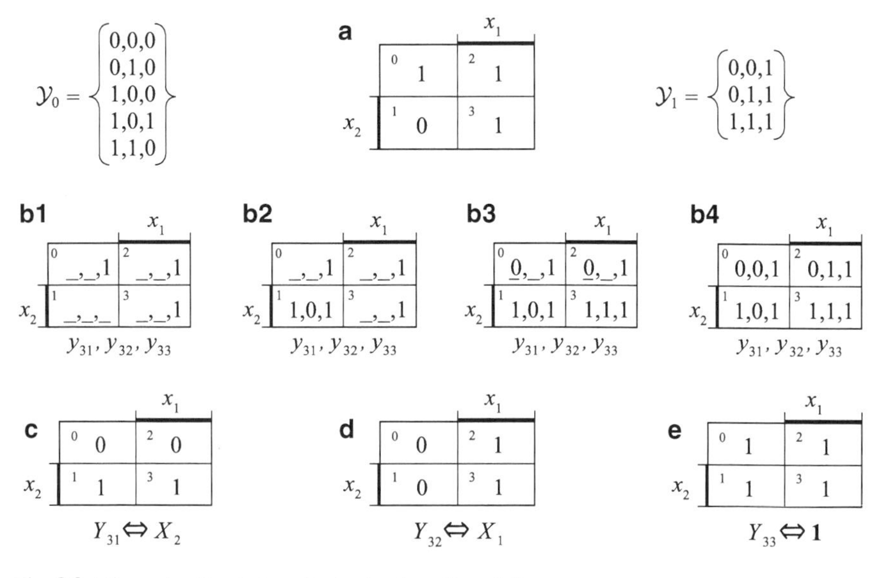

Fig. 8.8 Discussing the choice of n-tuples from $\mathcal{Y}_0$ and $\mathcal{Y}_1$

[1] In this paragraph (b), (c), etc., refer to sub figures of Fig. 8.7.

[2] In this paragraph (a), (b1), ..., (e) refer to the thus specified sub figures of Fig. 8.8.

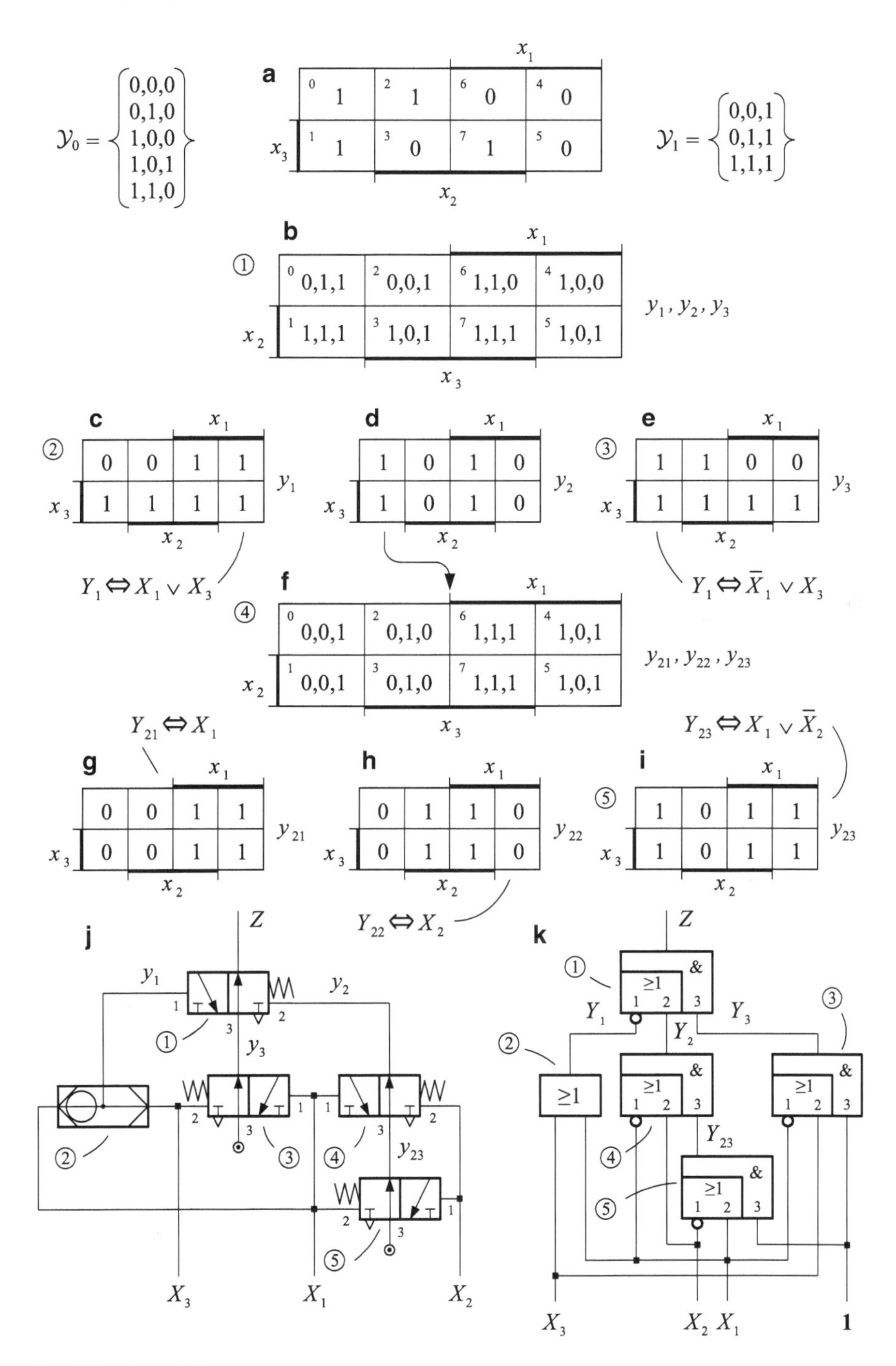

Fig. 8.9 Example 2

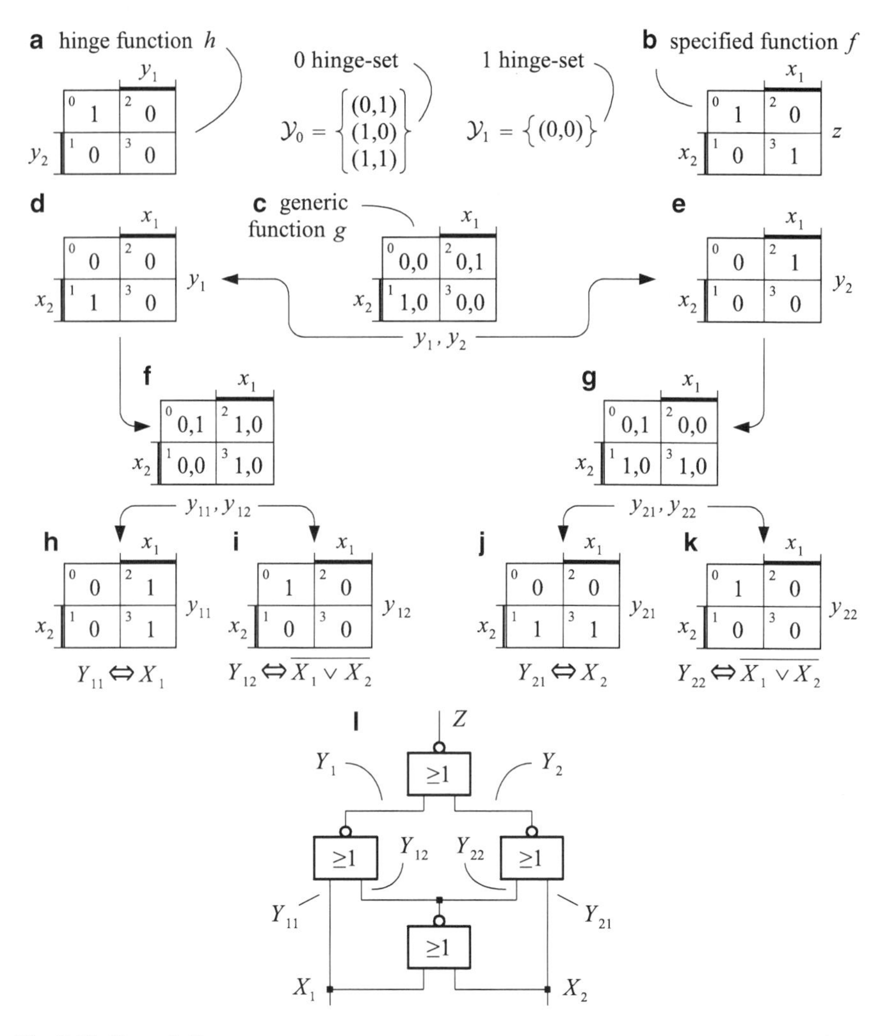

Fig. 8.10 Example 3

we have no choice but to begin with a K-map for the generic function as the one shown in (b1). If we could place a 1 in the third position of cell number 1 of (b1), we would be able to express Y_{33} as TRUTH. The triple to be inscribed in cell 1 of (b1) must be taken from $\mathcal{Y}_0$ (why?). There is only one such triple which has 1 as a last element, $(1, 0, 1)$, which brings us to (b2). Now having a 1 in position 1 of cell 1, it would be advantageous to also have a 1 in position 1 of cell 3. We could then describe Y_{31} as an IDENTITY (i.e., $Y_{31} \Leftrightarrow X_2$)—assuming, of course, that we could place 0 s in the first positions of cell 0 and 2. These aspirations are expressed in (b3) where the second position of cell 3 by necessity becomes 1. This leads to the desire to describe Y_{32} as an IDENTITY, i.e., as $Y_{32} \Leftrightarrow X_1$, which necessitates us to fill the K-map as shown in (b4). This triple-output K-map is then split into the single-output K-maps of the last row of Fig. 8.8.

8.6 Composing a Circuit: Example 2

Suppose we want to compose a circuit specified in Fig. 8.9a. As a hinge function let us accept either an OR or the composite function $Z \Leftrightarrow (\overline{Y}_1 \vee Y_2)Y_3$ of the NC-valve used in the previous section. After the explanations of that section, I leave you to pondering the possible solution shown in Fig. 8.9. The following calculation, which models the circuit of Fig. 8.9k, verifies the composed circuit because the result of the calculation can be read directly from the K-map of Fig. 8.9a.

$$
\begin{aligned}
Z &\Leftrightarrow (\overline{Y}_1 \vee Y_2)Y_3 \\
&\Leftrightarrow \big(\overline{X_1 \vee X_3} \vee (\overline{X}_1 \vee X_2)Y_{23}\big)(\overline{X}_1 \vee X_3) \\
&\Leftrightarrow \big(\overline{X}_1\overline{X}_3 \vee (\overline{X}_1 \vee X_2)(X_1 \vee \overline{X}_2)\big)(\overline{X}_1 \vee X_3) \\
&\Leftrightarrow (\overline{X}_1\overline{X}_3 \vee X_1X_2 \vee \overline{X}_1\overline{X}_2)(\overline{X}_1 \vee X_3) \\
&\Leftrightarrow \overline{X}_1\overline{X}_1\overline{X}_3 \vee \overline{X}_1X_1X_2 \vee \overline{X}_1\overline{X}_1\overline{X}_2 \vee \overline{X}_1\overline{X}_3X_3 \vee X_1X_2X_3 \vee \overline{X}_1\overline{X}_2X_3 \\
&\Leftrightarrow \overline{X}_1\overline{X}_3 \vee \overline{X}_1\overline{X}_2 \vee X_1X_2X_3 \vee \overline{X}_1\overline{X}_2X_3 \\
&\Leftrightarrow \overline{X}_1\overline{X}_3 \vee \overline{X}_1\overline{X}_2 \vee X_1X_2X_3
\end{aligned}
$$

8.7 Composing a Circuit: Example 3

To finish on a simpler note, you might like to see how to obtain the NOR circuit of the EQUIVALENCE shown in Fig. 8.1b. If so, follow the development laid out in Fig. 8.10.

Part II
Latches

Beginning with this division we enter the field of **sequential circuits**, circuits that have a memorising ability to various degrees. The simplest of these are the **latches**. To understand how memorisation works, we examine it in its elementary form—the latch. Latches prove to be the building blocks of all sequential circuits.

The theory of latches presented in Chap. 9 is general in that it is not restricted to feedback circuits. Most importantly, it contains a *time-independent* definition of a memory function which relies heavily on the concept of **well-behavedness**. It lays the basis for specifying a memory function in a reduced K-map, and for developing evaluation formulas. Chapter 10 applies the time-independent definition of the preceding chapter to **latches with feedback**. Here we discuss the all-important concept of **memorisation hazards**, and how to avoid them: The conventional method being the introduction of an inertial delay in the feedback, the ideal method, on the other hand, being to design the feedback as a **pre-established loop**. When designing latches, minimisation takes a back seat to reliability. Chapter 11 contains a complete discussion of all latches with two inputs and two outputs—the so-called elementary latches—presenting for these a unified set of graphic symbols. Chapter 12 introduces **latch composition**, stating how to realise a given latch on the basis of another. Its theory and application is covered extensively.

Chapter 9
Basic Theory of Latches*

To be able to develop an abstract concept of a latch, we need something from which to abstract. Section 9.1 introduces us to a number of devices and a circuit that intuitively have in common a property we call memorisation. Devices with this property are referred to as latches. Section 9.2 presents a formal concept of memorisation discussing it quite thoroughly. This somewhat unconventional definition was first proposed in Vingron (2004) in a slightly different form. This section, with its introduction of *setting*, *resetting*, and *memorising subsets* of input events, and the specification of a memory function in a reduced K-map, poses the prime part of this chapter. In Sect. 9.3 we start the logical description of latches (or memory) by introducing *inclusions* and *exclusions*. Section 9.4 introduces and develops the *basic memory-evaluation formulas* for latches. A generalisation of these formulas is presented in Sect. 9.5. The application of both the *basic memory-evaluation formulas* and the *generalised memory-evaluation formulas* is an extensive topic, and therefore covered in its own right in the next chapter. Conversely, the present chapter contains no applications.

9.1 What Is a Latch?

Informally stated, a **latch** is the simplest binary switching device or circuit with memorising capability. Latches come in all kinds of technologies and in many functional variants. The spool valve of Fig. 9.1a has a memory function: The valve retains the spool's position caused by a previous pressure pulse on one of the pilot lines x_1 or x_2. A different memory function is realised by the relay circuit of Fig. 9.1b: Briefly closing contact x_1 activates the *coil* y, its state of activation being retained by the coil-activated *contact*, y^*. The thyristor circuit of Fig. 9.1c poses an example for yet a third memory realisation and function: To make the thyristor fire, we not only need a supply voltage x_1 but also a pulse on the gate voltage x_2; the thyristor, then remaining open memorises the prior presence of a gate pulse.

S.P. Vingron, *Logic Circuit Design*, DOI 10.1007/978-3-642-27657-6_9,

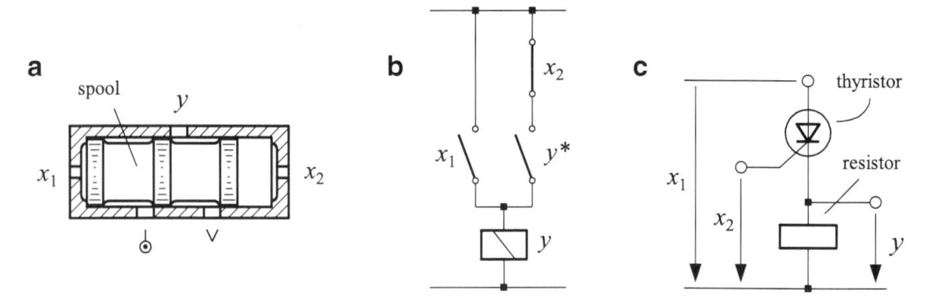

Fig. 9.1 Examples of latches

The memorising capability of latches is usually, but incorrectly, attributed to feedback, as exists, for instance, in the relay circuit of Fig. 9.1b. But, firstly, not every circuit in which feedback is incorporated has memorising capability, and, secondly, memorisation is not tied to feedback as the devices of Fig. 9.1a, c show. It is clearly necessary to explain memorisation independently of any technological realisation. Towards this end, consider the subsequent *events graph* of what is called a sampling circuit or sampling latch. As a latch, this circuit is called **D-latch** (i.e., *data latch*), and is *primus interparis* among the elementary latches (to be discussed later on).

A **sampling latch** has two inputs and one output. The prime input, the **data input** x_D, is **sampled**, i.e., its value is passed on to the output, y, during usually brief time intervals which are defined by the second input x_C, the **clock** or **control input**, being 1; between sampling, i.e., while x_C is 0, the output y retains (memorises) the last sampled value. This verbal specification is satisfied by the input–output behaviour depicted in the events graph of Fig. 9.2.

Before continuing, it seems advisable to reread what was said in Sect. 1.2 on the time dependent representation of binary signals in a **timing diagram**, and especially Sect. 1.3 where the concept of the **events graph** was introduced. Naming the latter representation a *graph* (and not a *diagram*) stresses the time independence of the signals: Neither is their duration relevant, nor is there any skew or inertia. The events graph describes only the input–output *behaviour* of a device or circuit, but in no

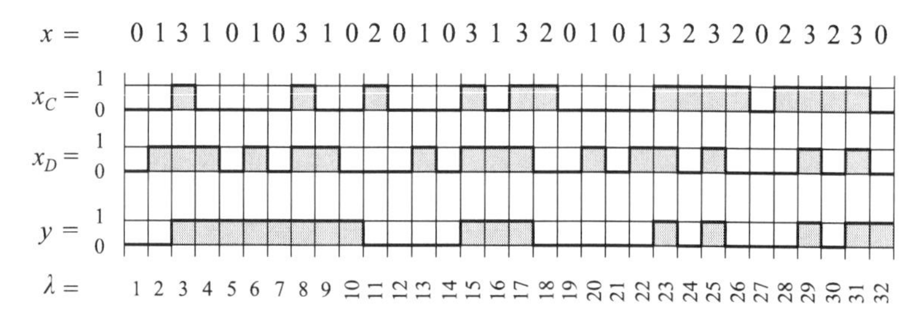

Fig. 9.2 An **events graph** depicting the input–output behaviour of the sampling circuit, or **D-latch**

way does it imply how to realise the behaviour. Next, let us gauge the information contained in the events graph of Fig. 9.2.

In drawing an events graph one tries to model the verbal specification thinking of the events graph as a well-behaved timing diagram. The location $\lambda = 3$ shows how $x_D = 1$ is sampled for the first time. The output remains 1 until $x_D = 0$ is sampled which is the case in the location $\lambda = 11$. As before, the output retains the zero value until $x_D = 1$ is sampled (which is the case at $\lambda = 15$) when the output again becomes one. I encourage you to continue reading the graph till you have verified it.

Now, consider the effect each input event $x \in \{0, 1, 2, 3\}$ has on the output y. Whenever $x = 3$ the output $y = 1$. Input events with this property are said to **set** the output. Whenever $x = 2$ the output $y = 0$. Input events with this property are said to **reset** the output. The input events $x = 0$ and $x = 1$, on the other hand, sometimes cause the output to be 1, and sometimes to be 0. Nevertheless, we can state unequivocally which output value, 0 or 1, to expect—namely, the preceding output value. An input event x, in any location λ, which causes an output value numerically identical to the previous one (to the output value in location $\lambda - 1$), is called a **memorising** input-event. A circuit or switching device that has at least one setting, one resetting, and one memorising input event—and no others—is called a latch. Please do now take the time to verify the setting, resetting, and memorising properties of the input events in Fig. 9.2.

9.2 The Memory Function

Latch refers to a device or circuit realised in a certain technology. The formal description of a latch, independent of any technology, is called a **memory** or **memory function**. In this section we concern ourselves with this formal aspect. A memorising input event x of a latch allows the output y_x to be (either) 0 or 1, which we can state as $y_x \in \{0, 1\}$. To coherently express *all* output values y_x (whether caused by setting, resetting, or memorising input events) as elements of a set, we use the following

Definition. A **memory** is a surjection

$$f : \mathcal{E} \mapsto \{\{0\}, \{1\}, \{0, 1\}\}.$$

Its **output**, y_x, is explained as an element of the value of the function $f(x)$, i.e., $y_x \in f(x)$ with $x \in \mathcal{E}$. The output is **well-behaved** meaning that

$$f(x(\lambda)) = \{0, 1\} \Rightarrow y_{x(\lambda)} = y_{x(\lambda-1)}$$

where $\mathcal{L} = \{1, 2, \ldots\}$ is an infinite index set. (9.1)

The memory definition (9.1) was inspired by Eilenberg and Elgot's (1970) definition of a relation, and first put forth in Vingron (1979) where a time parameter was used instead of an index set. The memory function defined in (9.1) is clearly a hybrid between a relation and a function. We now discuss this definition in detail.

A **surjection** is a function such that *each* element of the co-domain is a value of the function, i.e., is an image. The domain $\mathcal{E} = \{0, 1, 2, \ldots, 2^n - 1\}$ is, in our context, the set of the decimal equivalents $x = \sum_1^n x_i \cdot 2^{n-i}$ of the input events $(x_1, x_2, \ldots, x_n)$. For a combinational circuit the value of the function is the circuit's output: Not so for a memory function f where the images $f(x)$ are *sets* of output values so that the output y_x itself is an element of these sets, $y_x \in f(x)$.

The output values caused by setting or resetting input events are per definition well-behaved. The output value $y_{x(\lambda)}$ initiated by a *memorising* input event $x(\lambda)$—one that is mapped to $\{0, 1\}$—might switch spontaneously, or oscillate between 0 and 1 if it were not well-behaved.[1]It is valid to split an input event $x(\lambda)$, occupying a location λ of the index set $\mathcal{L}$, into an ordered pair of successive and *equal* input events $\big(x(\lambda - 1), x(\lambda)\big)$. In all such cases the circuit, being well-behaved, requires the output to be invariant, effectively inhibiting spontaneous switching.

The principle of being well-behaved, $f\big(x(\lambda)\big) = \{0, 1\} \Rightarrow y_{x(\lambda)} = y_{x(\lambda-1)}$, does not specify whether the input event $x(\lambda - 1)$ which initiates the prior output $y_{x(\lambda-1)}$ equals $x(\lambda)$, or not. If they differ, and if $x(\lambda)$ maps to $\{0, 1\}$ then the present output value $y_{x(\lambda)}$ *retains* the previous value $y_{x(\lambda-1)}$. Thus the principle of being well-behaved not only safe-guards the output of a memory function against spontaneous changes of the output value, it also ensures memorisation whenever an input event maps to $\{0, 1\}$.

Specifying a Memory Function in a K-Map

As a memory function f maps the input events $x \in \mathcal{E}$ to one of the sets $\{0\}$, $\{1\}$, or $\{0, 1\}$, it would seem that a K-map whose every cell contains one of these sets would be an appropriate representation—see Fig. 9.3a. But such a representation does not take into account that the memory function must be well-behaved, meaning that $f\big(x(\lambda)\big) = \{0, 1\} \Rightarrow y_{x(\lambda)} = y_{x(\lambda-1)}$. The correct representation, therefore, is to use a K-map the cells of which contain the output $y_x \in f(x)$, and not the function's value $f(x)$ itself—see Fig. 9.3b. The thus obtained K-map is a **reduced K-map** (see Sect. 5.5) and as such can be drawn in its standard form shown in Fig. 9.3c.

---

[1]A memory function is so very different to a conventional function because its output is *not* the image of the function. The image of a memory function is one of the *sets* $\{0\}, \{1\}, \{0, 1\}$ while the output of the memory function is an element of one of these sets. This only causes a problem when an input event x is present that is mapped to $\{0, 1\}$. Of course, y being an element of $\{0, 1\}$ it must be either 0 or 1. Yet, without further information it is impossible to say whether $y = 0$ or $y = 1$. In fact, the value of y could switch spontaneously between the values 0 and 1 without violating the condition $y \in \{0, 1\}$.

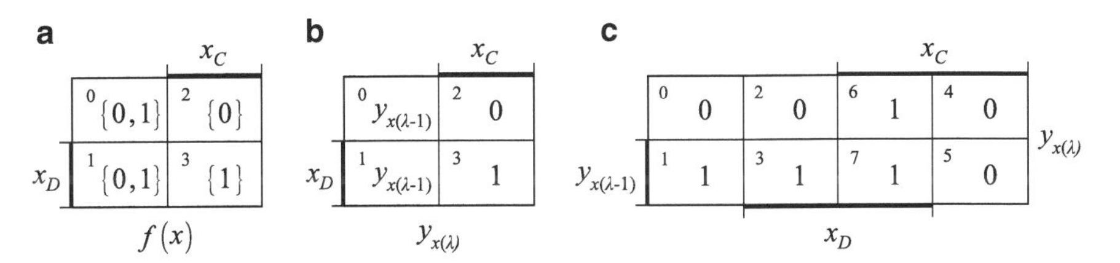

Fig. 9.3 Expressing the input–output behaviour of the sampling circuit of Fig. 9.2 in a K-map

A memory function f partitions its domain $\mathcal{E}$ into a **setting**, **resetting**, and **memorising** equivalent, $\mathcal{E}_1$, $\mathcal{E}_0$, and $\mathcal{E}_y$, respectively, whereby the memorising equivalent $\mathcal{E}_y$ is explicitly defined via well-behavedness:

$$\begin{aligned}
\mathcal{E}_1 &:= \{x(\lambda) | f(x(\lambda)) = \{1\}\}, \\
\mathcal{E}_0 &:= \{x(\lambda) | f(x(\lambda)) = \{0\}\}, \\
\mathcal{E}_y &:= \{x(\lambda) | (f(x(\lambda)) = \{0,1\}) \rightarrow (y_{x(\lambda)} = y_{x(\lambda-1)})\}.
\end{aligned} \tag{9.2}$$

These are easy to visualise in the reduced K-map of a memory function. Those cells containing 1s comprise the set of setting input events. Those containing 0s comprise the set of resetting cells. And those containing the symbol for the previous output, $y_{x(\lambda-1)}$, comprise the set of memorising input events.

Reading the Reduced K-Map of a Memory Function

The technical term **reading** (the reduced K-map of a memory function) refers to the process of developing an events graph for the memory function from its reduced K-map. Take, for example, the reduced K-map of the sampling circuit repeated in Fig. 9.4. To develop the events graph, one chooses a sequence of input events. We choose the same sequence of input events as in Fig. 9.2, which for your convenience is also repeated in Fig. 9.4. Now, as we proceed with the reading process, I encourage you to draw the output values into the events graph of Fig. 9.4 as they are obtained.

In the initial location ($\lambda = 1$) we take the input event x to be zero ($x = 0$). According to the K-map we have started with a memorising event; not knowing the preceding output value ($y_{x(\lambda-1)}$) we take the liberty of assuming it to be 0 so that the present value must also be 0—and this value is entered for y in location 1 of the events graph. In the second location ($\lambda = 2$) the input event is one ($x = 1$) which, according to the K-map, is again a memorising one. But this time we know the output of the previous input event (the previous output) to be 0 so that the present output must also be 0. The input event $x = 3$ at $\lambda = 3$ is a setting input event

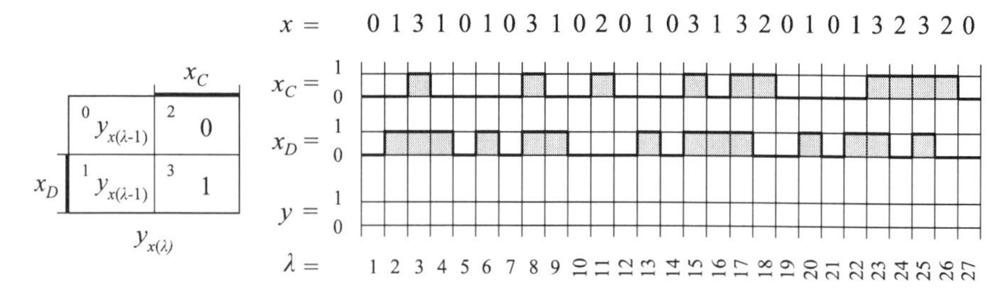

Fig. 9.4 Reading the reduced K-map of the sampling circuit according to the chosen sequence of input events stated in the events graph

(see the K-map) so $y = 1$ is entered into the events graph in location 3. Presumably these few explanations are sufficient to understand the reading procedure. Just this once, please do take the time to complete the events graph for this example. If you have any doubts, note that you must come up with the events graph of Fig. 9.2.

Before leaving this section, consider the required **infinity of the index set** $\mathcal{L}$, the set of locations λ. The definition of memory implies that whenever an input event occurs in a succession of input events of an events graph, and be the sequence infinitely long, each input event maintains its membership to one and the same subset $\mathcal{E}_1$, $\mathcal{E}_0$, $\mathcal{E}_y$, and never will an input event occur which does not belong to one of the setting, resetting, or memorising subsets. Thus, when adding a further input event to an events graph of a memory function, the required infinity of the index set $\mathcal{L}$ ensures that the input event added remains setting if its prior occurrences were setting, resetting if they were resetting, and memorising if they were memorising.

9.3 Introducing Inclusions and Exclusions

The evaluation formulas for memory functions are somewhat more complicated than those for combinational circuits, (5.19). Introducing special logic functions, or propositional forms, allows us to formulate the evaluation formulas for memory functions more succinctly. These special logic functions are called **inclusions**, $U(x)$, and **exclusions**, $V(x)$, and are defined as the abbreviations:

$$\boxed{\begin{array}{ll} U_1(x(\lambda)) :\Leftrightarrow x(\lambda) \in \mathcal{E}_1, & V_1(x(\lambda)) :\Leftrightarrow x(\lambda) \notin \mathcal{E}_1, \\ U_0(x(\lambda)) :\Leftrightarrow x(\lambda) \in \mathcal{E}_0, & V_0(x(\lambda)) :\Leftrightarrow x(\lambda) \notin \mathcal{E}_0, \\ U_y(x(\lambda)) :\Leftrightarrow x(\lambda) \in \mathcal{E}_y, & V_y(x(\lambda)) :\Leftrightarrow x(\lambda) \notin \mathcal{E}_y. \end{array}} \tag{9.3}$$

As set forth in Sect. 5.4 (at least for $x \in \mathcal{E}_1$), the right sides of the above expressions can be calculated as shown in the right sides of the following formulas. In these (and subsequently), the arguments $x(\lambda)$ of $U(\lambda)$ and $V(\lambda)$ are omitted for space reasons and for better readability.

$$
\begin{array}{|ll|}
\hline
U_1 \Leftrightarrow \bigvee_{\varphi:\mathcal{K}_\varphi \in \mathcal{K}(\mathcal{E}_1)} C_\varphi(x(\lambda)), & V_1 \Leftrightarrow \bigwedge_{\varphi:\mathcal{K}_\varphi \in \mathcal{K}(\mathcal{E}_1)} D_\varphi(x(\lambda)), \\
U_0 \Leftrightarrow \bigvee_{\varphi:\mathcal{K}_\varphi \in \mathcal{K}(\mathcal{E}_0)} C_\varphi(x(\lambda)), & V_0 \Leftrightarrow \bigwedge_{\varphi:\mathcal{K}_\varphi \in \mathcal{K}(\mathcal{E}_0)} D_\varphi(x(\lambda)), \\
U_y \Leftrightarrow \bigvee_{\varphi:\mathcal{K}_\varphi \in \mathcal{K}(\mathcal{E}_y)} C_\varphi(x(\lambda)), & V_y \Leftrightarrow \bigwedge_{\varphi:\mathcal{K}_\varphi \in \mathcal{K}(\mathcal{E}_y)} D_\varphi(x(\lambda)). \\
\hline
\end{array}
\tag{9.4}
$$

Remember that the cover $\mathcal{K}(\mathcal{E}_1)$ is a set of arbitrarily chosen K-sets whose union is $\mathcal{E}_1$, e.g., $\mathcal{K}(\mathcal{E}_1) = \{\mathcal{K}_{\varphi(1)}, \cdots, \mathcal{K}_{\varphi(L)}\}$ with $\mathcal{E}_1 = \mathcal{K}_{\varphi(1)} \cup \cdots \cup \mathcal{K}_{\varphi(L)}$, and that $\varphi: \mathcal{K}_\varphi \in \mathcal{K}(\mathcal{E}_1)$ is read: *those φ for which $\mathcal{K}_\varphi \in \mathcal{K}(\mathcal{E}_1)$*. Of course, you will usually be using the largest K-sets, i.e., prime sets, not just any K-sets, to obtain simple logic expression.

When working with inclusions, U, and exclusions, V, you will frequently want to fall back on the following summary of their interrelations:

$$
\begin{array}{|rr|}
\hline
U_1 \vee U_0 \vee U_y \Leftrightarrow \mathbf{1}, & V_1 V_0 V_y \Leftrightarrow \mathbf{0}, \\
U_1 \vee U_0 \Leftrightarrow V_y, & V_1 V_0 \Leftrightarrow U_y, \\
U_1 \vee U_y \Leftrightarrow V_0, & V_1 V_y \Leftrightarrow U_0, \\
U_0 \vee U_y \Leftrightarrow V_1, & V_0 V_y \Leftrightarrow U_1. \\
U_1 U_0 \Leftrightarrow \mathbf{0}, & V_1 \vee V_0 \Leftrightarrow \mathbf{1}, \\
U_1 U_y \Leftrightarrow \mathbf{0}, & V_1 \vee V_y \Leftrightarrow \mathbf{1}, \\
U_0 U_y \Leftrightarrow \mathbf{0}, & V_0 \vee V_y \Leftrightarrow \mathbf{1}. \\
\hline
\end{array}
\tag{9.5}
$$

Proving these formulas is simple. First consider $U_1 \vee U_0 \vee U_y \Leftrightarrow \mathbf{1}$. As $\mathcal{E}_1 \cup \mathcal{E}_0 \cup \mathcal{E}_y = \mathcal{E}$ we can write $(x \in \mathcal{E}_1) \vee (x \in \mathcal{E}_0) \vee (x \in \mathcal{E}_y) \Leftrightarrow (x \in \mathcal{E})$ which proves $U_1 \vee U_0 \vee U_y \Leftrightarrow \mathbf{1}$ because $x \in \mathcal{E} \Leftrightarrow \mathbf{1}$.

To prove $U_1 \vee U_0 \Leftrightarrow V_y$ we reason as follows:

$$
\begin{aligned}
\mathcal{E}_1 \cup \mathcal{E}_0 &= \mathcal{E} \setminus \mathcal{E}_y, \\
x \in (\mathcal{E}_1 \cup \mathcal{E}_0) &\Leftrightarrow x \in (\mathcal{E} \setminus \mathcal{E}_y) \\
(x \in \mathcal{E}_1) \vee (x \in \mathcal{E}_0) &\Leftrightarrow (x \in \mathcal{E}) \wedge (x \notin \mathcal{E}_y) \\
U_1 \vee U_0 &\Leftrightarrow \mathbf{1} \wedge V_y \\
U_1 \vee U_0 &\Leftrightarrow V_y
\end{aligned}
$$

Finally, $U_1 U_0 \Leftrightarrow \mathbf{0}$ follows directly from $\mathcal{E}_1$ and $\mathcal{E}_0$ being mutually disjoint.

9.4 Basic Memory Evaluation-Formulas

An evaluation formula states when the output of a switching function is 1, and when it is 0. Expressing this for a memory function is somewhat more intricate than for a combinational function (see (5.19)). To simplify the arguments, let us only ask when a memory's output is 1. Referring to the partitions $\mathcal{E}_1$, $\mathcal{E}_0$, and $\mathcal{E}_y$ of (9.2) we note: The output $y(\lambda)$ is obviously 1 in the presence of a setting input event $x \in \mathcal{E}_1$. But $y(\lambda)$ can also be 1 when a memorising input event $x \in \mathcal{E}_y$ is present, assuming that the prior output value, $y(\lambda - 1)$, was 1. Expressing these two cases formally we get:

Case 1: A setting input event is present—$x \in \mathcal{E}_1$. The setting input event $\mathcal{E}_1 := \{x(\lambda) | f(x(\lambda)) = \{1\}\}$ can equivalently be expressed as

$$x(\lambda) \in \mathcal{E}_1 \Leftrightarrow f(x(\lambda)) = \{1\}.$$

As the output $y_{x(\lambda)}$ is an element of $f(x(\lambda))$, the latter in our case being the one-element set $\{1\}$, the output $y_{x(\lambda)}$ itself is 1 allowing us to substitute $f(x(\lambda)) = \{1\}$ by $y_{x(\lambda)} = 1$ or, better still, by the logic variable $Y_{x(\lambda)}$. We can now rewrite $x(\lambda) \in \mathcal{E}_1 \Leftrightarrow f(x(\lambda)) = \{1\}$ as

$$U_1(x(\lambda)) \Leftrightarrow Y_{x(\lambda)}. \tag{9.6}$$

Case 2: A memorising input event is present—$x \in \mathcal{E}_y$. The definition of the memorising input event, $\mathcal{E}_y$ of (9.2), can be expressed as

$$x(\lambda) \in \mathcal{E}_y \Leftrightarrow (f(x(\lambda)) = \{0, 1\}) \rightarrow (y_{x(\lambda)} = y_{x(\lambda-1)}). \tag{*}$$

Successively, we transform each of the propositional forms $x(\lambda) \in \mathcal{E}_y$, $f(x(\lambda)) = \{0, 1\}$, and $y_{x(\lambda)} = y_{x(\lambda-1)}$ into logical expression. The first is trivial:

$$x(\lambda) \in \mathcal{E}_y \Leftrightarrow U_y.$$

The output $y_{x(\lambda)}$ being an element of $f(x(\lambda))$, which itself equals $\{0, 1\}$, allows us to deduce as follows:

$$f(x(\lambda)) = \{0, 1\} \Leftrightarrow y_{x(\lambda)} \in \{0, 1\} \Leftrightarrow y_{x(\lambda)} = 0 \ \vee \ y_{x(\lambda)} = 1 \Leftrightarrow \overline{Y}_{x(\lambda)} \vee Y_{x(\lambda)} \Leftrightarrow \mathbf{1}.$$

We transform the third propositional form, $y_{x(\lambda)} = y_{x(\lambda-1)}$, by using the logic variables $Y_{x(\lambda)} \Leftrightarrow y_{x(\lambda)} = 1$ and $Y_{x(\lambda-1)} \Leftrightarrow y_{x(\lambda-1)} = 1$. Using a truth table, it is easy to prove

$$y_{x(\lambda)} = y_{x(\lambda-1)} \Leftrightarrow Y_{x(\lambda)} \leftrightarrow Y_{x(\lambda-1)}.$$

The above logical form, (*), of the memorising input events can now be rewritten

$$U_y \Leftrightarrow \mathbf{1} \rightarrow (Y_{x(\lambda)} \leftrightarrow Y_{x(\lambda-1)}),$$

$$U_y \Leftrightarrow Y_{x(\lambda)} \leftrightarrow Y_{x(\lambda-1)},$$

$$U_y \Leftrightarrow Y_{x(\lambda)} Y_{x(\lambda-1)} \vee \overline{Y}_{x(\lambda)} \overline{Y}_{x(\lambda-1)}.$$

ANDing each side of the above formula with $Y_{x(\lambda-1)}$ we obtain

$$U_y Y_{x(\lambda-1)} \Leftrightarrow Y_{x(\lambda)} \underbrace{Y_{x(\lambda-1)} Y_{x(\lambda-1)}}_{Y_{x(\lambda-1)}} \vee \overline{Y}_{x(\lambda)} \underbrace{\overline{Y}_{x(\lambda-1)} Y_{x(\lambda-1)}}_{\mathbf{0}},$$

$$U_y Y_{x(\lambda-1)} \Leftrightarrow Y_{x(\lambda)} Y_{x(\lambda-1)} \vee \mathbf{0},$$

$$U_y Y_{x(\lambda-1)} \Leftrightarrow Y_{x(\lambda)} Y_{x(\lambda-1)}. \tag{9.7}$$

This is as far as Case 2 takes us, for unlike conventional mathematics, in mathematical logic we are not allowed to cancel $Y_{x(\lambda-1)}$ in (9.7).

Combining Cases 1 and 2 to obtain the evaluation formula. We obtain the value of the output variable $Y_{x(\lambda)}$ by combining the (9.6) and (9.7) via OR, i.e., ORing their left sides and right sides separately:

$$Y_{x(\lambda)} \Leftrightarrow U_1 \qquad (9.6\ldots)$$

$$\underline{Y_{x(\lambda)} Y_{x(\lambda-1)}} \Leftrightarrow \underline{U_y Y_{x(\lambda-1)}} \qquad (9.7\ldots)$$

$$Y_{x(\lambda)} \vee Y_{x(\lambda)} Y_{x(\lambda-1)} \Leftrightarrow U_1 \vee U_y Y_{x(\lambda-1)}$$

$$Y_{x(\lambda)}(\mathbf{1} \vee Y_{x(\lambda-1)}) \Leftrightarrow U_1 \vee U_y Y_{x(\lambda-1)}$$

$$Y_{x(\lambda)} \Leftrightarrow U_1 \vee U_y Y_{x(\lambda-1)}$$

This is the first of the memory evaluation formulas. From it we can develop the others with the help of the inclusion and exclusion formulas summarised in (9.5). We start by developing the exclusion form:

$$\begin{aligned} Y_{x(\lambda)} &\Leftrightarrow U_1 \vee U_y Y_{x(\lambda-1)} \\ &\Leftrightarrow V_0 V_y \vee V_1 V_0 Y_{x(\lambda-1)} \\ &\Leftrightarrow V_0\big(V_y \vee V_1 Y_{x(\lambda-1)}\big) \\ &\Leftrightarrow V_0 \underbrace{\big(V_y \vee V_1\big)}_{\mathbf{1}} \big(V_y \vee Y_{x(\lambda-1)}\big) \\ Y_{x(\lambda)} &\Leftrightarrow V_0\big(V_y \vee Y_{x(\lambda-1)}\big). \end{aligned}$$

Lastly, we invert the obtained evaluation formulas:

$$\overline{Y}_{x(\lambda)} \Leftrightarrow \overline{U_1 \vee U_y Y_{x(\lambda-1)}}, \qquad \overline{Y}_{x(\lambda)} \Leftrightarrow \overline{V_0(V_y \vee Y_{x(\lambda-1)})}$$

$$\Leftrightarrow \overline{U}_1 \wedge \overline{(U_y Y_{x(\lambda-1)})}, \qquad \Leftrightarrow \overline{V}_0 \vee \overline{(V_y \vee Y_{x(\lambda-1)})}$$

$$\Leftrightarrow \overline{U}_1(\overline{U}_y \vee \overline{Y}_{x(\lambda-1)}), \qquad \Leftrightarrow \overline{V}_0 \vee (\overline{V}_y \wedge \overline{Y}_{x(\lambda-1)})$$

$$\overline{Y}_{x(\lambda)} \Leftrightarrow V_1(V_y \vee \overline{Y}_{x(\lambda-1)}). \qquad \overline{Y}_{x(\lambda)} \Leftrightarrow U_0 \vee U_y \overline{Y}_{x(\lambda-1)}.$$

These so-called **basic evaluation formulas** are now gathered:

$$\boxed{\begin{aligned} Y_{x(\lambda)} &\Leftrightarrow U_1 \vee U_y Y_{x(\lambda-1)} \Leftrightarrow V_0(V_y \vee Y_{x(\lambda-1)}), \\ \overline{Y}_{x(\lambda)} &\Leftrightarrow U_0 \vee U_y \overline{Y}_{x(\lambda-1)} \Leftrightarrow V_1(V_y \vee \overline{Y}_{x(\lambda-1)}). \end{aligned}} \tag{9.8}$$

9.5 Generalised Memory Evaluation-Formulas

In this section we put forth a generalisation of memory evaluation-formulas, again leaving their application for the next chapter where their worth is demonstrated. The **generalised memory evaluation formulas** stating when the output is 1 are:

$$\boxed{\begin{aligned} Y_{x(\lambda)} &\Leftrightarrow U_1 \vee (U_{1S} \vee U_y) Y_{x(\lambda-1)} \Leftrightarrow V_0(V_{0S} V_y \vee Y_{x(\lambda-1)}), \\ \overline{Y}_{x(\lambda)} &\Leftrightarrow U_0 \vee (U_{0S} \vee U_y) \overline{Y}_{x(\lambda-1)} \Leftrightarrow V_1(V_{1S} V_y \vee \overline{Y}_{x(\lambda-1)}). \end{aligned}} \tag{9.9}$$

These formulas use arbitrary subsets, $\mathcal{E}_{1S}$ and $\mathcal{E}_{0S}$, of the setting and resetting equivalents:

$$\emptyset \subseteq \mathcal{E}_{1S} \subseteq \mathcal{E}_1, \qquad \emptyset \subseteq \mathcal{E}_{0S} \subseteq \mathcal{E}_0, \tag{9.10}$$

their respective inclusions and exclusions being written as

$$U_{1S} :\Leftrightarrow x \in \mathcal{E}_{1S}, \qquad U_{0S} :\Leftrightarrow x \in \mathcal{E}_{0S}. \tag{9.11a}$$

$$V_{1S} :\Leftrightarrow x \notin \mathcal{E}_{1S}, \qquad V_{0S} :\Leftrightarrow x \notin \mathcal{E}_{0S}. \tag{9.11b}$$

Most certainly you will have no difficulty in proving the following relationships:

$$\boxed{\begin{aligned} U_1 \vee U_{1S} &\Leftrightarrow U_1, & V_1 V_{1S} &\Leftrightarrow V_1, \\ U_0 \vee U_{0S} &\Leftrightarrow U_0, & V_0 V_{0S} &\Leftrightarrow V_0, \\ U_1 U_{1S} &\Leftrightarrow U_{1S}, & V_1 \vee V_{1S} &\Leftrightarrow V_{1S}, \\ U_0 U_{0S} &\Leftrightarrow U_{0S}, & V_0 \vee V_{0S} &\Leftrightarrow V_{0S}. \end{aligned}} \tag{9.12}$$

Using these formulas we derive $Y_{x(\lambda)} \Leftrightarrow U_1 \vee \big(U_{1S} \vee U_y\big)Y_{x(\lambda-1)}$, (9.9), from $Y_{x(\lambda)} \Leftrightarrow U_1 \vee U_y Y_{x(\lambda-1)}$, (9.8), in the following way:

$$\begin{aligned}
Y_{x(\lambda)} &\Leftrightarrow U_1 \vee U_y Y_{x(\lambda-1)} \\
&\Leftrightarrow U_1 \vee U_{1S} \vee U_y Y_{x(\lambda-1)} \\
&\Leftrightarrow U_1 \vee U_{1S}\mathbf{1} \vee U_y Y_{x(\lambda-1)} \\
&\Leftrightarrow U_1 \vee U_{1S}\big(\mathbf{1} \vee Y_{x(\lambda-1)}\big) \vee U_y Y_{x(\lambda-1)} \\
&\Leftrightarrow U_1 \vee U_{1S} \vee U_{1S} Y_{x(\lambda-1)} \vee U_y Y_{x(\lambda-1)} \\
Y_{x(\lambda)} &\Leftrightarrow U_1 \vee \big(U_{1S} \vee U_y\big) Y_{x(\lambda-1)}
\end{aligned}$$

The next chapter is entirely devoted to the practical application of the memory evaluation formulas, (9.9), in developing feedback latches. The methods developed there, and the feedback latches themselves, are the backbone to designing asynchronous sequential circuits as Division Three shows.

Chapter 10
Designing Feedback Latches*

The memory evaluation-formulas of the previous chapter are the backbone to designing *feedback* latches (such as the relay circuit of Fig. 9.1b). Yet, feedback can only be realised if we can show that the prior output $Y_{(\lambda-1)}$ may always (i.e., for all input events) be chosen to equal the present output $Y_{(\lambda)}$—for then and only then can we connect the two leads, thus realising feedback. The necessary analysis is carried out in Sect. 10.1 leading to modified evaluation formulas adapted to designing feedback latches. Section 10.2 points out a sensitive topic in the input–output behaviour of feedback latches—their susceptibility to hazards. Contrary to combinational circuits, an otherwise briefly erroneous output value can be perpetuated causing the latch to malfunction. There are two distinctly different ways to inhibit the occurrence of hazards. Section 10.3 discusses the cruder of these methods—the introduction of a delay in the feedback line. A more refined approach is taken up in Sect. 10.4 in which the design and working of pre-established feedback signals is discussed. Lastly, Sect. 10.5 looks into the possibility of minimising latches.

10.1 Feedback Evaluation-Formulas

We use a basic memory evaluation-formula to determine the dependence of the output $Y_{x(\lambda)}$ in a given location λ on the output $Y_{x(\lambda-1)}$ of the previous location $\lambda - 1$. The formula of our choice is $Y_{x(\lambda)} \Leftrightarrow U_1 \vee U_y Y_{x(\lambda-1)}$, (9.8). As a memory function partitions its domain $\mathcal{E}$ into the mutually disjoint and non-empty subsets $\mathcal{E}_1$, $\mathcal{E}_0$, and $\mathcal{E}_y$, (9.2), we must consider the above memory evaluation-formula for the following three cases:

Case 1: Take $x(\lambda)$ to be a setting input-event so that $U_1 \Leftrightarrow x(\lambda) \in \mathcal{E}_1 \Leftrightarrow \mathbf{1}$, $U_0 \Leftrightarrow x(\lambda) \in \mathcal{E}_0 \Leftrightarrow \mathbf{0}$, and $U_y \Leftrightarrow x(\lambda) \in \mathcal{E}_y \Leftrightarrow \mathbf{0}$. Then the memory's output becomes
$Y_{x(\lambda)} \Leftrightarrow U_1 \vee \left(U_y \wedge Y_{x(\lambda-1)}\right) \Leftrightarrow \mathbf{1} \vee \left(\mathbf{0} \wedge Y_{x(\lambda-1)}\right) \Leftrightarrow \mathbf{1} \vee \mathbf{0} \Leftrightarrow \mathbf{1}$.

S.P. Vingron, *Logic Circuit Design*, DOI 10.1007/978-3-642-27657-6_10,

Case 2: Take $x(\lambda)$ to be a resetting input-event so that $U_1 \Leftrightarrow x(\lambda) \in \mathcal{E}_1 \Leftrightarrow \mathbf{0}$, $U_0 \Leftrightarrow x(\lambda) \in \mathcal{E}_0 \Leftrightarrow \mathbf{1}$, and $U_y \Leftrightarrow x(\lambda) \in \mathcal{E}_y \Leftrightarrow \mathbf{0}$. Then the memory's output becomes
$Y_{x(\lambda)} \Leftrightarrow U_1 \vee (U_y \wedge Y_{x(\lambda-1)}) \Leftrightarrow \mathbf{0} \vee (\mathbf{0} \wedge Y_{x(\lambda-1)}) \Leftrightarrow \mathbf{0} \vee \mathbf{0} \Leftrightarrow \mathbf{0}$.

Case 3: Take $x(\lambda)$ to be a memorising input-event so that $U_1 \Leftrightarrow x(\lambda) \in \mathcal{E}_1 \Leftrightarrow \mathbf{0}$, $U_0 \Leftrightarrow x(\lambda) \in \mathcal{E}_0 \Leftrightarrow \mathbf{0}$, and $U_y \Leftrightarrow x(\lambda) \in \mathcal{E}_y \Leftrightarrow \mathbf{1}$. Then the memory's output becomes
$Y_{x(\lambda)} \Leftrightarrow U_1 \vee (U_y \wedge Y_{x(\lambda-1)}) \Leftrightarrow \mathbf{0} \vee (\mathbf{1} \wedge Y_{x(\lambda-1)}) \Leftrightarrow \mathbf{0} \vee Y_{x(\lambda-1)} \Leftrightarrow Y_{x(\lambda-1)}$.

In the first two cases, when $x(\lambda)$ is either a setting or resetting input-event, the output $Y_{x(\lambda)}$ in no way depends on $Y_{x(\lambda-1)}$, while $Y_{x(\lambda)}$ and $Y_{x(\lambda-1)}$ are equivalent, $Y_{x(\lambda)} \Leftrightarrow Y_{x(\lambda-1)}$, when $x(\lambda)$ is a memorising input-event. We may therefore require the equivalence between $Y_{x(\lambda)}$ and $Y_{x(\lambda-1)}$ for all λ of the infinite index set $\mathcal{L}$ of locations without contradicting Cases 1 and 2:

$$\boxed{Y_{x(\lambda)} \Leftrightarrow Y_{x(\lambda-1)}} \tag{10.1}$$

We call this the **postulate of feedback**. Using this postulate, we substitute $Y_{x(\lambda)}$ for $Y_{x(\lambda-1)}$ in $Y_{x(\lambda)} \Leftrightarrow U_1 \vee U_y Y_{x(\lambda-1)}$ obtaining (in unabridged notation)

$$Y_{x(\lambda)} \Leftrightarrow U_1(x(\lambda)) \vee U_y(x(\lambda))Y_{x(\lambda)}. \tag{10.2}$$

For better readability we revert to the abridged notation, used so far, where the argument $x(\lambda)$ is dropped. The basic memory evaluation formulas then take on the following form for the feedback latches:

$$\boxed{\begin{aligned} Y &\Leftrightarrow U_1 \vee U_y Y \Leftrightarrow V_0(V_y \vee Y), \\ \overline{Y} &\Leftrightarrow U_0 \vee U_y \overline{Y} \Leftrightarrow V_1(V_y \vee \overline{Y}). \end{aligned}} \tag{10.3}$$

We call the above formulas the **basic feedback evaluation-formulas**.

A similar analysis as above, of the generalised memory evaluation-formulas, (9.9), enables us to develop *generalised feedback evaluation-formulas*. The analysis necessary requires us to subdivide Case 1:

Case 1a: Take $x(\lambda)$ to be a setting input-event which also belongs to an arbitrarily chosen subset $\emptyset \subseteq \mathcal{E}_{1S} \subseteq \mathcal{E}_1$ so that $U_1 \Leftrightarrow x(\lambda) \in \mathcal{E}_1 \Leftrightarrow \mathbf{1}$, $U_{1S} \Leftrightarrow x(\lambda) \in \mathcal{E}_{1S} \Leftrightarrow \mathbf{1}$, $U_0 \Leftrightarrow x(\lambda) \in \mathcal{E}_0 \Leftrightarrow \mathbf{0}$, and $U_y \Leftrightarrow x(\lambda) \in \mathcal{E}_y \Leftrightarrow \mathbf{0}$. Then the memory's output becomes
$Y_{x(\lambda)} \Leftrightarrow U_1 \vee (U_{1S} \vee U_y)Y_{x(\lambda-1)} \Leftrightarrow \mathbf{1} \vee (\mathbf{1} \vee \mathbf{0})Y_{x(\lambda-1)} \Leftrightarrow \mathbf{1} \vee Y_{x(\lambda-1)} \Leftrightarrow \mathbf{1}$.

Case 1b: Take $x(\lambda)$ to be a setting input-event, but assume it not to belong to an arbitrarily chosen subset $\emptyset \subseteq \mathcal{E}_{1S} \subseteq \mathcal{E}_1$ so that $U_1 \Leftrightarrow x(\lambda) \in \mathcal{E}_1 \Leftrightarrow \mathbf{1}$, $U_{1S} \Leftrightarrow x(\lambda) \in \mathcal{E}_{1S} \Leftrightarrow \mathbf{0}$, $U_0 \Leftrightarrow x(\lambda) \in \mathcal{E}_0 \Leftrightarrow \mathbf{0}$, and $U_y \Leftrightarrow x(\lambda) \in \mathcal{E}_y \Leftrightarrow \mathbf{0}$. Then the memory's output becomes
$Y_{x(\lambda)} \Leftrightarrow U_1 \vee \big(U_{1S} \vee U_y\big)Y_{x(\lambda-1)} \Leftrightarrow \mathbf{1} \vee \big(\mathbf{0} \vee \mathbf{0}\big)Y_{x(\lambda-1)} \Leftrightarrow \mathbf{1} \vee \mathbf{0} \Leftrightarrow \mathbf{1}$.

Now following the above reasoning, we get the **generalised feedback evaluation-formulas**:

$$\boxed{\begin{aligned} Y &\Leftrightarrow U_1 \vee \big(U_{1S} \vee U_y\big)Y \Leftrightarrow V_0\big(V_{0S}V_y \vee Y\big), \\ \overline{Y} &\Leftrightarrow U_0 \vee \big(U_{0S} \vee U_y\big)\overline{Y} \Leftrightarrow V_1\big(V_{1S}V_y \vee \overline{Y}\big). \end{aligned}} \tag{10.4}$$

10.2 Design and Memorisation Hazards

Feedback circuits are designed according to either the inclusive (evaluation) formula $Y \Leftrightarrow U_1 \vee \big(U_{1S} \vee U_y\big)Y$, or the exclusive (evaluation) formula $Y \Leftrightarrow V_0\big(V_{0S}V_y \vee Y\big)$, (10.4). In the former instance we need to choose an arbitrary subset $\mathcal{E}_{1S}$ of all setting input-events $x \in \mathcal{E}_1$, while in the latter case we must choose a subset $\mathcal{E}_{0S} \in \mathcal{E}_0$ of resetting input-events. In this section we take these subsets to be empty, i.e., we assume that $\mathcal{E}_{1S} = \emptyset$ and $\mathcal{E}_{0S} = \emptyset$, which means we will be using the evaluation formulas in their *basic* forms $Y \Leftrightarrow U_1 \vee U_y Y$, and $Y \Leftrightarrow V_0\big(V_y \vee Y\big)$, (10.3). You will subsequently see that latches designed along these lines are the most hazard prone. In fact, the more ideal the switching elements (i.e., the faster they switch and the quicker they react), the greater the danger of the circuit malfunctioning. Studying this phenomenon is important to know how to avoid it. Designing latches with a minimal number of switching devices is of marginal importance—designing hazard-free latches is all-important.

Basic Design of a Latch

'*Basic*' in the above title refers to the use of the *basic* evaluation formulas $Q \Leftrightarrow U_1 \vee U_q Q$, and $Q \Leftrightarrow V_0\big(V_q \vee Q\big)$, (10.3). To demonstrate the design procedure, we use the example shown in Fig. 10.1 where we design the *inclusive* form of the D-latch. We will normally assume the latch we wish to design to be specified by a reduced K-map, see Fig. 10.1a for our example. We first calculate the circuit's inclusive formula, $Q \Leftrightarrow CD \vee \overline{C}\,Q$, as shown in the figure, the setting inclusion, U_1, being CD, the memorising inclusion, U_q, being $\overline{C}$. The right side of the inclusion formula, the expression $CD \vee \overline{C}\,Q$, allows us to draw the interim circuit naming its output Q due to $Q \Leftrightarrow CD \vee \overline{C}\,Q$.

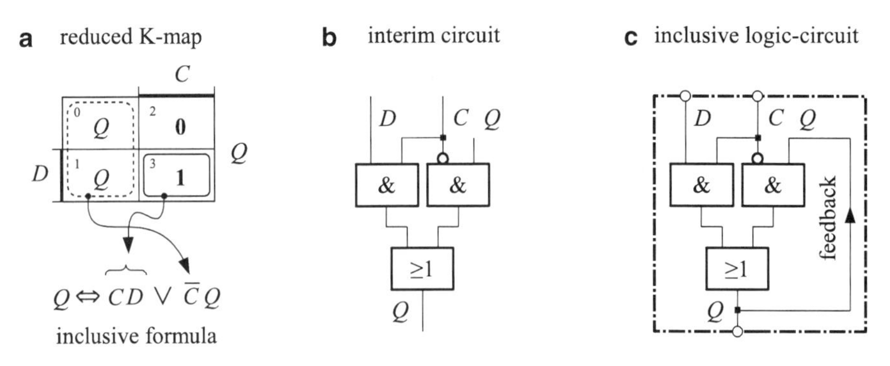

Fig. 10.1 Designing the D-latch in its basic disjunctive form

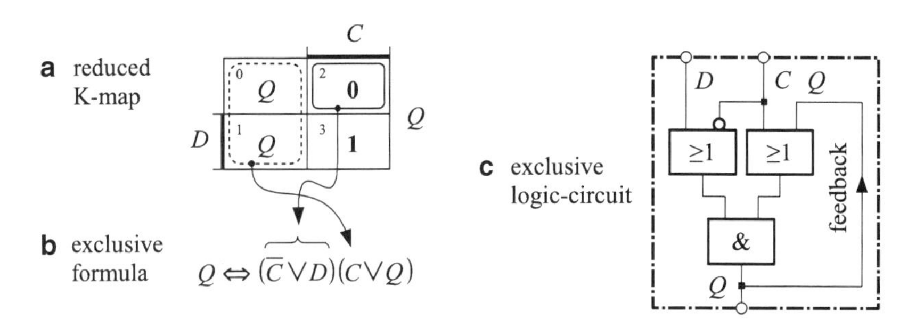

Fig. 10.2 Designing the D-latch in its basic conjunctive form

The value of the *input* Q is obtained by connecting the *output* Q to the *input* Q thus creating a **feedback** as shown in Fig. 10.1c. The frame around this circuit is only drawn to emphasize that the latch has only *two* independent input signals, namely C and D. The third input, Q, is not independent and is called the **internal state** of the latch. The logic circuit derived is that of the TTL D-latch 74SL75.

It is equally simple to design the logic circuit in its exclusive form, as Fig. 10.2 shows. The resetting exclusion, V_0, is obtained from the reduced K-map as $\overline{C} \vee D$, while the memorising exclusion V_q is C. The basic exclusive formula, $Q \Leftrightarrow V_0(V_q \vee Q)$, (10.3), then becomes $Q \Leftrightarrow (\overline{C} \vee D)(C \vee Q)$. This formula leads to the circuit of Fig. 10.2c. We next take a close look at the possibility of these circuits malfunctioning.

Occurrence of Memorisation Hazards

A **memorisation hazard** refers to the danger of a latch *malfunctioning* when switching from any input event (whether a setting, resetting or even memorising) to a memorising input-event (a *new* memorising input-event in the latter case). In this

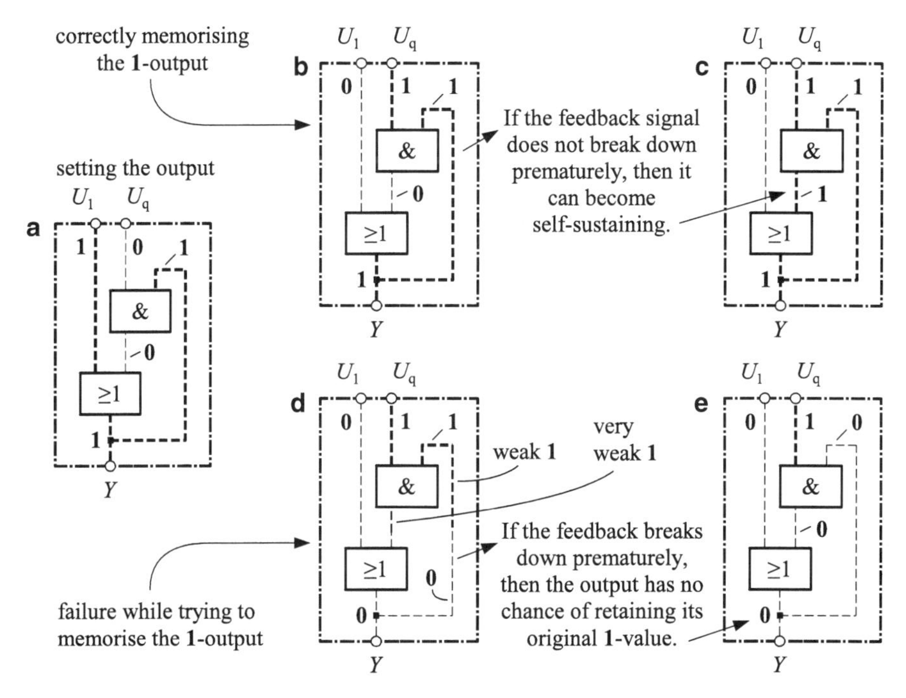

Fig. 10.3 Visualising the proper functioning of a basic disjunctive latch (*top row*), and the malfunctioning of the latch (*bottom row*)

context, **malfunctioning** means that the output *changes its value* when the (new) memorising input-event occurs instead of retaining the value.

The logic circuit in Fig. 10.3 is a graphical representation of $Y \Leftrightarrow U_1 \vee U_q Y$, (10.3). Figure 10.3a shows the signal flow in the presence of a setting input-event (in which case $U_1 \Leftrightarrow \mathbf{1}$, $U_0 \Leftrightarrow \mathbf{0}$, and $U_q \Leftrightarrow \mathbf{0}$). Note that the feedback signal is interrupted at the AND-gate. **For the latch to work properly**—see Fig. 10.3b—when switching to a memorising input-event ($U_1 \Leftrightarrow \mathbf{0}$, $U_0 \Leftrightarrow \mathbf{0}$, and $U_q \Leftrightarrow \mathbf{1}$) we must assume that the feedback signal remains **1** although the U_1-input to the OR (which causes the feedback signal to be **1**) has dropped to **0**. Both inputs to the AND now being **1**, the AND fires, thereby sustaining the **1**-signal in the feedback loop, see Fig. 10.3c.

A more realistic scenario—see Fig. 10.3d— is to assume that the **1**-signal in the feedback will start breaking down once U_1 becomes **0** this making it uncertain if the energy supplied by a weakening **1**-output of the AND-gate suffices to sustain or rebuild a **1**-signal in the feedback loop. In this case it is quite likely that the feedback loop will break down so that the output erroneously becomes **0**. In the following sections we concentrate on how to avoid this malfunctioning.

Having seen the strong likelihood of the basic disjunctive latch malfunctioning when a **1**-output is to be memorised, it is notable that **memorising a 0-output never fails**, i.e., there is no danger of the latch malfunctioning in this case. This

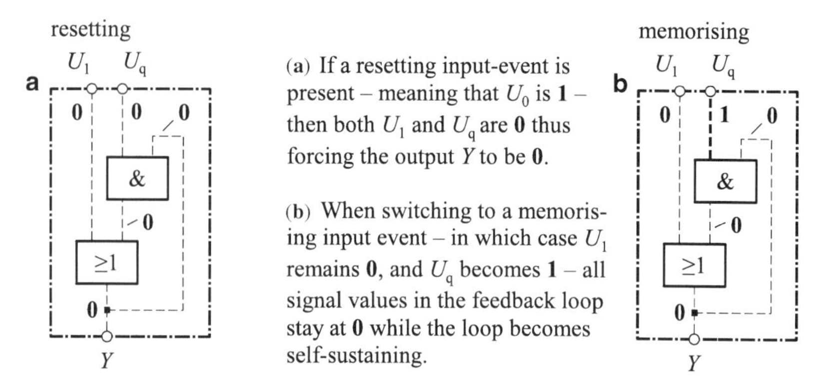

Fig. 10.4 A basic disjunctive latch never malfunctions when memorising a **0**-output

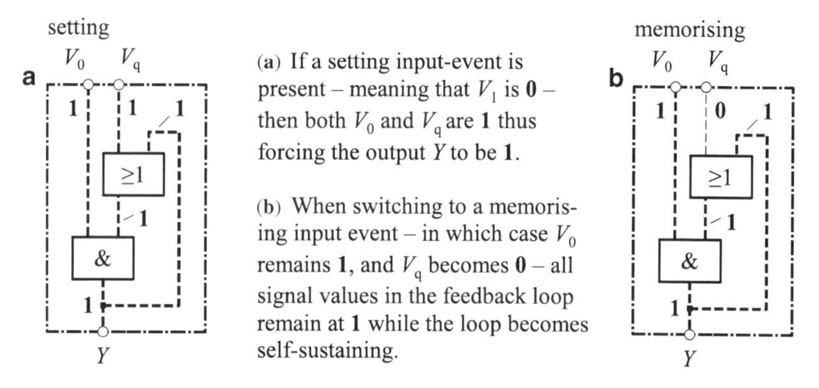

Fig. 10.5 A basic conjunctive latch never malfunctions when memorising a **1**-output

is visualised and explained in Fig. 10.4, and I strongly recommend pondering the figure.

The basic conjunctive latch of Fig. 10.5 which is designed according to $Y \Leftrightarrow V_0(V_y \vee Y)$, (10.3), is equally important. **It never malfunctions when a 1-output is to be memorised**. This is pictured and explained in Fig. 10.5, a figure I strongly encourage you to ponder.

On the other hand, the basic conjunctive latch shows its weakness when trying to memorise a **0**-output. The top row of Fig. 10.6 shows and explains the signal flow that must prevail for memorisation to be successful. The bottom row explains and pictures the signal flow that leads to the circuit malfunctioning. Again, I leave it to you to ponder this figure.

Designing feedback latches so that memorisation hazards are avoided as far as possible is the prime design objective, the reason being that if an error occurs it is not transient—it is permanent. Achieving these design goals is the topic of the next two sections.

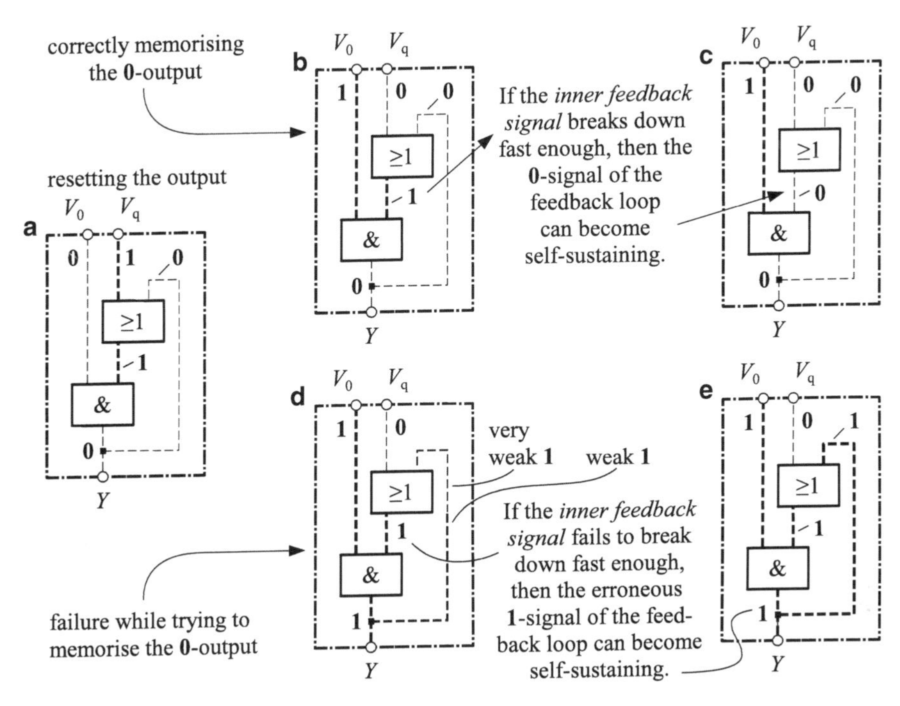

Fig. 10.6 Visualising the proper functioning of a basic conjunctive latch (*top row*), and the malfunctioning of the latch (*bottom row*)

10.3 Delayed Feedback

The switching device that gave rise to switching circuitry was the electro-magnetic relay developed by the Scottish–American scientist **Josef Henry** in 1835. Intrinsic to these devices is the inertia and delay with which the relay's contacts react to a change in the coil's state of energisation. This inertia and delay was—and mostly still is—taken to be an essential part of every feedback circuit, and is modelled by **installing an ideal inertial delay in the feedback**. The behaviour of the ideal inertial delay is pictured in Fig. 10.7. The leading and trailing edges of an input signal are delayed equally long by the constant delay-time Δt. *Inertial* refers to pulses and troughs shorter than the delay time Δt having no effect on the output. In this respect you can view a delay element as an *energy buffer* compensating brief and/or runt salients.

To see the effect of an inertial delay we first use the basic disjunctive form of the D-latch designed in Fig. 10.1 and shown in Fig. 10.8 with an inertial delay in the feedback. As basic disjunctive latches notoriously have difficulties memorising a **1**-output, we examine this in forcing a **1**-output by setting the latch. Figure 10.8a shows the signal flow for the setting input-event $C \Leftrightarrow$ **1** and $D \Leftrightarrow$ **1**, which specifies cell 3 in the K-map of Fig. 10.1a. Note that the **1**-signal in the feedback

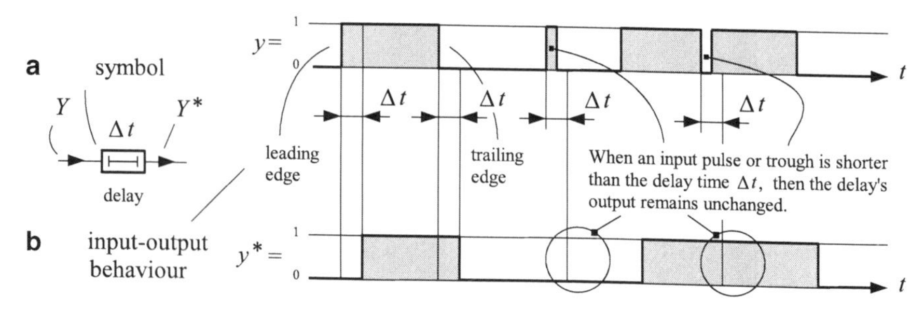

Fig. 10.7 Input–output behaviour of the ideal inertial-delay

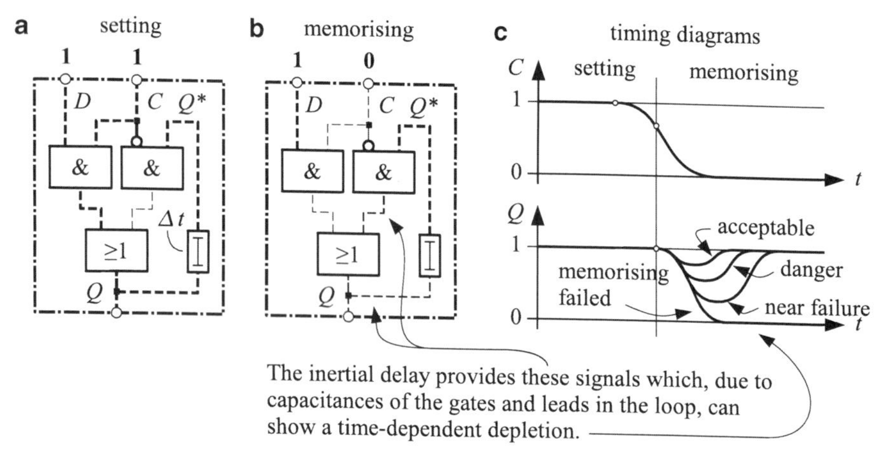

Fig. 10.8 Studying the effect of an inertial delay in the feedback of the basic disjunctive form of the D-latch

loop is interrupted between the loop's AND-gate and the OR-gate—just as is the case for the setting condition of the general circuit of Fig. 10.3a.

Switching to the memorising input-event $C \Leftrightarrow \mathbf{0}$ and $D \Leftrightarrow \mathbf{1}$ forces the output of the AND $(D \wedge C)$ to **0** thereby depriving the output-OR of its setting input signal. To compensate for this, the inertial delay must send its **1**-signal, Q^*, through the now open AND, $(\overline{C} \wedge Q^*)$, of the feedback loop, through the output-OR, and back to the delay—all hopefully fast enough to make the feedback loop self-sustaining. In real-world circuits this can, to a varying degree, lead to a time-dependent depletion of the output signal as schematically pictured in Fig. 10.8c, this showing that the inertial delay does not guarantee the proper functioning of memorisation.

An inertial delay in electronic circuits is realised by an IDENTITY, or two NEGATIONS in series. A classical example is the TTL D-latch 74SL75 shown in Fig. 10.9.

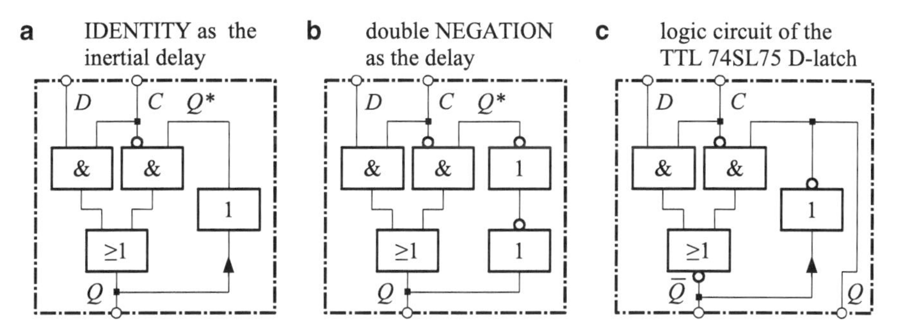

Fig. 10.9 The design of the TTL D-latch 74SL75

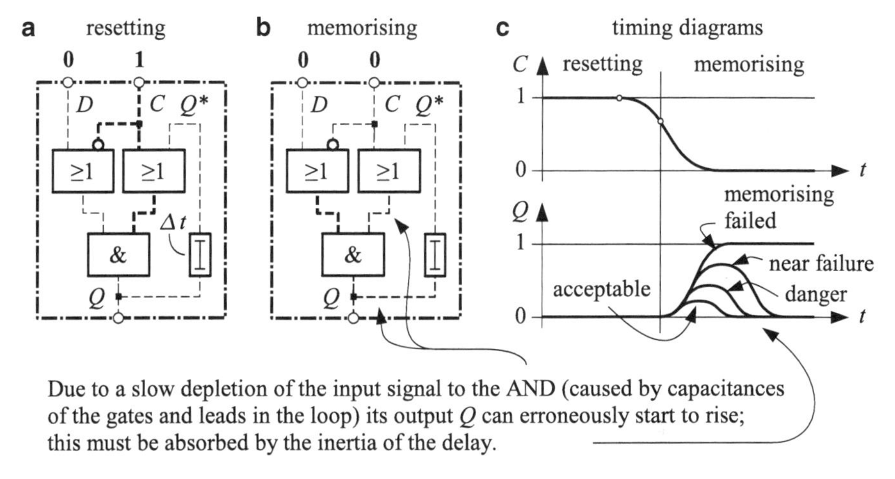

Fig. 10.10 Studying the effect of an inertial delay in the feedback of the basic conjunctive form of the D-latch

Not surprisingly, basic conjunctive latches are prone to malfunction when trying to memorise a **0**-output. This is pictured and discussed in Fig. 10.10 which, before continuing, you might want to ponder.

The more energy is stored in the delay of the feedback, and the longer the delay is able to dissipate this energy into the feedback loop, the smaller the time-dependent deviation of the value of the latch's output signal from its correctly memorised value. On the other hand, this has an adverse influence on the frequency with which input events may change while still causing correct output values (these are so-called **slow latches**). As you can see, employing a delay in the feedback to ensure the proper functioning of a feedback latch is not very effective. It is crude in that it has no theoretical basis and is a purely intuitive attempt to solve the problem of a memorisation hazard. A far better approach is shown in the next section.

10.4 Pre-established Feedback

The generalised feedback evaluation-formulas make a design method possible which proves ideal in avoiding memorisation hazards. Let us start with the inclusive form, $Y \Leftrightarrow U_1 \vee (U_{1S} \vee U_y)Y$, (10.4), used to develop *disjunctive* feedback latches. U_{1S} is the inclusion, the disjunctive logic expression, of an arbitrary subset $\mathcal{E}_{1S}$ of the setting input events $\mathcal{E}_1$, meaning $\emptyset \subseteq \mathcal{E}_{1S} \subseteq \mathcal{E}_1$. The case to be looked into in the context of avoiding memorisation hazards is when $\mathcal{E}_{1S} = \mathcal{E}_1$. As in the previous section, we demonstrate the effect for the D-latch, see Fig. 10.11. Do note the dashed loop in this figure: It represents the union $\mathcal{E}_1 \cup \mathcal{E}_q$ of the setting with the memorising input events. It is important to calculate its inclusion, let us call it U_{1q}, by using *prime sets*.

Summarising, we design disjunctive feedback latches according to the formula:

$$\boxed{Y \Leftrightarrow U_1 \vee U_{1q}Y \qquad \text{with} \qquad U_{1q}(x) \Leftrightarrow x \in (\mathcal{E}_1 \cup \mathcal{E}_q)} \tag{10.5}$$

and U_{1q} is expressed by using prime sets. **Circuits designed along these lines are completely free of memorisation hazards** as we will see. From the reduced K-map of Fig. 10.11 we can read the inclusions $U_1 \Leftrightarrow CD$ and $U_{1q} \Leftrightarrow \overline{C} \vee D$ so that $Y \Leftrightarrow U_1 \vee U_{1q}Y \Leftrightarrow CD \vee (\overline{C} \vee D)Y$ this leading to the logic circuit of Fig. 10.11.

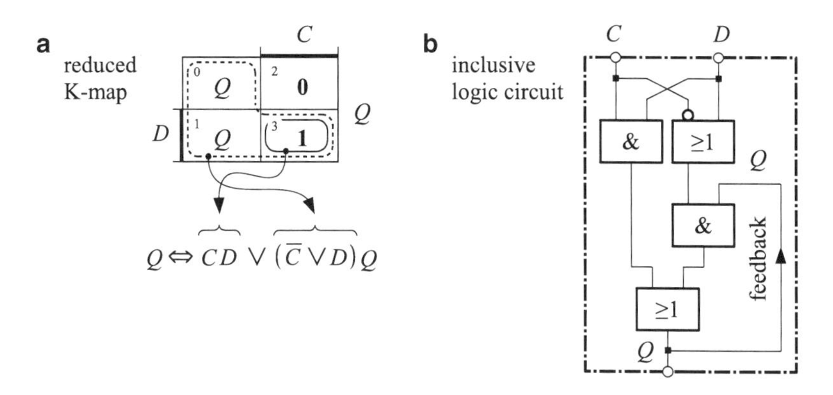

Fig. 10.11 Designing a D-latch according to $Y \Leftrightarrow U_1 \vee U_{1q}Y$, (10.5)

This being a *disjunctive* feedback-latch it *per se* never malfunctions when memorising a **0**-output. On the other hand, disjunctive latches (at least *basic* disjunctive latches) are prone to malfunction when trying to store a **1**-output signal (see, for example, Fig. 10.8a with its interrupted feedback signal). But consider the signal flow for a setting input-event for the new latch of Fig. 10.11—the signal flow being depicted in Fig. 10.12b: You will notice that the feedback loop carries an *uninterrupted **1**-signal throughout*. We call a loop with this property a **pre-established feedback-loop**. Its worth becomes evident when switching to a memorising input-event, see Fig. 10.12c: Although the setting inclusion, $C \wedge D$,

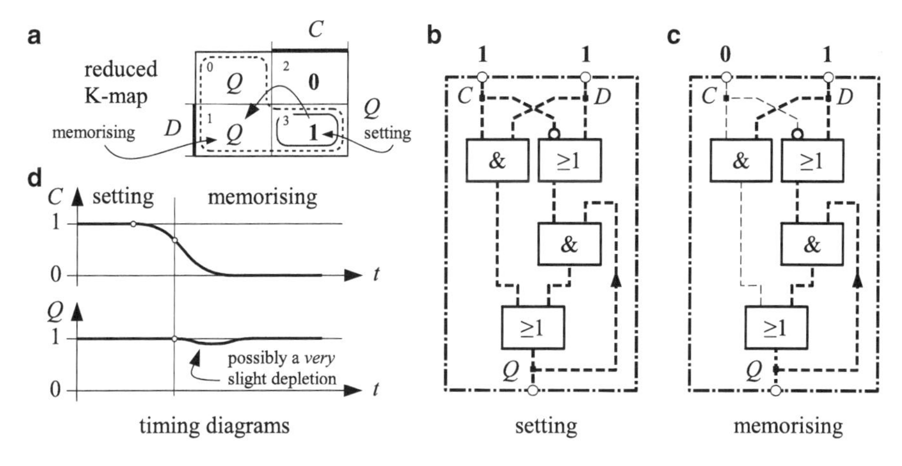

Fig. 10.12 The pre-established feedback ensures that a **1**-output of the disjunctive D-latch is stored faultlessly

no longer supplies the feedback loop with a **1**-signal via the output-OR, the loop already is and remains self-sustaining. The desired transition from the setting to a memorising input-event is shown in the reduced K-map of Fig. 10.12a. The timing diagram of Fig. 10.12d shows the flawless memorisation of the output signal. A possible minor depletion, as indicated in the figure, might occur if the input-OR $\overline{C} \vee D$ were to produce a **1**-hazard, or the output-OR a **0**-hazard.

Most engineers are less accustomed to designing *conjunctive* latches, as these use exclusions V_S and not inclusions U_S. To design a conjunctive latch completely free of memorisation hazards, we choose $\mathcal{E}_{0S}$ to be $\mathcal{E}_0$ when adapting $Y \Leftrightarrow V_0\big(V_{0S} V_y \vee Y\big)$, (10.4), and thus use the formula:

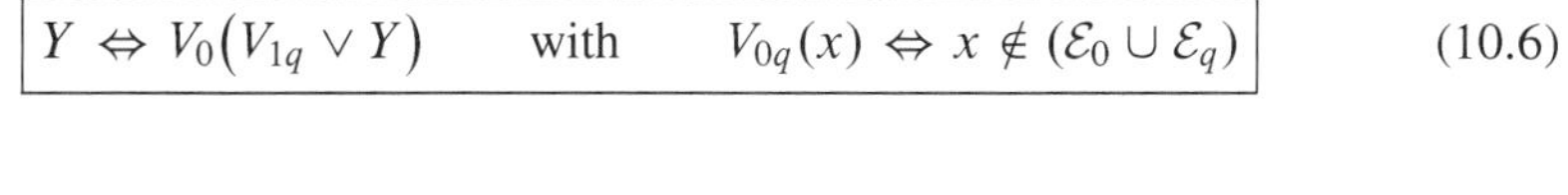

$$Y \Leftrightarrow V_0\big(V_{1q} \vee Y\big) \qquad \text{with} \qquad V_{0q}(x) \Leftrightarrow x \notin (\mathcal{E}_0 \cup \mathcal{E}_q) \tag{10.6}$$

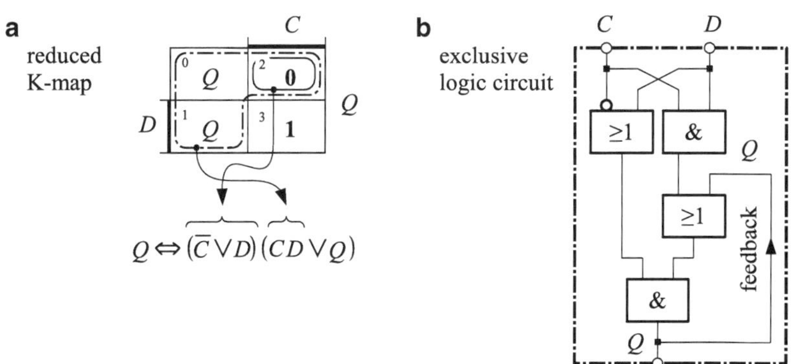

Fig. 10.13 Designing a D-latch according to $Y \Leftrightarrow V_0\big(V_{1q} \vee Y\big)$, (10.6)

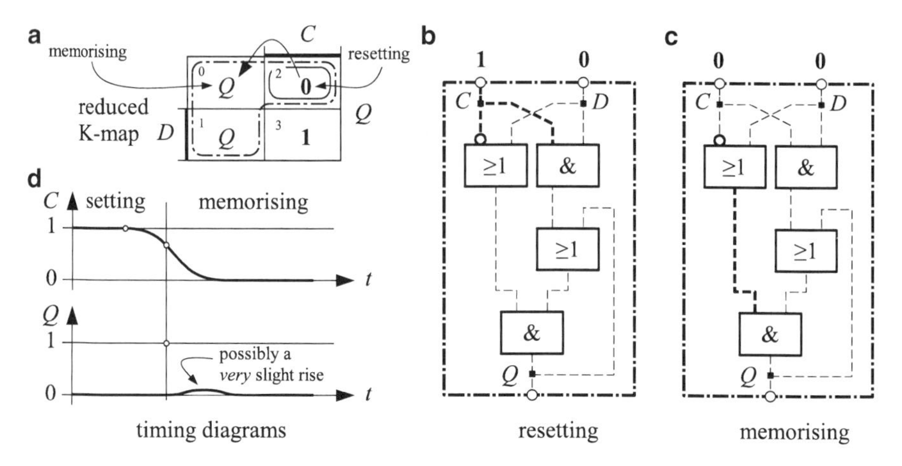

Fig. 10.14 The pre-established **0**-feedback ensures that a **0**-output of the conjunctive D-latch is stored faultlessly

where V_{0q} is assumed free of single-input hazards. From the reduced K-map of Fig. 10.13 we can read the exclusions $V_0 \Leftrightarrow \overline{C} \vee D$ and $V_{0q} \Leftrightarrow CD$ so that $Y \Leftrightarrow V_0\left(V_{1q} \vee Y\right) \Leftrightarrow (\overline{C} \vee D)(CD \vee Y)$ this leading to the logic circuit of Fig. 10.13.

This being a *conjunctive* feedback-latch it *per se* never malfunctions when memorising a **1**-output. On the other hand, conjunctive latches (at least *basic* conjunctive latches) are prone to malfunction when trying to store a **0**-output signal (see, for example, Fig. 10.10 with its interrupted **0**-feedback signal). This is not the case for the latch designed in Fig. 10.13. As you can see studying the Fig. 10.14b, c, **the 0-signal in the feedback loop is pre-established when the latch is reset**, and remains unaltered when switching to a memorising input event thus ensuring that the output signal also remains unchanged, meaning that no memorisation hazard occurs.

10.5 Minimisation

Minimising *combinational* circuits opens the door to the occurrence of hazards, i.e., of transient, spurious output-values. Similarly, minimising feedback-latches enables memorisation hazards to occur. These pose a serious danger because they can quickly cause an erroneous output signal and thus a malfunctioning of the latch. If you choose to design a minimised latch, it is recommended and sometimes necessary to insert an inertial delay in the feedback to reduce the likelihood of a memorisation hazard.

Let us detail the above considerations for the example of Fig. 10.15. Say, we want to design a *disjunctive* latch. To do so, we must use the inclusive formula

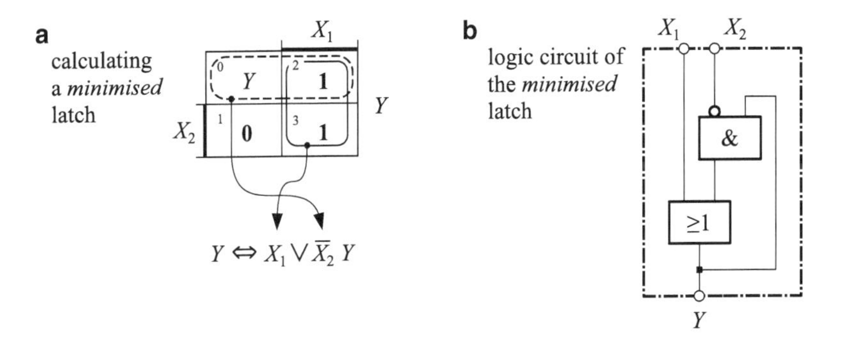

Fig. 10.15 Designing a minimised feedback circuit for a given K-map

$Y \Leftrightarrow U_1 \vee (U_{1S} \vee U_y)Y$, (10.4), choosing $\mathcal{E}_{1S}$ such that $\mathcal{E}_{1S} \cup \mathcal{E}_y$ is a prime set. The prime set to choose, being anchored on $\mathcal{E}_y = \{0\}$, is $\mathcal{E}_{1S} \cup \mathcal{E}_y = \{0, 2\}$ as represented by the dashed loop in Fig. 10.15a. Note that one does not choose (an arbitrary) subset $\mathcal{E}_{1S}$, one chooses a prime set, one anchored on one or more memorising input events, the chosen prime sets determining the subset $\mathcal{E}_{1S}$. From Fig. 10.15a we read $U_1 \Leftrightarrow X_1$ and $\mathcal{E}_{1S} \cup \mathcal{E}_y \Leftrightarrow \overline{X}_2$ so that $Y \Leftrightarrow U_1 \vee (U_{1S} \vee U_y)Y \Leftrightarrow X_1 \vee \overline{X}_2 Y$. This formula is depicted in Fig. 10.15b. You might like to note that this is the logic form of the relay circuit of Fig. 9.1b.

This circuit will not malfunction when memorising the **1**-output caused by input event 2, but it can very well malfunction when memorising the **1**-output initiated by input event 3. To avoid this latter malfunctioning some authors (for this and other reasons) recommend not using this circuit when a transition from the input event (1,1) to (0,0) is possible.

Chapter 11
Elementary Latches

This chapter presents an overview of the elementary latches, first classifying them, and then developing a system of latch symbols. The few latch symbols standardised in IEC 117–15 or IEEE 91–1984 simply don't cover the problem.

An **elementary latch** is a latch with two inputs and one output. There are 36 such latches, as compared with 10 combinational functions of two inputs and one output. It seems wise to discard 8 of the 36 latches as too risky to be technically feasible, still leaving us with 28 elementary latches. Any of these can appear in the realisation of sequential circuits—how many and which ones depends on the specification of the circuit. The elementary latches are classified in Sect. 11.1.

Two kinds of latch symbols are introduced. The first models any of the 36 elementary latches, but most conveniently the 28 technically feasible ones. By its definition, a symbol of this kind always represents two mutually inverted elementary latches. These symbols are introduced in Sect. 11.2 and detailed for all elementary latches in Sects. 11.3 and 11.4.

The symbols used for elementary latches with two inverted outputs are not capable of describing the behaviour of the so-called Eccles-Jordan latches. These are the oldest, most common and best known latches making it imperative to have latch symbols which represent them. The Eccles-Jordan latch is introduced and discussed quite thoroughly in Sect. 11.5 while its symbols are developed and presented in Sect. 11.6.

In the final section, Sect. 11.7, we take a look at the standard latch symbols documenting the five latches they stand for.

11.1 Classification of Elementary Latches

A latch can be described by an events table. The header of the events table for an elementary latch, see Fig. 11.1, contains the two binary (numeric) inputs x_1 and x_2. The table has four columns which we number by the decimal equivalent, x, of the input event (x_1, x_2). The columns are *weighted* from left to right by decreasing powers

S.P. Vingron, *Logic Circuit Design*, DOI 10.1007/978-3-642-27657-6_11,

$x_1 =$	0	0	1	1	
$x_2 =$	0	1	0	1	
$x \ldots$	0	1	2	3	... input event x
$(2^2-1)-x \ldots$	3	2	1	0	... complement of x
$2^{(2^2-1)-x} \ldots$	2^3	2^2	2^1	2^0	... *weight* of $y(x)$
$y_{s,r}(x) =$	$y(0)$	$y(1)$	$y(2)$	$y(3)$	... output value $y(x)$

Fig. 11.1 The full header of a events table for two inputs

of 2. The output values $y(0), y(1), y(2), y(3)$ assigned to the four columns stand for one of the symbols 0, 1, or q, i.e., $y(x) \in \{0, 1, q\}$. As each symbol 0, 1, or q must occur at least once in any given row, one of the symbols must occur exactly *twice*. A row containing two 1s is said to describe a **predominantly setting latch**. A row with two 0s describes a **predominantly resetting latch** while a **predominantly memorising latch** is described by a row in which the symbol q occurs twice.

Each latch $f_{s,r}$ and its output $y_{s,r}(x)$ is specified by *two* indices. The first index, denoted as s and called the **set index**, describes the position of 1s (i.e., the columns containing the 1s), while the second index, the so-called **reset index** r describes the columns containing 0s. They are determined as follows:

$$s = \sum_{x=0}^{2^2-1} s_x \cdot 2^{(2^2-1)-x}, \qquad \begin{cases} s_x = 1 & \text{if } y(x) = 1, \\ s_x = 0 & \text{if } y(x) \neq 1. \end{cases} \tag{11.1}$$

$$r = \sum_{x=0}^{2^2-1} r_x \cdot 2^{(2^2-1)-x}, \qquad \begin{cases} r_x = 1 & \text{if } y(x) = 0, \\ r_x = 0 & \text{if } y(x) \neq 0. \end{cases} \tag{11.2}$$

The **Predominantly Memorising** latches are summarised in the two tables of Fig. 11.2. They are referred to by the abbreviation **PQ-latches** (instead of PM-latches) because it is common to employ the letter Q (or q) for the memorising internal variable. The latches in equal rows are mutually inverted. Using the set and reset indices, we can state that **two latches are negated** if the set and reset indices

$x_1 =$	0	0	1	1	
$x_2 =$	0	1	0	1	
$x =$	0	1	2	3	
$y_{2,1}(x) =$	q	q	1	0	a
$y_{4,1}(x) =$	q	1	q	0	a
$y_{4,2}(x) =$	q	1	0	q	c
$y_{8,1}(x) =$	1	q	q	0	d
$y_{8,2}(x) =$	1	q	0	q	e
$y_{8,4}(x) =$	1	0	q	q	e

$x_1 =$	0	0	1	1	
$x_2 =$	0	1	0	1	
$x =$	0	1	2	3	
$y_{1,2}(x) =$	q	q	0	1	b
$y_{1,4}(x) =$	q	0	q	1	b
$y_{2,4}(x) =$	q	0	1	q	c
$y_{1,8}(x) =$	0	q	q	1	d
$y_{2,8}(x) =$	0	q	1	q	f
$y_{4,8}(x) =$	0	1	q	q	f

Fig. 11.2 The **Predominantly Memorising** latches (**PQ-latches**)

Predominantly setting latches

$x_1 =$	0	0	1	1	
$x_2 =$	0	1	0	1	
$x =$	0	1	2	3	
$(y_{6,1}(x) =)$	q	1	1	0	
$y_{10,1}(x) =$	1	q	1	0	a1
$y_{12,1}(x) =$	1	1	q	0	a1
$y_{5,2}(x) =$	q	1	0	1	b1
$(y_{9,2}(x) =)$	1	q	0	1	
$y_{12,2}(x) =$	1	1	0	q	c1
$y_{3,4}(x) =$	q	0	1	1	b1
$(y_{9,4}(x) =)$	1	0	q	1	
$y_{10,4}(x) =$	1	0	1	q	c1
$y_{3,8}(x) =$	0	q	1	1	d1
$y_{5,8}(x) =$	0	1	q	1	d1
$(y_{6,8}(x) =)$	0	1	1	q	

Predominantly resetting latches

$x_1 =$	0	0	1	1	
$x_2 =$	0	1	0	1	
$x =$	0	1	2	3	
$(y_{1,6}(x) =)$	q	0	0	1	
$y_{1,10}(x) =$	0	q	0	1	a0
$y_{1,12}(x) =$	0	0	q	1	a0
$y_{2,5}(x) =$	q	0	1	0	b0
$(y_{2,9}(x) =)$	0	q	1	0	
$y_{2,12}(x) =$	0	0	1	q	c0
$y_{4,3}(x) =$	q	1	0	0	b0
$(y_{4,9}(x) =)$	0	1	q	0	
$y_{4,10}(x) =$	0	1	0	q	c0
$y_{8,3}(x) =$	1	q	0	0	d0
$y_{8,5}(x) =$	1	0	q	0	d0
$(y_{8,6}(x) =)$	1	0	0	q	

Fig. 11.3 The **Predominantly Setting and Resetting** latches (**PSR-latches**). The functions in parentheses lead to **technically risky circuits** as explained in Fig. 11.4

of one latch equal, respectively, the reset and set indices of the other:

$$\boxed{f_{s,r} = \overline{f}_{r,s}} \tag{11.3}$$

Latches marked with equal alphabetic letters 'a' to 'f' in Fig. 11.2 are **transposed latches**—latches you obtain by interchanging the input variables to the circuits.

The **Predominantly Setting and Resetting** latches (the **PSR-latches**) are summarised in the two tables of Fig. 11.3. Here too, the latches in equal rows are complementary. The transposed latches that are predominantly setting are marked 'a1' to 'd1', those of the predominantly resetting latches are marked 'a0' to 'd0'.

Those latches—$f_{6,1}$, $f_{9,2}$, $f_{9,4}$, $f_{6,8}$, and $f_{1,6}$, $f_{2,9}$, $f_{4,9}$, $f_{8,6}$—whose outputs $y_{s,r}(x)$ are written in parentheses lead to **risky circuits** as explained for the examples $f_{6,1}$ and $f_{1,6}$ of Fig. 11.4.

Consider the predominantly resetting latch of Fig. 11.4a. Its characteristic property is that the only setting input event, cell 3, is *not* adjacent to the only memorising input event in cell 0. To memorise the output value 1, caused by the setting

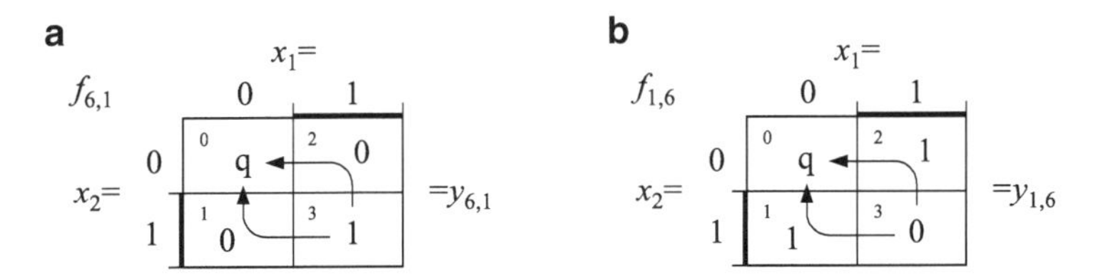

Fig. 11.4 Technically risky elementary latches

input event $(x_1, x_2) = (1, 1)$, both input variables must change their values *simultaneously* from 1 to 0. If simultaneousness is not achieved, switching from $(x_1, x_2) = (1, 1)$ to $(x_1, x_2) = (0, 0)$ does not occur directly, rather $(x_1, x_2) = (1, 1)$ switches via $(x_1, x_2) = (0, 1)$ to $(x_1, x_2) = (0, 0)$ if x_1 changes its value first (lower path), or $(x_1, x_2) = (1, 1)$ switches via $(x_1, x_2) = (1, 0)$ to $(x_1, x_2) = (0, 0)$ if x_2 changes its value first (upper path). The input events $(x_1, x_2) = (0, 1)$ and $(x_1, x_2) = (1, 0)$ are both *resetting* input events and when *previous* to the present and memorising input event $(x_1, x_2) = (0, 0)$ for long enough cause the output value of 0 they invoke to be memorised, and not the value of 1 invoked by the setting input event $(x_1, x_2) = (1, 1)$.

11.2 Symbols for Elementary Latches*

Gates were introduced as graphic symbols that represent logic connectives or functions such as NEGATION, AND, OR. An elementary **latch symbol** is a graphic symbol which represents one of the 36 elementary latches listed in the previous section. In practice, we restrict our interest to the 28 technically feasible elementary latches, i.e., omitting the 8 technically risky ones. Gates and latch symbols thus stand for an *idealised* behaviour that can be *approximated* by physical realisations.

The basic latch symbol is shown in Fig. 11.5. The symbol has two inputs and two outputs. Their dependence on the inputs is defined such that the outputs are mutually inverted. To define the dependence of the outputs on the inputs, it is necessary to introduce **local variables** A and B for the two inputs. Then, the space adjoining a certain output contains the list of those input events (of the local input variables A and B) which set that output, and reset the other one.

These symbols have the following typical properties:

Property 1: The output variables are inverted. This aspect (among others) is illustrated in Fig. 11.6. Speaking of two inverted outputs of a single latch, is not quite as straight forward as referring to inverted outputs of a combinational circuit.

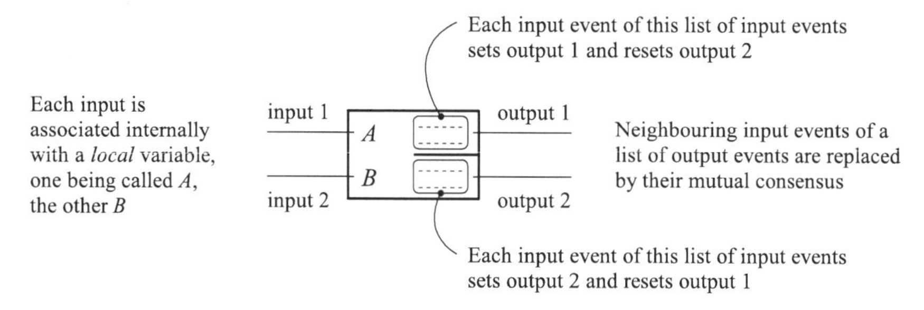

Fig. 11.5 Introducing the latch symbol for an elementary latch with two inverted outputs

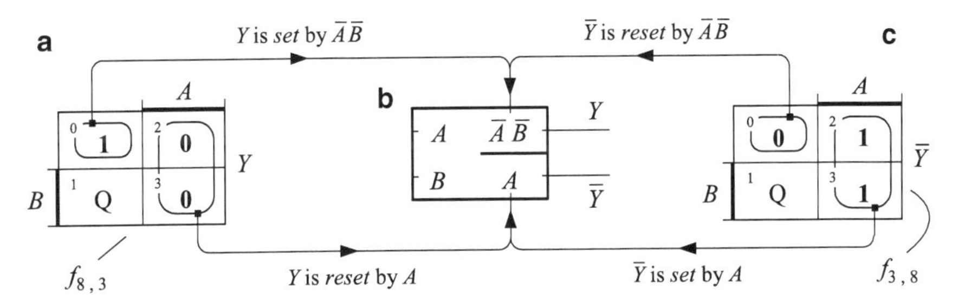

Fig. 11.6 Drawing the latch symbol for the inverted functions $f_{8,3}$ and $f_{3,8}$

The reason is this: When *powering up* a latch (when switching its power supply on), the outputs must be inverted even if the input event present is a memorising one. As we shall see, this requirement will force us to abandon certain feedback designs of a latch with two inverted outputs.

Let us first detail how to express the behaviour of an elementary latch by a latch symbol, using as an example the latch $f_{8,3}$ whose K-map is given in sub-figure (a), and whose output is referred to as Y. The output Y is *set* by the input event $\overline{A}\ \overline{B}$ which is therefore assigned to the space in the latch symbol adjoining output Y. In this case, the list of setting input events consists of a single input event, the input event $\overline{A}\ \overline{B}$.

The output Y of $f_{8,3}$, sub-figure (a), is *reset* by either of the input events $A\overline{B}$ or AB, i.e., by their consensus $A \Leftrightarrow A\overline{B} \vee AB$. So, instead of entering the list $A\overline{B}$, AB into the space of the latch symbol dedicated for the resetting input events of Y, we inscribe it with the resetting consensus, A, as shown in sub-figure (b).

Now take a look at the inverted function $f_{3,8}$, sub-figure (c), whose output $\overline{Y}$ is of course inverted to the output Y of $f_{8,3}$, sub-figure (a). Here A is the setting consensus of $\overline{Y}$, and $\overline{A}\ \overline{B}$ its resetting input event—entries already entered into the latch symbol of sub-figure (b).

Property 2: A latch symbol with a local variable (say, A) which is fed an inverted global input-signal, see Fig. 11.7a, is equivalent to a latch symbol where the global input-signal is *not* inverted (see Fig. 11.7b), but where the associated local variable *is* inverted in the setting and resetting output lists.

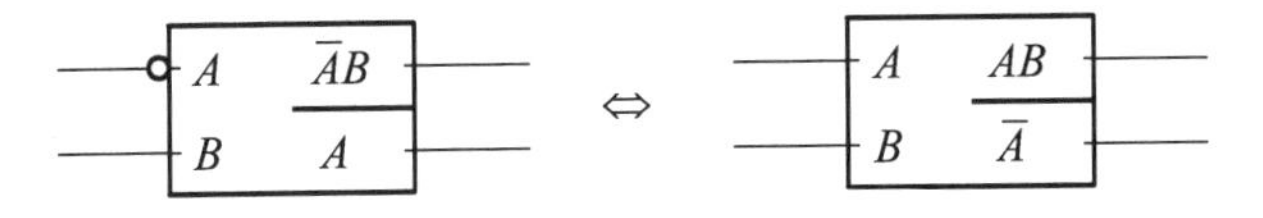

Fig. 11.7 Inverting an input

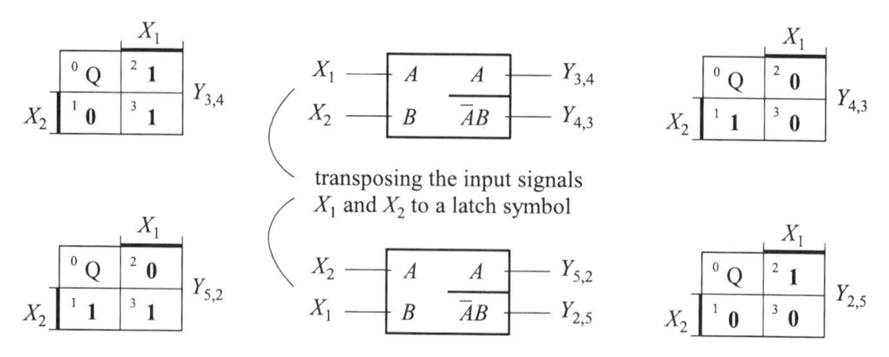

Fig. 11.8 Transposing (interchanging) the inputs

Property 3: Latches whose inputs are transposed (interchanged)—see the two latch symbols of Fig. 11.8—are described by K-maps whose inscribed symbols 0, 1, q are transposed, i.e., mirrored across the main diagonal.

11.3 Predominantly Memorising Latches*

The latch symbols of the **Predominantly Memorising latches** (the **PQ-latches**[1]) are contained in Fig. 11.9 together with the K-maps defining their behaviour. The latch in the first row, the one with the output $Y_{1,2}$, is called the **D-latch**. It plays an important role in switching theory, as it is used to *sample* a data signal, and to *synchronise* a data signal with a clock signal. Referring to these uses, the local inputs are denoted D (for the data signal) and C (for the clock signal), instead of A and B as is otherwise common. *Sampling* was defined in Sect. 9.1, while *synchronising* is discussed in a later chapter. In regard to these applications the D-latch is *primus inter pares*—first among equals. This phrase also points to the fact that all latches are equivalent in that any single (technically feasible) latch can be used to realise all others (this topic is analysed in the following chapter on Latch Composition). The D-latch is later considered in quite some detail.

Note that the latch symbols in Fig. 11.10 are identical with those of Fig. 11.9, but that the (global) input signals are transposed (i.e., interchanged). Therefore the output signals describe different functions which is also shown in comparable K-maps.

The four distinct latch symbols of Fig. 11.9 (or of Fig. 11.10, if you prefer) cover all 12 predominantly memorising latches (as listed in Fig. 11.2). The symbols in the top rows of Figs. 11.9 and 11.10 represent the four functions $f_{1,2}$, $f_{2,1}$,

[1]As mentioned previously, the Predominantly Memorising Latches are referred to by the abbreviation PQ-latches (instead of PM-latches) because it is common to employ the letter Q (or q) for the memorising internal variable.

$f_{1,4}$, and $f_{4,1}$ while the bottom rows represent the functions $f_{4,8}$, $f_{8,4}$, $f_{2,8}$, and $f_{8,2}$. The two symbols in the two middle rows each represent only two (mutually inverted) functions—$f_{2,4}$, $f_{4,2}$, and $f_{8,1}$, $f_{1,8}$—these latches being invariant to the transposition of their input signals.

11.4 Predominantly Setting and Resetting*

The four latch symbols of Fig. 11.11 are capable of representing any of the 16 technically feasible **Predominantly Setting and Resetting latches** (**PSR-latches**) summarised in the table of Fig. 11.3. Each latch symbol in Fig. 11.11 represents two functions. Transposing the inputs to the symbols provides us with the next set of eight functions. The Figs. 11.11 and 11.12 picture all possible 16 PSR-latches.

Designing latches with the help of the evaluation formulas (10.4) leads to disjunctive and conjunctive latches. Both these latches are contained in Fig. 11.11. As a matter of interest, the thyristor in the circuit of Fig. 9.1c represents the conjunctive latch $Y_{1,12}$.

Designing feedback circuits whose behaviour is identical to that of PSR-latches is no trivial matter, and is therefore deferred to the next chapter where the synthesis of latches by composition is discussed. Here, we use an example to show the difficulties one can encounter. In Fig. 11.13 we attempt to develop circuits for the latch of sub-figure (b)—the PSR-latch in the first row of Fig. 11.11. To calculate feedback circuits, the memorising internal variable Q, used in Fig. 11.11, is replaced by the respective feedback signal, $Y_{3,4}$ and $Y_{4,3}$.

A first attempt to obtain a feedback circuit could be to calculate each output, $Y_{3,4}$ and $Y_{4,3}$, from the K-maps of sub-figures (a) and (c), respectively, combining the results to obtain the circuit of sub-figure (d). The grievous property of this circuit is that both output variables, $Y_{3,4}$ and $Y_{4,3}$, are **0** when the circuit powers up with both inputs being **0**. This common situation, in which the outputs are *not* mutually inverted, is unacceptable.

Sub-figure (e) shows a quite acceptable solution—one output, $Y_{4,3}$, is not designed as an independent latch, rather, it is simply the complement of the other output, $Y_{3,4}$. The only disadvantage is that the outputs $Y_{3,4}$ and $Y_{4,3}$ do not switch simultaneously when using real-world switching devices, the output $Y_{4,3}$ always lagging behind by the delay time of the inverter. While this might be acceptable in some situations, it is generally undesirable. We are thus still left with the need to develop a satisfactory solution—a problem to be taken quite seriously, and looked into in the next chapter.

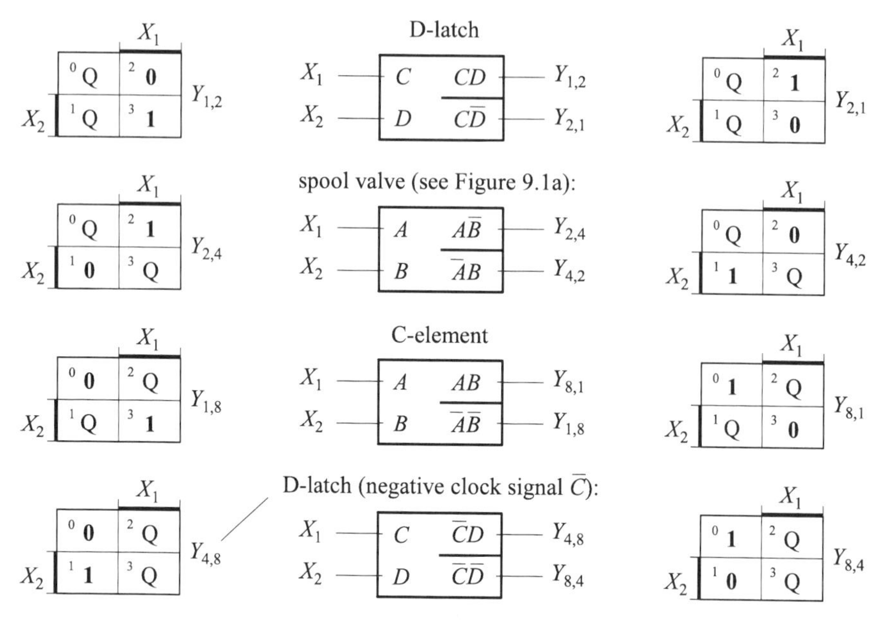

Fig. 11.9 Predominantly Memorising latches (PQ-latches)

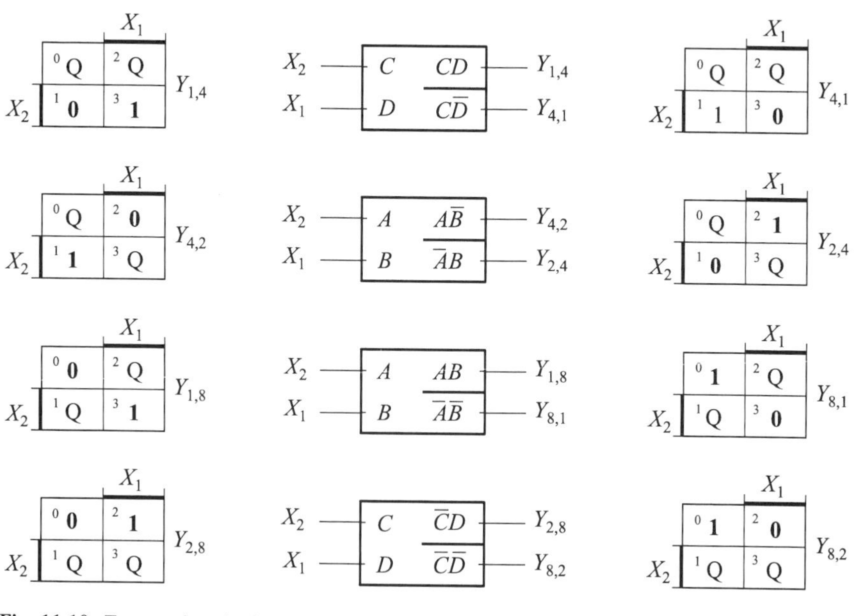

Fig. 11.10 Transposing the inputs to the **Predominantly Memorising latches** (**PQ-latches**) of Fig. 11.9

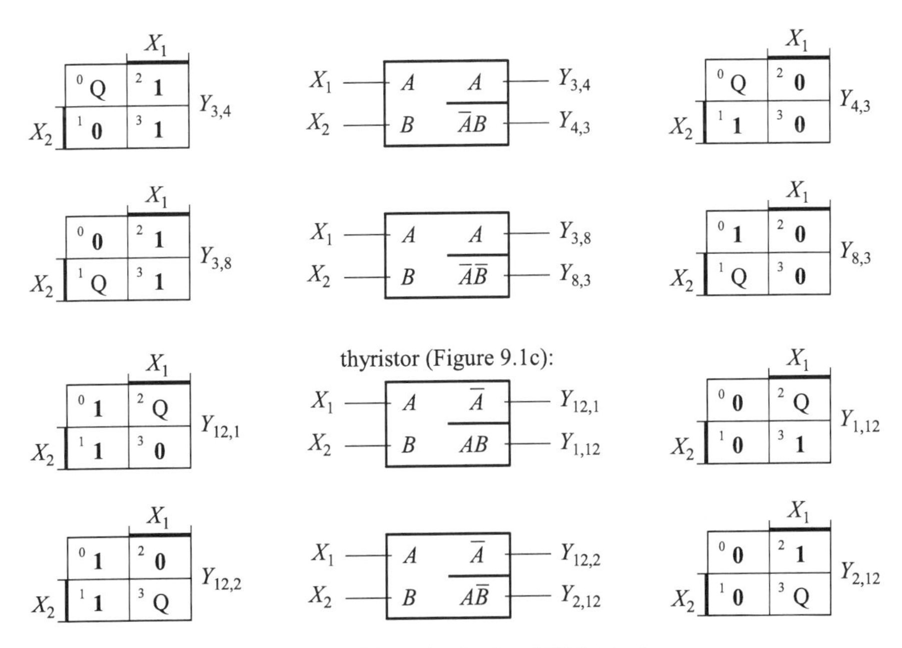

Fig. 11.11 Predominantly Setting and Resetting latches (PSR-latches)

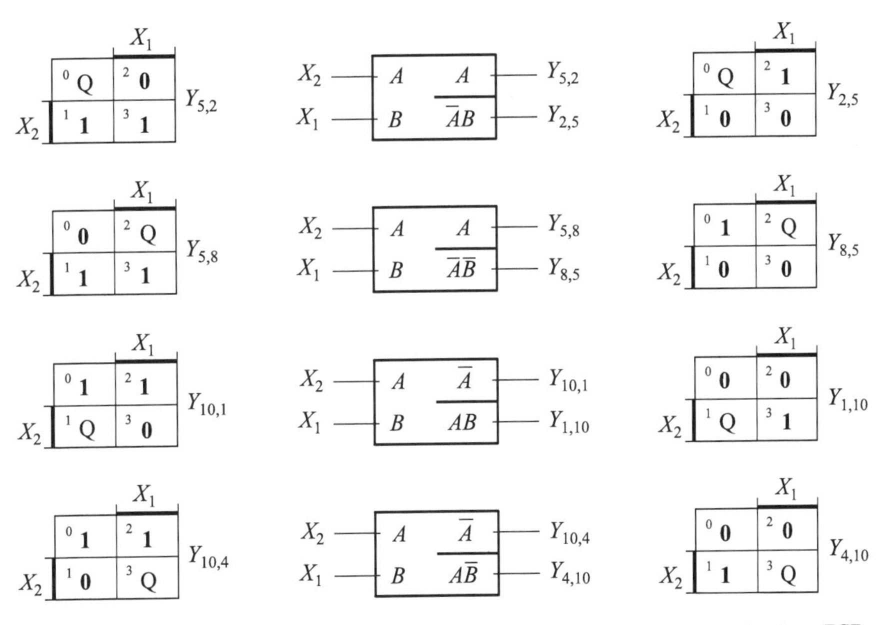

Fig. 11.12 Transposing the inputs to the **Predominantly Setting and Resetting latches** (**PSR-latches**) of Fig. 11.11

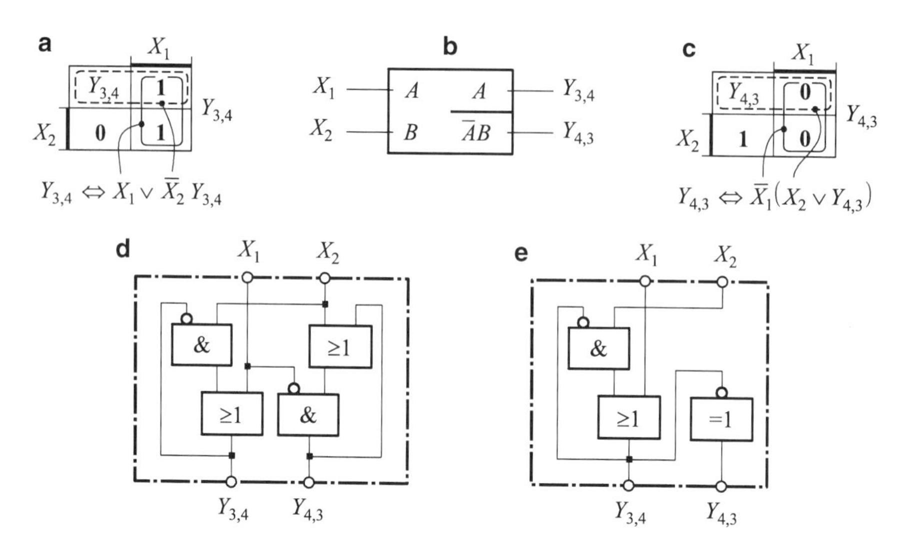

Fig. 11.13 Two unsatisfactory realisations—sub-figures (**d**) and (**e**)—of the latch defined in sub-figure (**b**)

11.5 Eccles-Jordan Latches—the Principle

In essence, Eccles-Jordan latches are cross-coupled NAND gates, or cross-coupled NOR gates. To demonstrate the principle properties, only the NANDs latch is used in this section. All said of the NANDs latch applies in like manner to the dual NORs latch.[2] The principle of the cross-coupled NANDs and NORs latches was established in 1919 (note the early date) by Eccles and Jordan[3] using electronic radio valves. These latches are still the most common in use today, not least because they prove invaluable in developing robust PSR-latches, and a robust D-latch.

The basic design of the cross-coupled NANDs latch is shown in Fig. 11.14a. Please note that *the feedback signals are not delayed*. The inputs to NAND-1 are thus X_1 and Z (Z, and not Z^*), while the inputs to NAND-2 are X_2 and Y (again note: Y, and not Y^*). This allows us to calculate the outputs Y and Z thus:

$$\begin{aligned} Y &\Leftrightarrow \overline{X_1 \wedge Z} & Z &\Leftrightarrow \overline{X_2 \wedge Y} \\ &\Leftrightarrow \overline{X_1 \wedge \overline{X_2 \wedge Y}} & &\Leftrightarrow \overline{X_2 \wedge \overline{X_1 \wedge Z}} \\ Y &\Leftrightarrow \overline{X_1} \vee X_2 Y & Z &\Leftrightarrow \overline{X_2} \vee X_1 Z \end{aligned} \tag{11.4}$$

---

[2]If this field is new to you, I invite you to rewrite this section for the cross-coupled NORs latch.

[3]Eccles, W. H., F. W. Jordan: '*A Method of Using Two Triode Valves in Parallel for Generating Oscillations*'. The Radio Review, vol. 1, pp. 80–83, November 1919. You can order a copy via the British Library: www.bl.uk

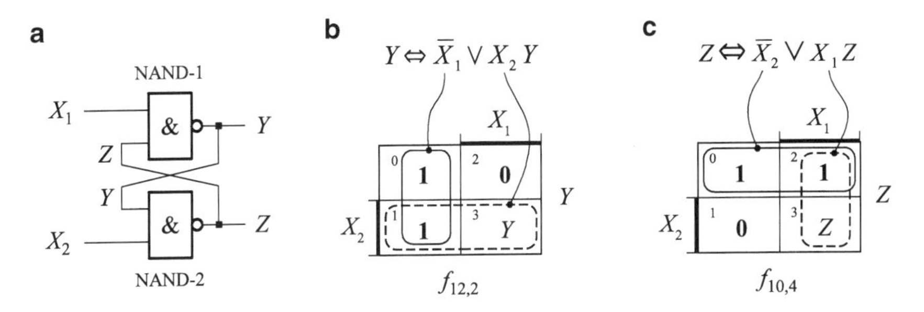

Fig. 11.14 Analysing the basic principle of the cross-coupled NAND-latch

These formulas are those of latches, and unequivocally describe the K-maps pictured in Fig. 11.14b, c and thereby the behaviour of the cross-coupled NANDs latch. We use this latch and the K-maps that describe its behaviour to demonstrate the general properties of the Eccles-Jordan latches—properties that are remarkable and ambivalent.

Property 1: An Eccles-Jordan latch has exactly one input event that produces identical output values for Y and Z. This input event is called the **common input event** as its effect is common to both outputs. In our example, the common input event is $\overline{X}_1\overline{X}_2$ as it sets both outputs. This is called a ***setting* common (input) event**. All Eccles-Jordan NANDs latches have a *setting* common event, whereas all Eccles-Jordan NORs latches have a *resetting* common event.

Property 2: The outputs of an Eccles-Jordan latch are *not* inverted due to the existence of a common input event.

Property 3: The **Eccles-Jordan latches have a property quite unacceptable**: Their **output values are undefined** when switching from the *common input event* to the *memorising input event*. The common input event is *never* adjacent to the memorising one, so that the values of both input variables must change simultaneously to achieve the direct transition from the common to the memorising input event. The signal flow for our example latch is shown in Fig. 11.15. The common input event, $(\mathbf{0}, \mathbf{0})$, forces both outputs to be **1** as well as the feedback signals. When switching to the memorising input event, $(\mathbf{1}, \mathbf{1})$, both NANDs, **in an ideal circuit**, force the outputs and the feedbacks to **0**, these forcing the outputs and the feedbacks to **1**, which forces the outputs back to **0**, and so on until one of the feedback signals wins the **race** to the NAND gates. The only certain way to avoid this race condition is to ensure that the common input event is never presented to the Eccles-Jordan latch. The theory of Latch Composition, covered in the next chapter, enables us to ensure this.

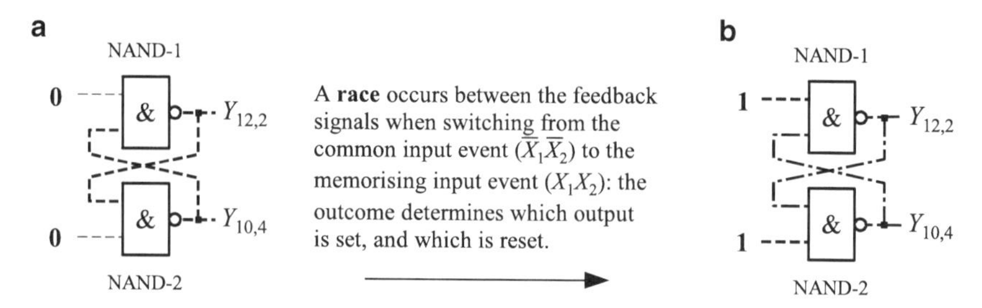

Fig. 11.15 The genesis of a race of the feedback signals in an Eccles-Jordan latch

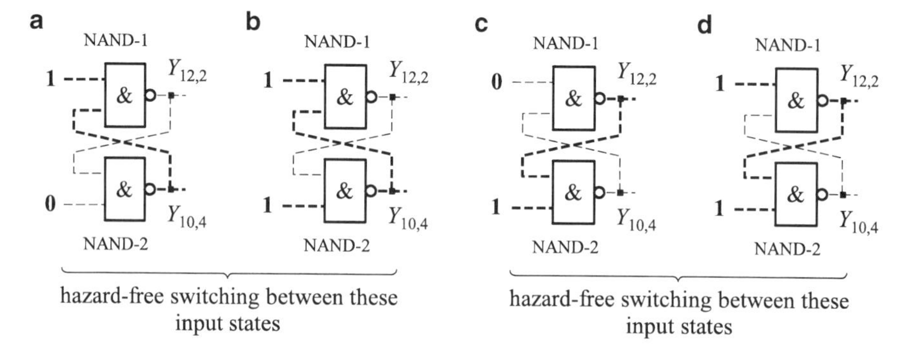

Fig. 11.16 Pre-established feedback loops when switching between the memorising input event and the adjacent setting and resetting input events

Property 4: Importantly and advantageously, **Eccles-Jordan latches are hazard-free** when switching between a setting or resetting input events (cells 1 or 2 in Fig. 11.14b, c) and the memorising input event(cell 3). The signal flow in the feedback loops is pre-established, and is not changed or influenced by the mentioned transitions. Specifically, switching back and forth between cells 1 and 3 leaves the outputs and the internal signal flow unchanged, as does switching between cells 2 and 3, see Fig. 11.16.

11.6 Eccles-Jordan Latches—Their Symbols*

The latch symbol for an Eccles-Jordan latch is designed along the lines of that for the elementary latches. Each input is associated with a local variable, usually named A and B, as in the introductory example-symbols of Fig. 11.17. Each output is associated with the input event which sets that output and resets the other

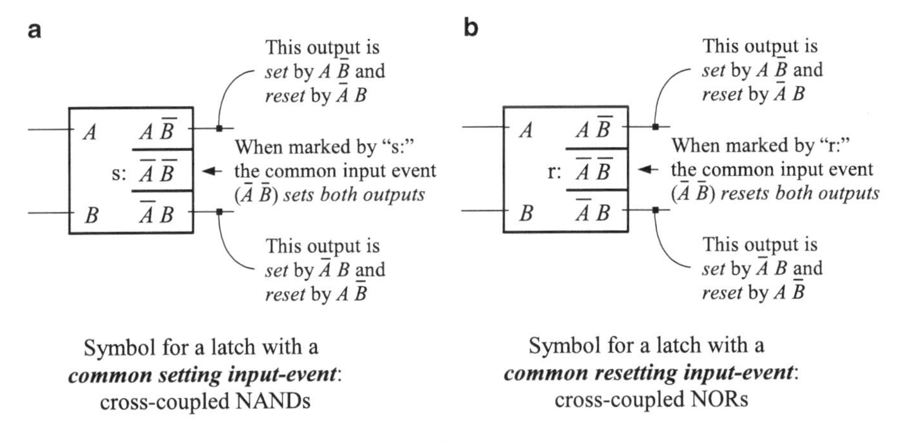

Fig. 11.17 The latch symbol for an Eccles-Jordan latch

one. In the example of Fig. 11.17, the input event $A\overline{B}$ sets the topmost output, and resets the bottommost one. Vice versa, $\overline{A}B$ sets the bottommost output, and resets the topmost one. So far, this is the same principle as for symbols of the elementary latches. But contrary to these, it is necessary to specify the common input event for an Eccles-Jordan latch. This is done in the cell between those assigned to the outputs. In both our example symbols of Fig. 11.17, the common input event is $\overline{A}\ \overline{B}$. But if the common input event is used to **set** both outputs, it is prefixed by **'s:'**, whereas it is prefixed with **'r:'** when used to **reset** both outputs.

All Eccles-Jordan latches are summarised in the tables of Figs. 11.18 and 11.19. The symbols of Fig. 11.18 are characterised as having common *setting* input events, and thus are realised by cross-coupled NAND gates. The functions realised by these symbols are specified by K-maps to the left of the symbols. Do take the time necessary to see how the **0**s and **1**s of a K-map lead to the associated input events in the latch symbol next to the K-maps. Then, make sure you understand how a latch symbol describes the signal flow in the Eccles-Jordan latch to the right of the symbol.

The local variables A and B, and the input events of them, are row-wise identical for the latch symbols in like rows of Figs. 11.18 and 11.19. The only difference between the symbols of these tables is the definition of the common input event. Figure 11.19 contains all Eccles-Jordan latches with a *resetting* common event, and are thus realised by cross-coupled NOR gates. A further comparison of like symbols of Eccles-Jordan latches with differing common input events shows that they describe inverted latches.

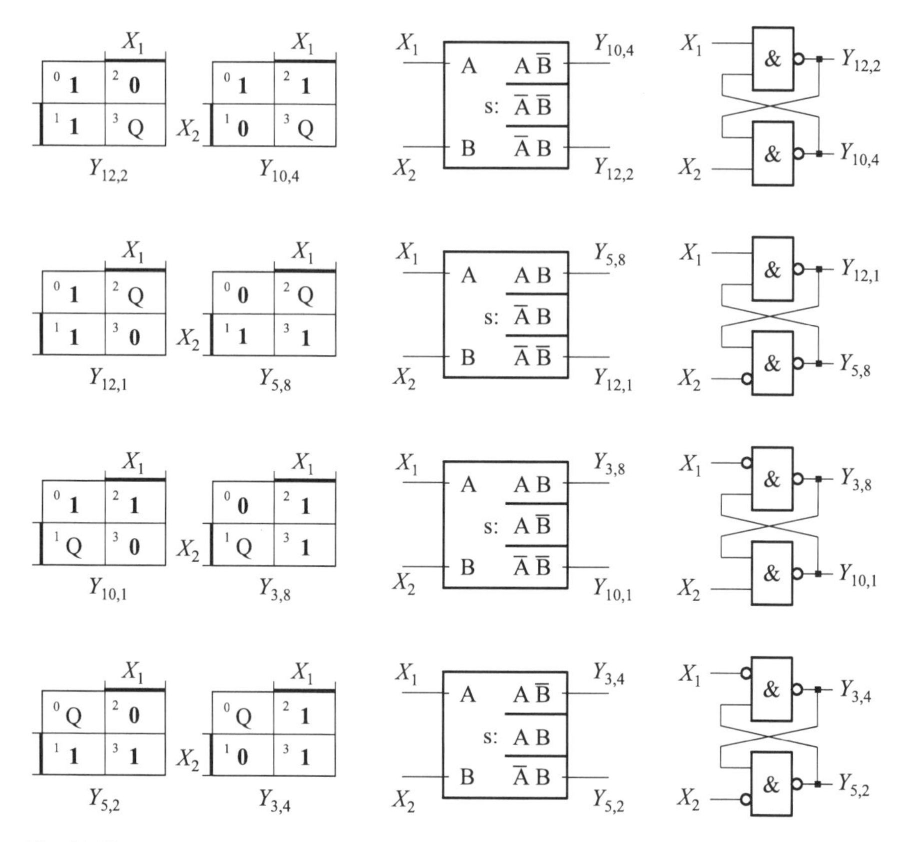

Fig. 11.18 The Eccles-Jordan NAND latches

11.7 Standard Symbols for Latches

The concepts of setting, resetting, and memorising *input events* are not present in conventional switching theory or in the IEC 117–15 standard, rather, only the notion of setting and resetting *input variables*, or **setting** and **resetting dependences** (as they are called in the IEC 117–15 standard) are used. This notion stems from the assumption that one input *sets* while the other *resets* the output, and that when neither is *active*, the output is *not changed* (abbreviated as 'nc', a paraphrase for 'memorised').

The latches to which the above concepts of the IEC 117–15 standard apply are usually referred to as **Set-Reset latches** (or **SR-latches**), and are summarised in Fig. 11.20, the K-maps of column 1 showing the latches' functions, column 2 their IEC 117–15 symbols (sub-figures (a) to (d)), and column 3 the equivalent latch symbols developed in the previous sections. In going along with the concept of *setting* and *resetting* inputs, we replace the internal variables A and B of our symbols in the rightmost column by the mnemonics S (for *setting*) and R

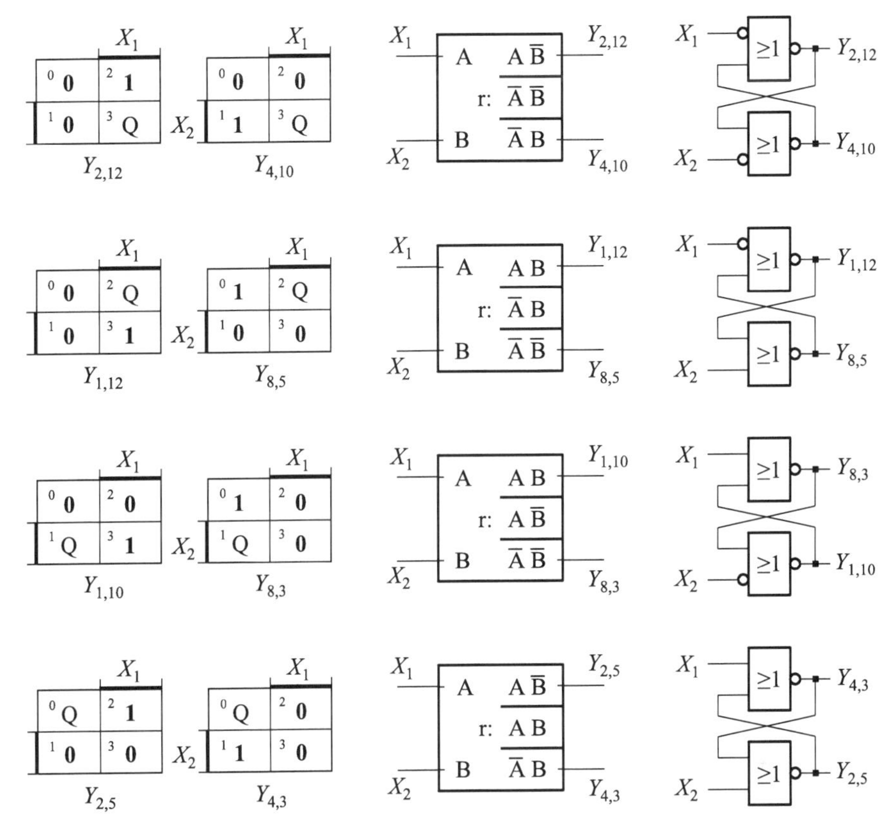

Fig. 11.19 The Eccles-Jordan NOR latches

(for *resetting*), respectively. The standard (IEC 117–15) symbols use so-called **qualifying symbols** or **labels** (here, the integers 1 or 2) common to an input and one or both outputs. An output thus associated with the setting input is *set* irrespective of the value assigned to the resetting input. Likewise, the output thus associated with R is *reset*, irrespective of the value of S.

The IEC 117–15 standard stipulates that all output lines of an element that is not subdivided (such as an IEC latch symbol) carry identical logic values except when otherwise indicated by associated qualifying symbols or labels. For instance, by this definition the bottom output lines of sub-figures (a) and (b), prior to being inverted, carry the same output signals as the respective top output lines—especially as both output lines are assigned the same labels (the integer 1), but not *due* to both output lines being assigned the same label. The identity of the output values can only be broken if each output line is assigned different labels as, e.g., in sub-figures (c) or (d).

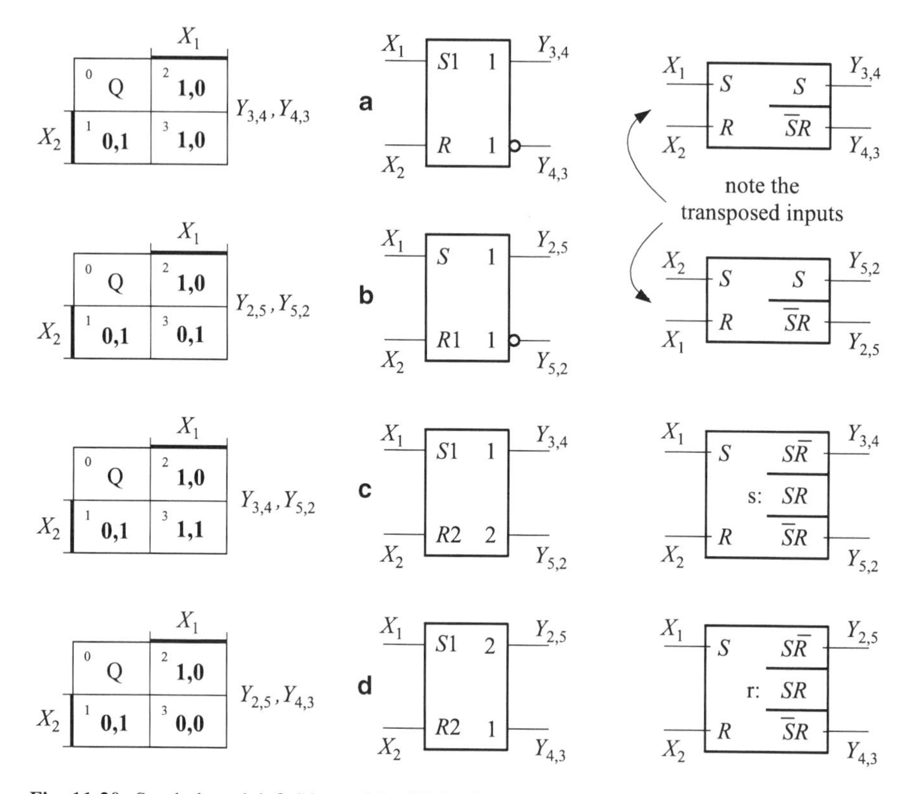

Fig. 11.20 Symbols and definitions of the **SR-latches**

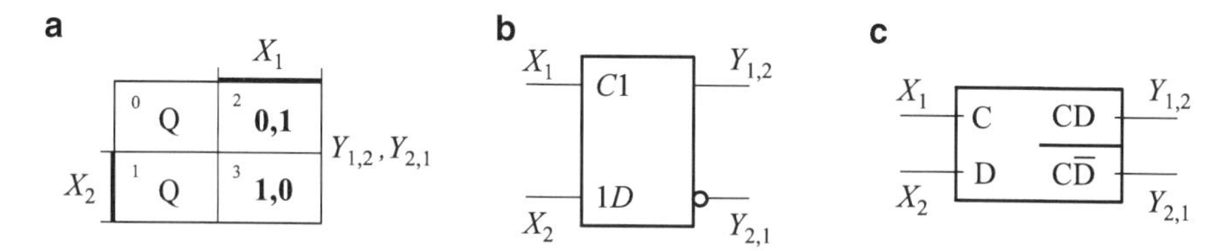

Fig. 11.21 Symbols and definitions of the **D-latches**

The **D-latch** is defined by the K-map of Fig. 11.21a, the standard symbol is given in sub-figure (b). The local variable C, used in sub-figure (b), is reserved as specifying a so-called **control dependency**. A control input is usually used to enable (or disable) other inputs (D, S, or R) of latches. The control input is assigned a label (in the case of the D-latch: the number 1). The *input* lines that are to be dependent on the control input are assigned the *label* of the control C and only then specified as D, S, or R inputs.

Chapter 12
Latch Composition*

Methodologically, the composition of latches is an extension to the composition of combinational circuits, discussed in Chap. 8, and is thus based on the catenation of functions. *Latch composition* allows us to realise a given (or specified) latch f by employing a chosen latch h as the hinge function, and developing for these the generic function g as a combinational circuit, a circuit called the *pre-logic* of h. The composition problem of latches boils down to finding, or developing, the combinational pre-logic g which satisfies the catenation $hg = f$ for the given memory functions f and h. The theory of latch composition is put forth quite extensively in the first section.

In Sect. 12.2 we point to the Eccles-Jordan latch as an optimal general purpose hinge latch. We do so by designing an optimal circuit for the D-latch, a latch that is the *primus inter paris* of latches: It alone (among all elementary latches) has a K-map that can be further reduced to one with a single input; in its own right it realises sampling, and in a D-flipflop it can be used for synchronisation. The D-latch, by the way, is a **predominantly memorising** latch—a PQ-latch—the only such latch of general interest.

A major application of latch composition is the optimal design of the **predominantly setting and resetting latches**—the PSR-latches of Fig. 11.3. You will remember that in Sect. 11.4 both attempts to develop a satisfactory circuit for the PSR-latch (see Fig. 11.13) were not all that successful. The optimal design of PSR-latches can only be achieved by latch composition. The realisations of the PSR-latches by NAND-gates and by NOR-gates are developed and summarised in Sect. 12.3—they are the prime single result of this chapter.

Gated latches, touched on in Sect. 12.4, are mainly mentioned to augment the standard latch-symbols Sect. 11.7. The input AND-gates to a gated latch have only a superficial similarity with a generic function of a composed latch.

S.P. Vingron, *Logic Circuit Design*, DOI 10.1007/978-3-642-27657-6_12,

12.1 Principle of Latch Composition

To visualise the problem of latch composition we shall consider in detail the example laid out in Fig. 12.1 referring, in this section, to its sub-figures simply as (a), (b), and (c). The symbol f denotes the latch (or memory function) to be composed; in our example we want to compose the D-latch shown in the K-map of (a). h is the hinge function on the basis of which f is to be composed; as shown in the K-map of (b), the hinge latch h used in our example is a set-dominant latch.

A memory function, as you will remember, maps a set of input events $\mathcal{E}$ into the set $\widehat{\mathcal{Q}} = \{\{0\}, \{1\}, \{0, 1\}\}$, see (9.1), and it is the graphical representation of this mapping that points the way to solving the composition problem. The *K-map* of the latch f we want to design is shown in sub-figure (a) together with the *mapping* for f. The correspondence between the *mapping* and the *K-map* is hopefully obvious, but if not, might be clarified by the sub-section 'Specifying a

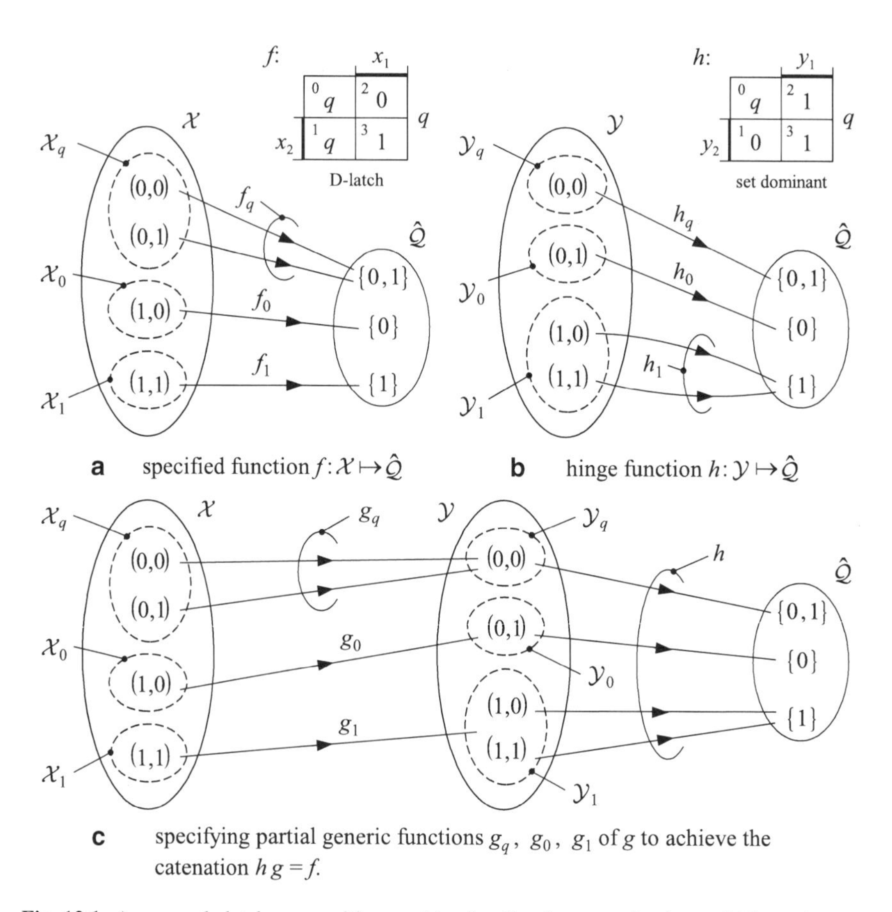

Fig. 12.1 An example latch-composition used to visualise the generalisations of this section

Memory Function in a K-map' in Sect. 9.2. As shown there (refer to Fig. 9.3a), the mapping of the function f could also have been expressed by a K-map. But such a representation is not easily applied to the *catenation* of functions. In a similar manner as for the function f, discussed above, sub-figure (b) depicts the K-map and the mapping for the hinge function h; do take a moment or so to check the correspondence between the K-map and the mapping for the hinge function.

As both given memory functions $f : \mathcal{X} \mapsto \widehat{\mathcal{Q}}$ and $h : \mathcal{Y} \mapsto \widehat{\mathcal{Q}}$ have the same co-domain $\widehat{\mathcal{Q}}$, we can—as indicated in sub-figure (c)—always find a function g so that the catenation hg equals f. Allow me to detail this: As f maps both input events $(0,0)$ and $(0,1)$ of $\mathcal{X}$ to $\{0,1\}$ of $\widehat{\mathcal{Q}}$—see sub-figure (a)—these input events must, in the catenation diagram of sub-figure (c), map to the input event $(0,0) \in \mathcal{Y}$, for this is the only input event which h maps to $\{0,1\}$. A like consideration necessitates us to map $(1,0) \in \mathcal{X}$ to $(0,1) \in \mathcal{Y}$ as the former input event, $(1,0)$, is mapped to $\{0\}$ by f, while $(0,1)$ is mapped to $\{0\}$ by h. An ambiguous situation arises when mapping $(1,1) \in \mathcal{X}$ to one of the input events $(1,0)$ or $(1,1)$ of $\mathcal{Y}$. As either mapping is correct, we can arbitrarily choose one of them (but of course: only one).

The above example will hopefully help you in visualising the generalisations introduced below. To obtain a generic function $g : \mathcal{X} \mapsto \mathcal{Y}$, we first partition its domain $\mathcal{X}$ and co-domain $\mathcal{Y}$ (as indicated in the sub-figures (a) and (b) of our example):

$$\mathcal{X}_0 := \{x | f(x) = \{0\}\}, \qquad \mathcal{Y}_0 := \{y | h(y) = \{0\}\}, \tag{12.1a}$$

$$\mathcal{X}_1 := \{x | f(x) = \{1\}\}, \qquad \mathcal{Y}_1 := \{y | h(y) = \{1\}\}, \tag{12.1b}$$

$$\mathcal{X}_q := \{x | f(x) = \{0,1\}\}, \qquad \mathcal{Y}_q := \{y | h(y) = \{0,1\}\}. \tag{12.1c}$$

These partitions are used as domains of the *partial functions*

$$f_0 : \mathcal{X}_0 \mapsto \{\{0\}\}, \qquad h_0 : \mathcal{Y}_0 \mapsto \{\{0\}\}, \tag{12.2a}$$

$$f_1 : \mathcal{X}_1 \mapsto \{\{1\}\}, \qquad h_1 : \mathcal{Y}_1 \mapsto \{\{1\}\}, \tag{12.2b}$$

$$f_q : \mathcal{X}_q \mapsto \{\{0,1\}\}, \qquad h_q : \mathcal{Y}_q \mapsto \{\{0,1\}\}. \tag{12.2c}$$

The domains of these partial functions, being partitions of $\mathcal{X}$ and $\mathcal{Y}$, respectively, are mutually disjoint,

$$\mathcal{X}_0 \cap \mathcal{X}_1 = \emptyset, \qquad \mathcal{Y}_0 \cap \mathcal{Y}_1 = \emptyset,$$

$$\mathcal{X}_0 \cap \mathcal{X}_q = \emptyset, \qquad \mathcal{Y}_0 \cap \mathcal{Y}_q = \emptyset,$$

$$\mathcal{X}_1 \cap \mathcal{X}_q = \emptyset, \qquad \mathcal{Y}_1 \cap \mathcal{Y}_q = \emptyset,$$

while their respective unions comprise the original domains $\mathcal{X}$ and $\mathcal{Y}$,

$$\mathcal{X}_0 \cup \mathcal{X}_1 \cup \mathcal{X}_q = \mathcal{X}, \qquad \mathcal{Y}_0 \cup \mathcal{Y}_1 \cup \mathcal{Y}_q = \mathcal{Y}.$$

As visualised in sub-figure (c) of our example, the partitions $\mathcal{X}_0$, $\mathcal{X}_1$, $\mathcal{X}_q$ and $\mathcal{Y}_0$, $\mathcal{Y}_1$, $\mathcal{Y}_q$ allow us to define the partial generic functions

$$\boxed{g_0 : \mathcal{X}_0 \mapsto \mathcal{Y}_0, \qquad g_1 : \mathcal{X}_1 \mapsto \mathcal{Y}_1, \qquad g_q : \mathcal{X}_q \mapsto \mathcal{Y}_q.} \tag{12.3}$$

From the above definitions, (12.1), (12.2), and (12.3), it follows that

$$\begin{array}{ccc}
f_0 \cap f_1 = \emptyset, & f_0 \cap f_q = \emptyset, & f_1 \cap f_q = \emptyset, \\
h_0 \cap h_1 = \emptyset, & h_0 \cap h_q = \emptyset, & h_1 \cap h_q = \emptyset, \\
g_0 \cap g_1 = \emptyset, & g_0 \cap g_q = \emptyset, & g_1 \cap g_q = \emptyset, \\
f_0 \cup f_1 \cup f_q = f, & h_0 \cup h_1 \cup h_q = h, & g_0 \cup g_1 \cup g_q = g.
\end{array}$$

But, most importantly, the following logical implications allow us to choose the generic functions g_0, g_1 and g_q:

$$\boxed{\begin{array}{lcl}
g_0 : \mathcal{X}_0 \mapsto \mathcal{Y}_0 & \Rightarrow & h_0\ g_0 = f_0, \\
g_1 : \mathcal{X}_1 \mapsto \mathcal{Y}_1 & \Rightarrow & h_1\ g_1 = f_1, \\
g_q : \mathcal{X}_q \mapsto \mathcal{Y}_q & \Rightarrow & h_q\ g_q = f_q.
\end{array}} \tag{12.4}$$

These implications are interpreted as follows : **(a)** Any *arbitrarily chosen* function g_0 from $\mathcal{X}_0 = \{x|f(x) = \{0\}\}$ to $\mathcal{Y}_0 = \{y|h(y) = \{0\}\}$ satisfies the catenation $h_0\ g_0 = f_0$; **(b)** any *arbitrarily chosen* function g_1 from $\mathcal{X}_1 = \{x|f(x) = \{1\}\}$ to $\mathcal{Y}_1 = \{y|h(y) = \{1\}\}$ satisfies the catenation $h_1\ g_1 = f_1$; **(c)** any *arbitrarily chosen* function g_q from $\mathcal{X}_q = \{x|f(x) = \{0,1\}\}$ to $\mathcal{Y}_q = \{y|h(y) = \{0,1\}\}$ satisfies the catenation $h_q\ g_q = f_q$.

The partial hinge-sets $\mathcal{Y}_0$, $\mathcal{Y}_1$, and $\mathcal{Y}_q$ are the resetting, setting, and memorising input events of the hinge latch $h : \mathcal{Y} \mapsto \{\{0\},\{1\},\{0,1\}\}$, respectively. They are best determined visually from the K-map of the hinge latch. Take the hinge function $h : \mathcal{Y} \mapsto \{\{0\},\{1\},\{0,1\}\}$ to have m binary inputs, $y_1, \ldots, y_m$, so that it is described by an m-dimensional K-map whose cells contain either 0, 1 or q.

(12.5)

Determining the partial hinge-sets $\mathcal{Y}_0$, $\mathcal{Y}_1$, and $\mathcal{Y}_q$ from the K-map of a chosen hinge function $h : \mathcal{Y} \mapsto \{\{0\},\{1\},\{0,1\}\}$:

(a) $\mathcal{Y}_0 = \{(y_1,\ldots,y_m)|h(y_1,\ldots,y_m) = \{0\}\}$ consists of the input events $(y_1,\ldots,y_m)$ of those cells containing the output symbol 0;

(b) $\mathcal{Y}_1 = \{(y_1,\ldots,y_m)|h(y_1,\ldots,y_m) = \{1\}\}$ consists of the input events $(y_1,\ldots,y_m)$ of those cells containing the output symbol 1;

(c) $\mathcal{Y}_q = \{(y_1,\ldots,y_m)|h(y_1,\ldots,y_m) = \{0,1\}\}$ consists of the input events $(y_1,\ldots,y_m)$ of those cells containing the output symbol q;

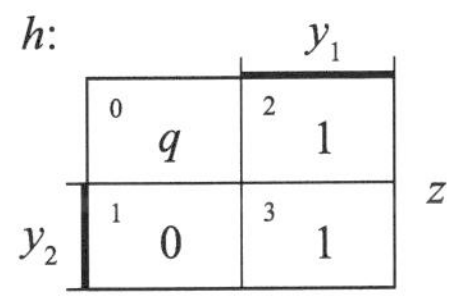

h:

		y_1
	$^0\ q$	$^2\ 1$
y_2	$^1\ 0$	$^3\ 1$

z $\quad \mathcal{Y}_0 = \{(0,1)\}, \qquad \mathcal{Y}_1 = \{(1,0),(1,1)\}, \qquad \mathcal{Y}_q = \{(0,0)\}$

Fig. 12.2 Illustrating the use of procedure (12.5) to obtain $\mathcal{Y}_0, \mathcal{Y}_1, \mathcal{Y}_q$ from h

To illustrate the above procedure for our example take a glance at the K-map of Fig. 12.2 taken from sub-figure (b). $\mathcal{Y}_0$ consists of the only input event $(0, 1)$ because this is the only cell in the K-map containing the output value 0. By a like argument, $\mathcal{Y}_q$ contains only $(0, 0)$. $\mathcal{Y}_1$, on the other hand, contains the cells $(1, 0)$ and $(1, 1)$ because both cell of the K-map have the output value 1 inscribed.

We now turn to developing the generic function $\underset{\frown}{g} : \mathcal{X} \mapsto \mathcal{Y}$. As both this function and the given memory function $f : \mathcal{X} \mapsto \widehat{\mathcal{Q}}$ have the same domain $\mathcal{X}$, they can both be described by identically organised n-dimensional K-maps for the external variables $x_1, \ldots, x_n$. But whereas the cells of the K-map for f contain the symbols 0, 1, or q, the cells of the K-map for g contain the input events $(y_1, \ldots, y_m)$ which are elements of resetting, setting, or memorising partitions $\mathcal{Y}_0$, $\mathcal{Y}_1$, or $\mathcal{Y}_q$ of $\mathcal{Y}$. The basic solution to the catenation problem, stated in (12.4), enables us to specify which input events $(y_1, \ldots, y_m)$ to inscribe into any given cell of the K-map of g. Specifically, this can be stated as in the

Procedure for developing the K-map of a generic function $g : \mathcal{X} \mapsto \mathcal{Y}$. Draw an *empty* K-map identical to the one for $f : \mathcal{X} \mapsto \widehat{\mathcal{Q}}$ the cells of which are entered as follows:

(a) If a cell x of the K-map for f contains the output symbol 0, then the same cell of the K-map for g contains an arbitrary element, i.e., input event $(y_1, \ldots, y_m)$, of $\mathcal{Y}_0$.
(b) If a cell x of the K-map for f contains the output symbol 1, then the same cell of the K-map for g contains an arbitrary element, i.e., input event $(y_1, \ldots, y_m)$, of $\mathcal{Y}_1$.
(c) If a cell x of the K-map for f contains the output symbol q, then the same cell of the K-map for g contains an arbitrary element, i.e., input event $(y_1, \ldots, y_m)$, of $\mathcal{Y}_q$.

(12.6)

To illustrate the use of procedure (12.6), we apply it to our running example as pictured in Fig. 12.3 to which the following explanations refer (the partitions $\mathcal{Y}_0, \mathcal{Y}_1, \mathcal{Y}_q$ of $\mathcal{Y}$ in sub-figure (b) are taken from Fig. 12.2). The output value 0 in cell $(x_1, x_2) = (1, 0)$ of the K-map of f requires us to enter an arbitrary element (y_1, y_2) of $\mathcal{Y}_0$ into cell $(x_1, x_2) = (1, 0)$ of the K-map of g. As $\mathcal{Y}_0 \in \{(0, 1)\}$ contains $(0, 1)$ as only element, it is this we must enter into cell $(x_1, x_2) = (1, 0)$ of g. But do note that it is customary to omit the parentheses, writing only 0,1 instead of $(0, 1)$. This makes it easier to interpret the K-map of g as containing two distinct outputs—y_1 and y_2.

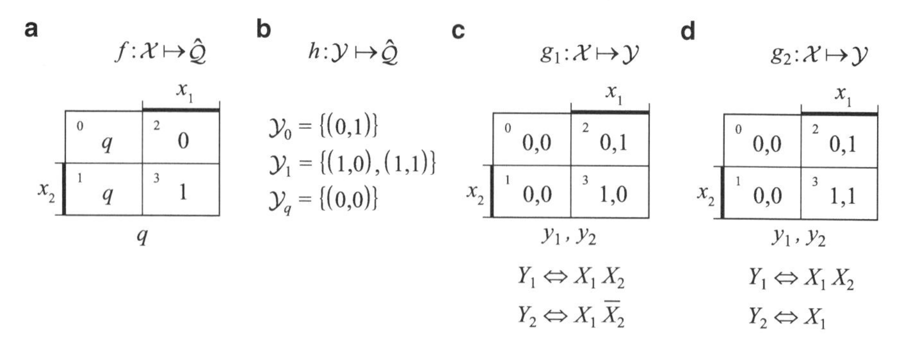

Fig. 12.3 Illustrating the use of procedure (12.6) to obtain a K-map for the pre-logic g

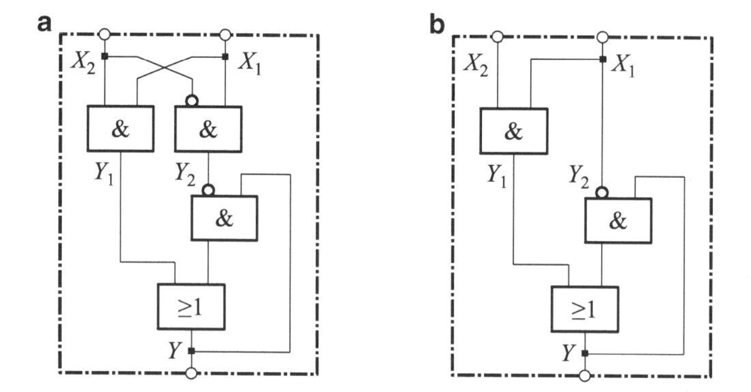

Fig. 12.4 Drawing the circuits for the catenation problem stated in Fig. 12.1 and whose pre-logic is developed in Fig. 12.3

Each cell of the K-map of f containing the variable q correspond to like cells in g where, independently from one another, each contains an arbitrary element (y_1, y_2) of $\mathcal{Y}_q \in \{(0,0)\}$. There only being one element to choose from, both cells $(0,0)$ and $(0,1)$ must contain the same value, i.e., (0,0).

Cell $(x_1, x_2) = (1,1)$ of the K-map of f contains a 1 so that the like cell in g must contain an arbitrary element (y_1, y_2) of $\mathcal{Y}_1 \in \{(1,0),(1,1)\}$. Choosing $(1,0)$ leads to a generic function we call g_1—see Fig. 12.3c—while the alternative choice, $(1,1)$, leads to g_2—see Fig. 12.3d.

To complete the running example of this section, we put forth in Fig. 12.4 the solutions for the catenated circuits when using a minimised hinge latch.

12.2 D-Latch Designs

There are two different ways of designing latches: either as a feedback circuit using the evaluation formulas (12.4), or by composition as discussed in the previous section. We compare both methods by applying them to the realisation of the D-latch of Fig. 11.21. Note that we want the circuits to have two outputs, and these to be inverted. Theoretically there is no reason to give preference to any of the 12 latches. Nevertheless, the D-latch is singular in a number of respects: (a) It governs *sampling* and *synchronising*; (b) it is used to define theoretical models (so-called *automata*) of *synchronous* circuits, and (c) it is the only latch whose K-map can be reduced to a 1-dimensional K-map (see Fig. 12.5b). The first two points are discussed later, the last one is mentioned below.

The Feedback Design of the D-Latch

For the following remarks please consult Fig. 12.5—the sub-figures of which are simply referred to as (a), (b), (c), and (d). We start the feedback design of the D-latch by specifying its K-map, e.g., as in (a). For future reference let me point out that the reduced K-Map of the D-latch, sub-figure (a), can be further reduced to obtain the *fully reduced* K-map of sub-figure (b). The hazard-free version of the logic formula for the D-latch, $Q \Leftrightarrow CD \vee (\overline{C} \vee D)Q$, is obtained from (a) as in the example of Sect. 10.4. From this logic formula we obtain the circuit of the D-latch shown in (c). Naturally, this circuit has only one output, this being the non-inverted output Q of the D-latch. If, in addition, we want an inverted output $\overline{Q}$, we can simply add an inverter as shown in (d). Although this circuit is viable, i.e., will work successfully, it has the *practical* drawback of $\overline{Q}$ being delayed with respect to Q (see Fig. 10.7). Although circuits of this kind are sometimes required, they are generally avoided in favour of the design discussed next.

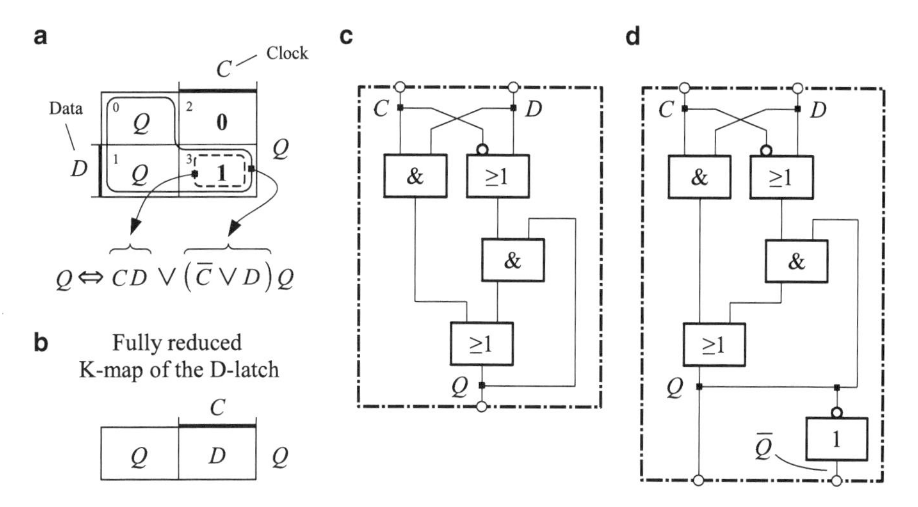

Fig. 12.5 The hazard-free feedback design of the D-latch

The Compositional Design of the D-Latch

When composing a latch, the first decision is what hinge latch to use. If we can avoid a race in the feedback, and enforce complementarity of its outputs, a good argument speaks in favour of a basic Eccles-Jordan latch as hinge latch: it is hazard-free and symmetrical so that its non-inverted and inverted outputs are (theoretically) likely to switch simultaneously.

The basic Eccles-Jordan NAND-latch is repeated in Fig. 12.6 so as to be able to determine its hinge sets $\mathcal{Y}_0$, $\mathcal{Y}_1$, and $\mathcal{Y}_Q$. If *any* input event ($\overline{Y}_1\overline{Y}_2$, $\overline{Y}_1 Y_2$, $Y_1\overline{Y}_2$, or $Y_1 Y_2$) can be presented to the hinge latch, then any transition from one to the other can occur: Most undesirably, the transition from the *common* input event $\overline{Y}_1\overline{Y}_2$ to the memorising input event $Y_1 Y_2$ could cause a race in the feedback. To avoid this situation, we decide to ensure that the Eccles-Jordan latch is not presented with the common input event $\overline{Y}_1\overline{Y}_2$. In other words, we are going to ensure that the inputs to the Eccles-Jordan latch are never simultaneously **0**. This has the desired side effect that the outputs are always inverted, i.e., are inverted for any occurring input events. Avoiding the common input event $\overline{Y}_1\overline{Y}_2$ is achieved by dropping it from $\mathcal{Y}_1$ which is indicated in Fig. 12.6b, c.

The actual composition is carried out in Fig. 12.7 to which we now refer. The latch to be composed, the D-latch, is represented by its K-map f. Having determined the hinge sets as $\mathcal{Y}_0 = \{(1,0)\}$, $\mathcal{Y}_1 = \{(0,1)\}$, and $\mathcal{Y}_Q = \{(1,1)\}$ the generic function is unequivocally determined as the K-map g. Evaluating the outputs Y_1 and Y_2 of the K-map leads to their stated formulas, $Y_1 \Leftrightarrow \overline{C\,D}$ and $Y_2 \Leftrightarrow \overline{C\,\overline{D}}$, and these, together with the basic Eccles-Jordan latch of Fig. 12.6a, allow us to draw the circuit of Fig. 12.7a.

For circuits intended for mass production—and the D-latch certainly belongs to this category—it is very worth while to reduce the number of their gates as far as possible. In our case, by transforming Y_2:

$$Y_2 \Leftrightarrow \overline{C\,\overline{D}} \Leftrightarrow \overline{\mathbf{0} \vee C\,\overline{D}} \Leftrightarrow \overline{C\,\overline{C} \vee C\,\overline{D}} \Leftrightarrow \overline{C\,(\overline{D} \vee \overline{C})} \Leftrightarrow \overline{C\,\overline{D\,C}} \Leftrightarrow \overline{C\,Y_1},$$

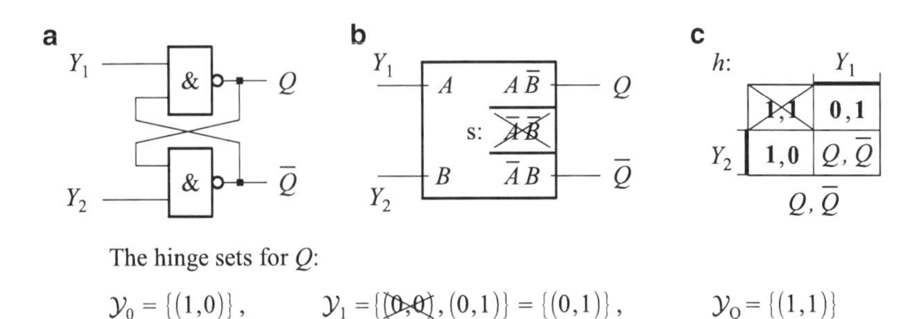

Fig. 12.6 The basic Eccles-Jordan NAND-latch and its hinge sets $\mathcal{Y}_0$, $\mathcal{Y}_1$, $\mathcal{Y}_Q$

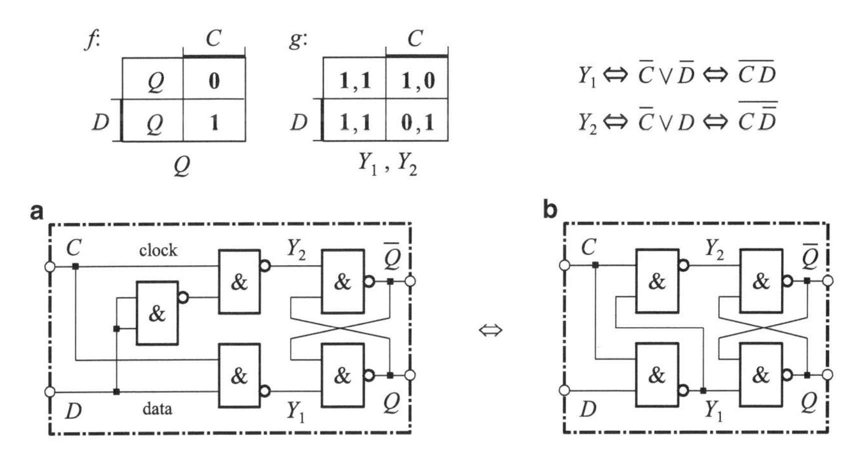

Fig. 12.7 Composing the D-latch on the basis of an Eccles-Jordan latch

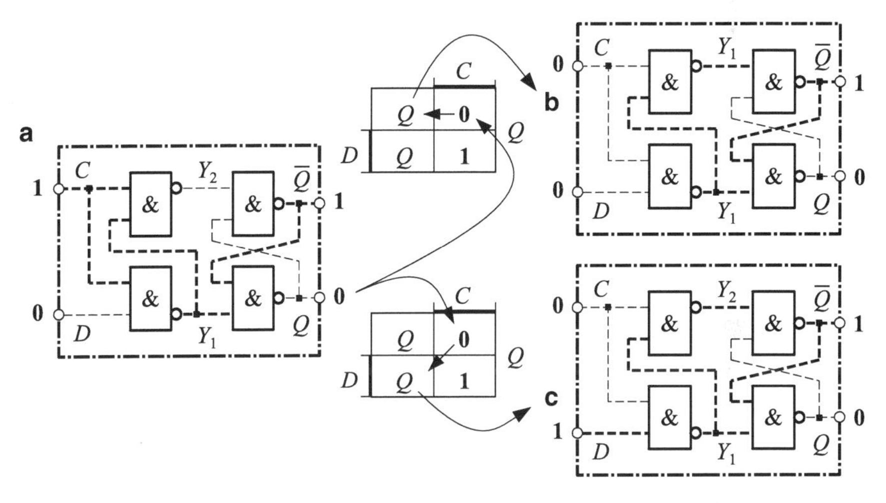

Fig. 12.8 Analysing the D-latch when the input event changes from $C\overline{D}$ to $\overline{C}\ \overline{D}$, (**a–b**), or from $C\overline{D}$ to $\overline{C}D$, (**a–c**)

we obtain the simpler circuit of Fig. 12.7b which is of course equivalent to that of Fig. 12.7a.

Analysing the Signal Flow in the Composed D-Latch

A partial analysis of the composed D-latch is shown in Fig. 12.8 to which this and the next paragraph refers. Explicitly, I shall only speak of the output Q, leaving it to you to verify that $\overline{Q}$ really is the inverse of Q. The top K-map indicates that we first study the circuit when the input event switches from $C\overline{D}$ (where Q is reset) to $\overline{C}\ \overline{D}$ (where Q is memorised). For the input event $C\overline{D}$, the signal flow within the D-latch is pictured in (a). Do note that it is $\overline{Q}$ that is forcibly set, and that Q is reset as a result. Now, when switching to the input event $\overline{C}\ \overline{D}$ we obtain the signal flow shown in (b). The feedback signals prove to have been pre-established because they have not changed.

On the other hand, when switching from the input event $C\overline{D}$ to $\overline{C}D$ (which requires that both input variables, C and D, *change their values simultaneously*) the internal signal flow as shown in (c) is established. Even in the case of this unlikely transition, the circuit functions flawlessly, the internal input signals Y_1 and Y_2 having been pre-established.

A like analysis as in Fig. 12.8 is needed to discuss the transitions from the setting input event CD to the memorising input events $\overline{C}\ \overline{D}$ and $\overline{C}D$. This, I would recommend you to undertake.

Finally, it should be noted that for none of the input events to the composed D-latch are the internal variables Y_1 and Y_2 of its internal Eccles-Jordan latch both **0** ensuring that no race occurs within the Eccles-Jordan latch.

12.3 PSR-Latches Using NAND or NOR-Gates

When introducing **Predominantly Setting and Resetting latches** (**PSR-latches**) in Sect. 11.4, pain was taken to point out (using the example of Fig. 11.13) that developing robust feedback circuits is not a trivial matter. In fact, the only viable approach is the design by composition as demonstrated and argued for the D-latch in the previous section.[1] The argument, there as here, is to use the Eccles-Jordan latch as hinge latch. Let us begin by designing circuits with NAND gates so that we must also use the NAND version of the basic Eccles-Jordan latch.

Figure 12.9 demonstrates how to develop the K-maps of the generic functions g associated with each PSR-latch, given the partial hinge-sets $\mathcal{Y}_0 = \{(1,0)\}$, $\mathcal{Y}_1 = \{(0,1)\}$, and $\mathcal{Y}_Q = \{(1,1)\}$ for the Eccles-Jordan NAND-latch.[2] The procedure used is put forth in (12.6), and I encourage you to make sure you know how the K-maps g of Fig. 12.9 were obtained using this procedure. Reading the K-maps g gives us the expressions for the pre-logic, Y_1 and Y_2. Implementing the formulas for the pre-logic leads to the NAND circuits of Fig. 12.10, our solutions for robust PSR-latches.

Even a cursory glance at the circuits for the PSR-NAND-latches of Fig. 12.10 shows that the simplest of these circuits, that for $f_{1,12}$, serves naturally as a hinge latch for the other circuits. The other latches ($f_{4,3}$, $f_{8,3}$, and $f_{2,12}$) are then obtained

[1]You might possibly want to take another look at the 'The Compositional Design of the D-Latch' in the previous section.

[2]These partial hinge-sets were developed in Fig. 12.6.

The hinge sets for the Eccles-Jordan NAND-latch (see Figure 29.6):

$\mathcal{Y}_0 = \{(1,0)\}$, $\mathcal{Y}_1 = \{\cancel{(0,0)}, (0,1)\} = \{(0,1)\}$, $\mathcal{Y}_Q = \{(1,1)\}$

$f_{4,3}$:

		X_1
	Q	0
X_2	1	0

$Y_{4,3}$

g:

		X_1
	1,1	1,0
X_2	0,1	1,0

Y_1, Y_2

$Y_1 \Leftrightarrow X_1 \vee \overline{X_2} \Leftrightarrow \overline{\overline{X_1}\, X_2}$

$Y_2 \Leftrightarrow \overline{X_1}$

$f_{8,3}$:

		X_1
	1	0
X_2	Q	0

$Y_{8,3}$

g:

		X_1
	0,1	1,0
X_2	1,1	1,0

Y_1, Y_2

$Y_1 \Leftrightarrow X_1 \vee X_2 \Leftrightarrow \overline{\overline{X_1}\, \overline{X_2}}$

$Y_2 \Leftrightarrow \overline{X_1}$

$f_{1,12}$:

		X_1
	0	Q
X_2	0	1

$Y_{1,12}$

g:

		X_1
	1,0	1,1
X_2	1,0	0,1

Y_1, Y_2

$Y_1 \Leftrightarrow \overline{X_1} \vee \overline{X_2} \Leftrightarrow \overline{X_1\, X_2}$

$Y_2 \Leftrightarrow X_1$

$f_{2,12}$:

		X_1
	0	1
X_2	0	Q

$Y_{2,12}$

g:

		X_1
	1,0	0,1
X_2	1,0	1,1

Y_1, Y_2

$Y_1 \Leftrightarrow \overline{X_1} \vee X_2 \Leftrightarrow \overline{X_1\, \overline{X_2}}$

$Y_2 \Leftrightarrow X_1$

Fig. 12.9 Calculating the pre-logic for Eccles-Jordan NAND latches to obtain the NAND-gate solutions of the PSR-latches

by inverting either or both of the inputs of $f_{1,12}$. This is also expressed symbolically in the left column of latch symbols, the four latch symbols being identical, and only their inputs being inverted as required.

The latch symbols in the right column are the individual symbols for each latch, symbols in which neither of the inputs are inverted. You might want to refer to Property 2 in Sect. 11.2 when comparing the left hand and right hand latch symbols in Fig. 12.10: You can switch between the left hand latch symbols and the right hand ones by inverting an input signal and its associated local (internal) variable. The outputs of the symbols have been arranged to coincide with those of the circuits when these are rotated clockwise by 90°.

The PSR-*NOR*-latches of Fig. 12.11 of course realise exactly the same functions as do the PSR-*NAND*-latches of Fig. 12.10. Allow me to encourage you to calculate these circuits along the same lines that led to the PSR-NAND-latches (first determining the hinge-sets $\mathcal{Y}_0$, $\mathcal{Y}_1$, and $\mathcal{Y}_Q$ for the Eccles-Jordan NOR-latch, then the generic functions g, and then designing the circuits). Comparing the *three latch*

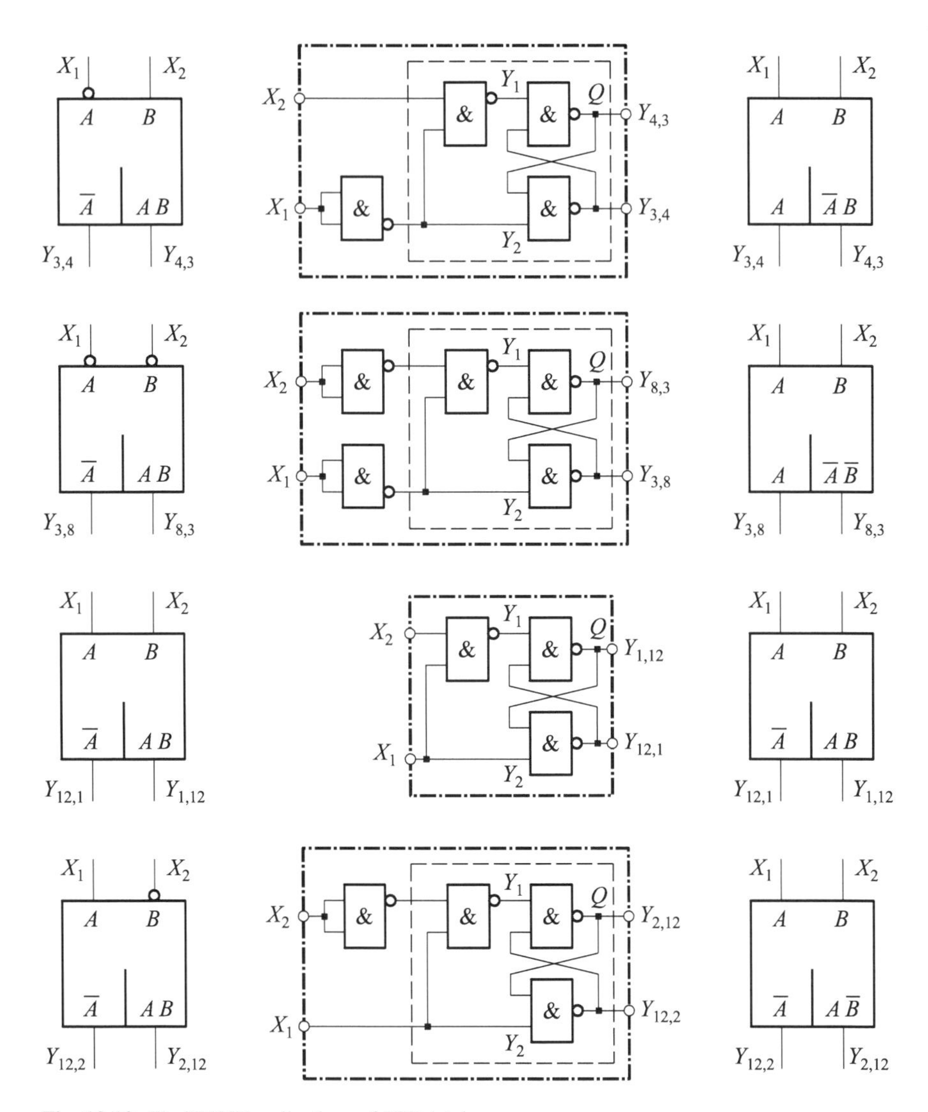

Fig. 12.10 The NAND realisations of PSR-latches

symbols in any given row of both figures *put side by side*, you will notice you can switch between any of these latch symbols by inverting an appropriate input signal and its associated local (internal) variable. The two-output K-map in any given row of such a combined figure describes the behaviour of each of the three latch symbols and of the NAND and NOR circuits of the row.

We now have realisations for all the standard SR-latch-symbols of Fig. 11.20. The top *two* SR-latches are both, e.g., realised by the NAND-latch or NOR-latch in the top row of Fig. 12.10 or Fig. 12.11, while the bottom two SR-latches are realised by Eccles-Jordan NAND and NOR-latches, respectively.

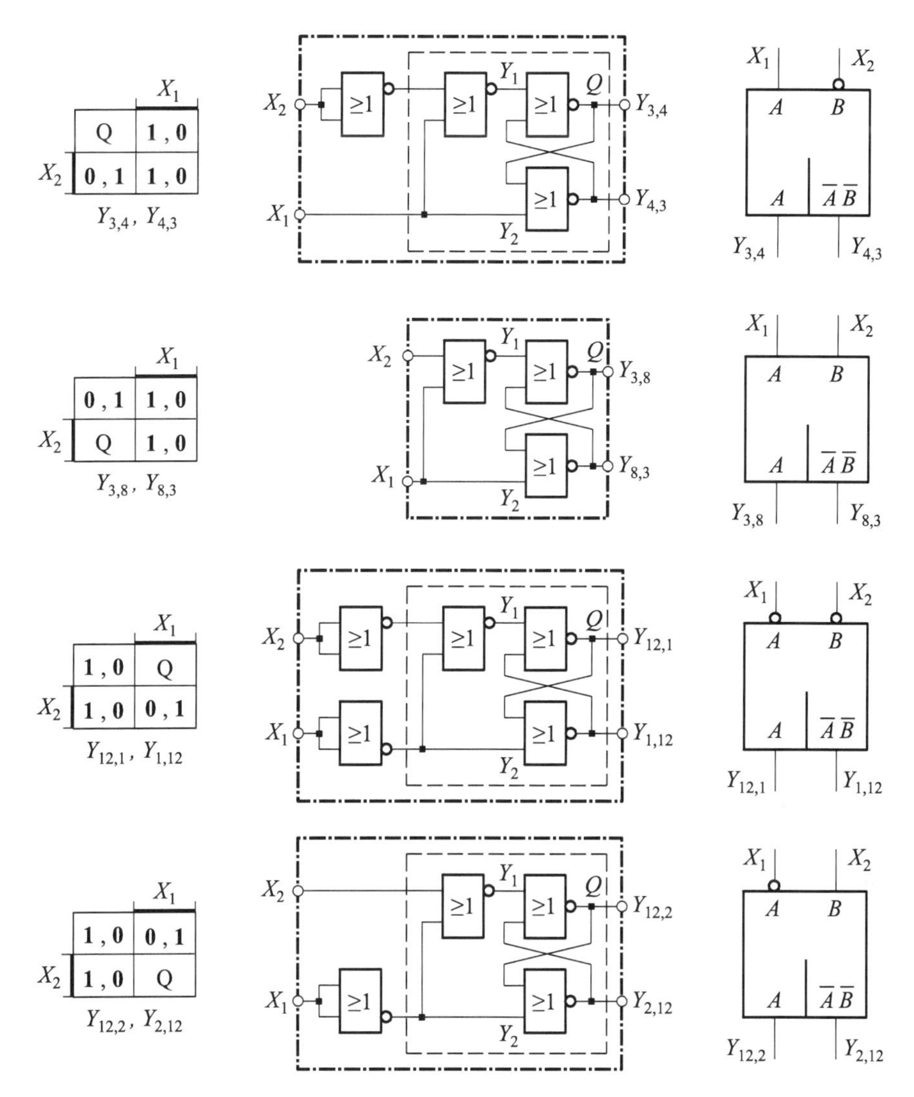

Fig. 12.11 The NOR realisations of PSR-latches

12.4 Synchronous Latch-Inputs

None of the theoretical knowledge on latches presented in this Division was known when the first electric and electronic computers were being developed in the late 1930s and early 1940s. But since 1919 the principle of the Eccles-Jordan latch was common knowledge, and of course the set-dominant relay circuit (see Fig. 2.11), and its dual, the reset-dominant relay circuit. Putting these memory devices to use within larger circuits was done intuitively, and frequently led to hazard prone circuits.

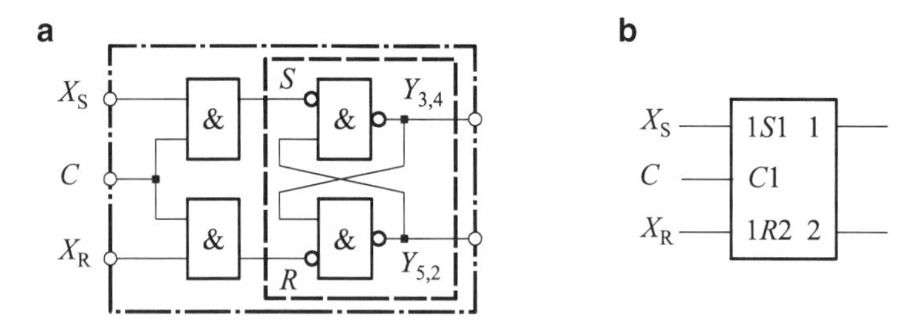

Fig. 12.12 Synchronising latch inputs X_S, X_R with a control signal C

The method employed to get to grips with the problems incurred was to ensure that one could determine when data was sent to the latches. This was done with the help of a control signal that, when active (usually, when its value was **1**), allowed data to be passed to the latch. The simple, if not simplistic way to achieve this was to AND the latch's input signal with the control signal as shown in the example of Fig. 12.12a.

When the AND-gates are incorporated into the latch design, the inputs, X_S and X_R, are referred to as **synchronous inputs**, and the latches informally as **gated latches**. Actually, synchronisation is a knottier problem than simply ANDing an input signal with a control signal. Synchronisation is looked into more generally in a later chapter. Gated latches are, to this day, so common that the IEC-standard 117–15 reserved the variable C as a control variable, calling it a **control dependency**, and meaning that the control variable is ANDed with other input signals: The controlled inputs are preceded with the label which succeeds the control variable, as shown in Fig. 12.12b.

Part III
Asynchronous Circuits

A **sequential circuit** has a memorising ability to a certain degree. One distinguishes between two types of sequential circuits: asynchronous and synchronous. An **asynchronous (sequential) circuits** employs input signals that are continuously present. A **synchronous (sequential) circuit** accepts an input signal only during a usually brief time interval when a (periodic) clock signal is 1.

The theory of sequential circuits, as developed by Huffman (1954), is the theory of **asynchronous circuits**, and these, and only these, are the circuits we are concerned with. You can find excellent coverages of the theory in the literature, among the best being Krieger (1969), and Dietmeyer (1971). Despite its remarkable achievements, Huffman's theory and its subsequent development (e.g., by Moore, Mealy, Unger, Tracey, to mention only a few of the pioneers), had and still has serious pitfalls.

The theory of asynchronous circuits presented in this division breaks with standard theory, the theory by Huffman and his disciples, except for the use of Huffman's flow table. The backbone to *describing*, *calculating*, and *verifying* a sequential circuit is the so-called **word-recognition tree** discussed in great detail with the help of many examples in Chap. 13; it was conceived by Vingron (1983). Using the word-recognition tree to calculate circuits depends completely on the theory of latches put forth in Division Two, in particular in Chap. 10. Huffman's flow table remains an indispensable tool. In Chap. 14 it is explained as a program governing the behaviour of a so-called sequential automaton. How to write such programs is discussed extensively. The prime and unsolved problem in Huffman's theory is finding a so-called state assignment for the flow table. This is a binary encoding of the rows of the flow table enabling us to determine latches which, working in unison, realise the circuit's memory. Chapter 15 presents an **algorithmic solution to the state assignment problem**—the algorithm being called iterative catenation. Chapter 17 contains a novel, algorithmic procedure for transforming a flow table into one with the least number of rows thereby making it possible to find the least number of latches (and the latches themselves) needed to realise the circuit—this process is called merging the flow table. The final chapter is concerned with the problem of verifying the input–output behaviour of sequential circuits, a problem generally taken to be unsolved, and by some, considered to be (at least economically) unsolvable. As Chap. 18 shows, the problem is not only solvable, the results obtained are in fact very economically applicable.

Chapter 13
Word-Recognition Tree*

A sequential circuit reacts to *sequences* of input events. A sequence of input events is called an **input word**. A sequential circuit is specified by the shortest and the least number of input words that determine the circuit's behaviour. As both these properties (the length of input words, and the number of input words) are finite, the number of memory elements needed to realise a sequential circuit is also finite. A sequential circuit can be specified by a tree that allows us to mark the effect input words have on the circuit's output. Such a tree allows us to pinpoint or recognise all the words needed to define the circuit—we thus call these trees **word-recognition trees**; they were devised by Vingron (1982). In this chapter we introduce the word-recognition trees by example, at the same time showing that they enable a direct derivation of the circuit.

13.1 Priority-AND

More often than not, the wording of a problem makes it obvious that the circuit to be designed is a sequential circuit as opposed to a combinational one. This is the case when for one and the same input event, the output sometimes is 0 and other times is 1. In the first example, for instance, the output can be either 0 or 1 when both inputs are 1:

Example 13.1 (Priority-AND). The output y is 1 when both inputs x_1 and x_2 are 1, x_1 having become 1 prior to x_2.

If not otherwise stated, all our examples will have two inputs and a single output. Where possible and reasonable a problem will be given a catchy name (e.g., Priority-AND) to help in recalling the problem when referring to it. In the above example, as in following ones, let it go without saying that we want to express the problem *formally*, and then *calculate* and *draw* the logic circuit.

The natural way to express a sequential problem formally is by way of a **word-recognition tree**, as the one shown in Fig. 13.1a. **Trees** conventionally grow

S.P. Vingron, *Logic Circuit Design*, DOI 10.1007/978-3-642-27657-6_13,

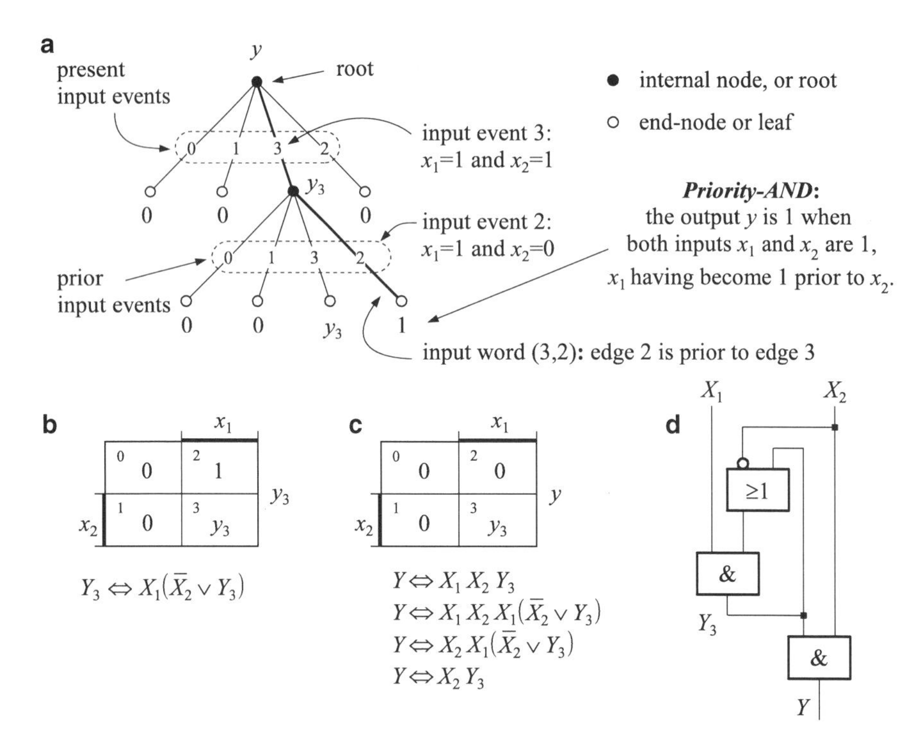

Fig. 13.1 Priority-AND: (**a**) tree representation; (**b, c**) K-maps of the elementary sub-trees and their calculation; (**d**) logic circuit

downward and consist of **nodes** (the circles) and directed **edges** (the connections from one node to another). One distinguishes between an edge that leads *to* a node, a so-called **input edge**, and one that leads *away from* a node, an **output edge**. Each node has exactly *one* input edge, with the exception of a special node called the **root** which has *no* input edge. Some nodes, referred to as **end nodes**, or **leaves**, *have no output edges*. Those nodes that *do have output edges* and an input edge are referred to as **internal nodes**. The root is a special node that is neither an internal node nor an end node.

The **edges of a word-recognition tree** are associated with input events $x \in \{0, 1, \ldots, 2^n - 1\}$. The number of output edges leading away from the root or from an internal node is 2^n, one for each input event. A succession of ℓ connected edges from the root to a leaf represents an *input word* $w(\ell)$. An **input word** $w(\ell)$ is an ℓ-tuple $(x^{(1)}, x^{(2)}, \ldots, x^{(\ell)})$ of input events its elements listed from root to leaf, such that the first input event, $x^{(1)}$, is the present input event, the second input even, $x^{(2)}$, is the prior one, the third, $x^{(3)}$, the pre-prior one, and so on.

Each node is identified by the input word, $w(\ell)$, its elements leading from the root to that node. The **node** $w(\ell)$ **of a word-recognition tree** is assigned an output value $y_{w(\ell)}$; this output value is induced by and exists simultaneously with the present input event $x^{(1)}$ when the prior input event was $x^{(2)}$, the pre-prior $x^{(3)}$, etc. The argument ℓ of a longest input word $w(\ell)$ is called the tree's **height**. The **complexity** of a sequential circuit increases primarily with the height of the word-recognition tree, and secondarily with the number of its internal nodes.

Drawing the word-recognition tree for the Priority-AND, shown in Fig. 13.1a, is strait forward: according to the problem's statement, the only node to be assigned an output value of 1 is the node $(3,2)$; all other nodes are assigned a zero output, with the exception of node $(3,3)$ whose output value, $y_{3,3}$, must equal, and therefore is assigned, the output value y_3. The claim that $y_{x,x} = y_x$ is of prime importance. You might like to argue its meaning and validity yourself—if so, interrupt your reading here ...

Quite generally, $y_{x^{(1)}}$ is an output value caused by a present input event $x^{(1)}$ *irrespective of those input events preceding it*; the output value $y_{x^{(1)},x^{(2)}}$ is also caused by the input event $x^{(1)}$, but only when $x^{(1)}$ is preceded by the input event $x^{(2)}$ which we call the prior input event (to $x^{(1)}$). Although the output value, y_x, of an internal node can vary (is either 0 or 1), we require a sequential circuit to be well-behaved, that is, the output may only change if the prior input event *differs* from the present one. If they do *not* differ, i.e., if $x^{(1)} = x$ and $x^{(2)} = x$, then **well-behavedness** requires that $y_x = y_{x,x}$.

Having now drawn the word-recognition tree of Fig. 13.1a, we turn to the **calculation of the circuit**. For this, the values and variables of the output nodes of each elementary sub-tree are transferred to a K-map whose output is the root variable of the sub-tree. Following this procedure, the elementary sub-tree emanating from the root y of sub-figure (a) corresponds to the K-map of sub-figure (c); the elementary sub-tree whose root is y_3 corresponds to the K-map of sub-figure (b). The evaluation of the K-maps is also shown in sub-figures (b) and (c). The formulas obtained, $Y_3 \Leftrightarrow X_1(\overline{X}_2 \vee Y_3)$ and $Y \Leftrightarrow X_2Y_3$, allow us to **draw the logic circuit** of sub-figure (d).

To see how the circuit works, I encourage you to check its signal flow, for instance, for the following sequence of input events: 0, 2, 3, 1, 3, 2, 3, 2, 0, 1, 3, 1, 0—the associated output values should be 0, 0, 1, 0, 0, 0, 1, 0, 0, 0, 0, 0, 0.

Next, consider a slightly more rigorous version of the above problem.

Example 13.2 (Expanded Priority-AND). The output y is 1 when both inputs x_1 and x_2 are 1 having changed their values from $(0,0)$ via $(1,0)$ to $(1,1)$.

Before taking a formal approach, I invite you to try and develop the circuit intuitively—a practice still quite common and very instructive. Don't give up too soon ....

In a formal approach, we start by drawing the word-recognition tree shown in Fig. 13.2a. Why are the **level-1 outputs** y_0, y_1, and y_2 zero? These are the output values caused when at least one input is 0, the problem implicitly stating that these outputs must be 0 because the output may only be 1 when *both* inputs are 1. Why does a sub-tree emanate from the level-1 node y_3? The output y_3, the output value we register when both inputs are 1, may be 0 or 1, according to its prior input events.

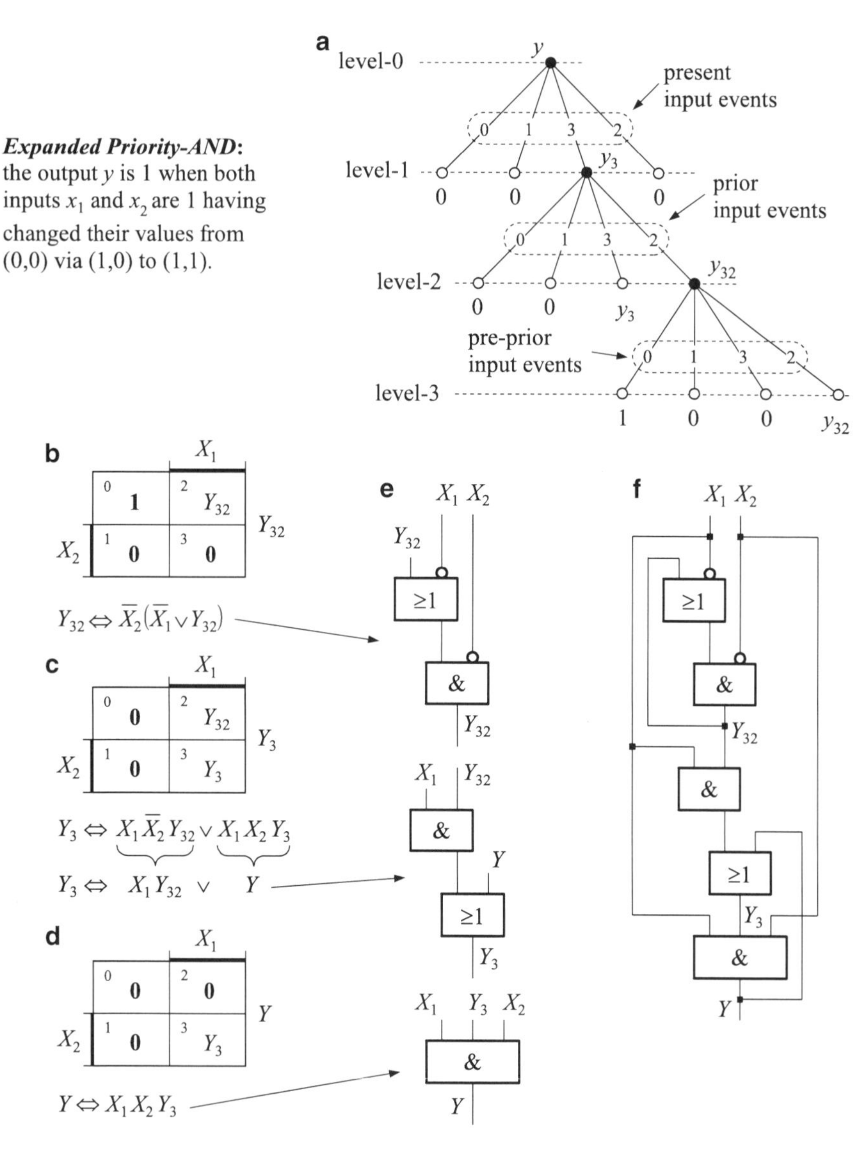

Fig. 13.2 Expanded Priority-AND: (**a**) tree representation; (**b, c, d**) K-maps of the elementary sub-trees and their calculation; (**e**) interim logic circuits; (**f**) final logic circuit

The value of y_3 for all possible prior input events is stated in a sub-tree emanating from node 3.

What values do we assign the **level-2 outputs** $y_{3,0}$, $y_{3,1}$, $y_{3,3}$, and $y_{3,2}$? The problem states that the output may only be 1 when the input event prior to (1,1) is (1,0). The outputs $y_{3,0}$ and $y_{3,1}$ have prior input events (0,0) and (0,1), respectively, and thus must be 0. $y_{3,2}$, as y_3 discussed in the above paragraph, is undetermined (can be either 0 or 1) its value depending on the pre-prior input event. To determine the value of $y_{3,2}$ we let a sub-tree emanate from node (3,2) allowing us to state the values of the outputs $y_{3,2,0}$, $y_{3,2,1}$, $y_{3,2,3}$, and $y_{3,2,2}$. As discussed for the word-recognition tree of the previous example—Fig. 13.1a—$y_{3,3} = y_3$.

The **level-3 outputs** $y_{3,2,0}$, $y_{3,2,1}$, and $y_{3,2,3}$ all have well determined values. $y_{3,2,0}$ expresses the output value caused by the input event 3 when it is preceded by the input event 2, itself preceded by the input event 0—according to the problems specification, $y_{3,2,0}$ is the only output to be 1. By this argument, the outputs $y_{3,2,1}$, and $y_{3,2,3}$ must be 0. In a like argument, that enabled us to write $y_{3,3} = y_3$, we now deduce that $y_{3,2,2} = y_{3,2}$. The general principle is this: **An output $y_{w(\ell)}$ caused by an input word $w(\ell) = (x^{(1)}, \ldots, x^{(\ell-2)}, x^{(\ell-1)}, x^{(\ell)})$ whose ultimate input event $x^{(\ell)}$, and penultimate input event $x^{(\ell-1)}$ are equal—$x^{(\ell-1)} = x^{(\ell)}$—is indistinguishable from the input word $y_{w(\ell-1)}$ invoked by the input word $w(\ell - 1) = (x^{(1)}, \ldots, x^{(\ell-2)}, x^{(\ell-1)})$.** This is a direct consequence of **well-behavedness**.

We next **transform each elementary sub-tree into a K-map**. If you want to draw the circuit with the signal flow going from top to bottom, it is helpful to start with the elementary sub-trees of the highest level, and continue with those nearer and nearer the root. Consider the K-map of Fig. 13.2b developed from sub-tree $y_{3,2}$, the sub-tree whose root is $y_{3,2}$. The root variable $y_{3,2}$ is the output variable $Y_{3,2}$ of the K-map. The numbers in the upper left corners of the cells are the input events, these also marking the edges of the sub-trees. The values and variables inscribed into the cells are those taken from corresponding nodes of the edges of sub-tree $y_{3,2}$. For instance, the **1** in cell 0 corresponds to the node value 1 of edge 0 of sub-tree $y_{3,2}$. You will now hopefully have no trouble creating the K-maps for Y_3 and Y of sub-figures (c) and (d).

When **evaluating the K-maps** note that all K-maps whose output variable is also inscribed into a cell of its K-map represents a feedback latch allowing us to calculate the latch by the generalised feedback-formulas (10.4). For our example, the evaluation of the K-maps is shown in sub-figures (b, c, d), the prime point of interest being the reduced form of Y_3. The logic circuits for the formulas of $Y_{3,2}$, Y_3, and Y are each drawn separately in sub-figure (e) making it easier to see how to **draw the final circuit**, Fig. 13.2f by connecting identical variables.

13.2 Two-Hand Safety Circuit

When actuating the ram of a press, it is a requirement that neither hand of the operator be in the pinch zone which is guaranteed if both hands are needed simultaneously in activating the press. This text is not the proper place to discuss how to mount the two activation push buttons. But it is clear that they cannot simply be ANDed; if they were, one of the activation buttons could be fastened allowing the press to be operated with one hand while simultaneously operating in the pinch zone with the other hand. Thus, the press may only be started when both activation buttons are depressed simultaneously, and the ram only moving down as long as both activation buttons are being depressed. A more concise formulation is this:

Example 13.3 (Two-Hand Safety Problem). The output y is 1 when both inputs x_1 and x_2 are 1 having changed simultaneously from $(0, 0)$ to $(1, 1)$.

In standard switching theory it is **deplorable practice** to allow two or more inputs to change their values *simultaneously*. The reason conventionally put forth is that it is nigh impossible to ensure a desired simultaneous change of input values. But the **counter argument** is that we want to be free to explicitly specify what reaction of the circuit to expect *if* certain input signals change their values simultaneously. In our discussion of sequential circuits we abide by this counter argument for the *synthesis* of circuits.

The design of the Two-Hand Safety Circuit starts by specifying its word-recognition tree, Fig. 13.3a, which hopefully poses no problem after the explanations in the previous section. Of the corresponding K-maps, sub-figures (b) and (c), the former is interesting, as it (not unexpectedly) represents a risky latch, the latch $f_{8,6}$. In drawing the circuit, it is advisable to first draw the interim circuits individually, sub-figure (d), and then to connect all equally named variables to

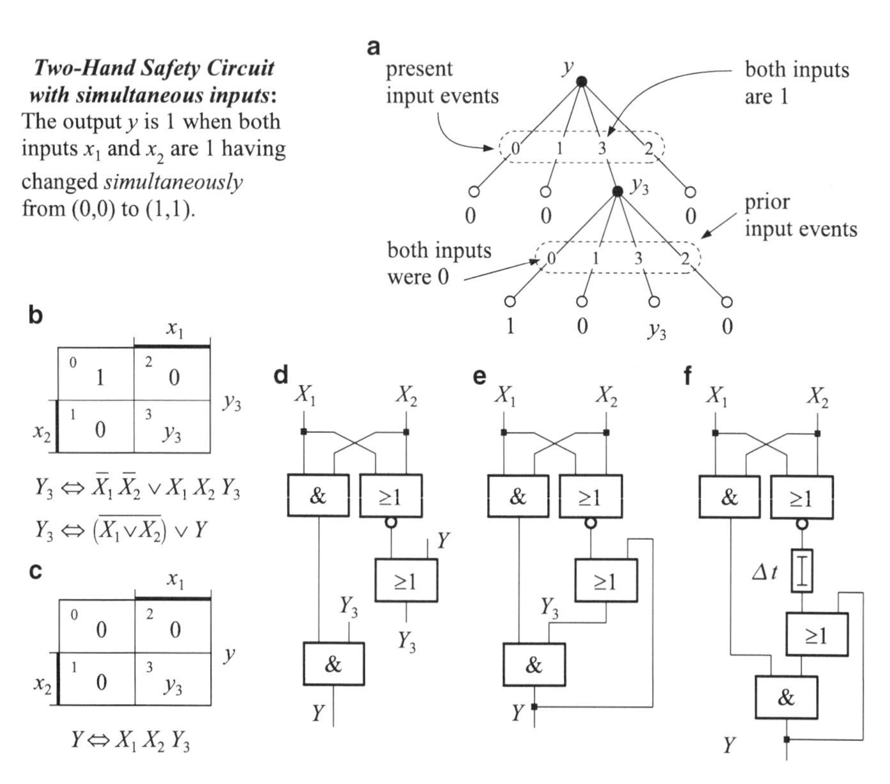

Fig. 13.3 Two-Hand Safety Circuit: (a) tree representation; (b, c) K-maps of the elementary sub-trees and their calculation; (d) interim logic circuits; (e) final logic circuit; (f) necessary delay Δt

obtain the final logic circuit, sub-figure (e). Theoretically and practically there is only a very meagre chance to activate the output of this circuit. A prerequisite for the circuit's applicability is the insertion of a delay as shown in sub-figure (f). If the input values of both x_1 and x_2 change from 0 to 1 within the time span Δt, the circuit accepts the change as simultaneous.

H.H.Glättli proposed a variant of the Two-Hand Safety Problem which we use here only as a design exercise. Consider the possibility of allowing the inputs to change successively, this leading to the following problem:

Example 13.4 (Expanded Two-Hand Safety Problem). The output y is 1 when both inputs x_1 and x_2 are 1 having changed their values

(α) simultaneously from (0,0) to (1,1), or
(β) successively from (0,0) via (0,1) to (1,1), or
(γ) successively from (0,0) via (1,0) to (1,1).

The problem is easily stated in a word-recognition tree, see Fig. 13.4. Compare the two elementary sub-trees $y_{3,1}$ and $y_{3,2}$. Realising that the output variable of a feedback latch is a **free variable** allows us to rename it as desired. By this token, the elementary sub-trees $y_{3,1}$ and $y_{3,2}$ and the latches they represent are identical, $y_{3,1} \Leftrightarrow y_{3,2}$. With this in mind, note that $Y_{3,2}$ in cell 2 of sub-figure (c) may be replaced by $Y_{3,1}$ greatly simplifying the result for Y_3. Developing the logic circuit in Fig. 13.4 is hopefully self-explaining.

13.3 D-Flipflop and T-Flipflop

Computer circuits are commonly designed as **synchronous sequential circuits**, circuits that only take notice of input signals the moment a clock or control signal switches on. An external input that can be fed into a computer circuit at any time (independently of a control signal) is referred to as being **asynchronous**.

Input signals are fed into a *synchronous* sequential circuit via a **synchronising circuit** called a **D-flipflop**. A visualisation of its behaviour is shown in Fig. 13.5 and can be explained verbally as follows (as the arrows show when x_C equals 1):

Example 13.5 (D-flipflop). The value of y_{DF} remains unchanged as long as x_C is 0, and equals the value x_D had prior to when x_C became 1.

The verbal explanation for the D-flipflop translates directly into the word-recognition tree of Fig. 13.6a. This tree is more conveniently expressed using *logic* variables (instead of numeric variables as was previously the case) and by expressing the input events assigned to the nodes as logic input variables or their negations. The first part of the verbal definition, 'the value of y_{DF} remains unchanged as long as x_C is 0', lets us mark the node of edge $\overline{X}_C$ as Y_{DL}. The second part of the verbal definition, '[the value of y_{DF}] equals the value x_D had prior to when x_C became 1', requires the node of the word $(X_C, \overline{X}_C)$ to be marked with the value of X_D.

Expanded Two-Hand Safety Circuit
The output y is 1 *iff* both inputs x_1 and x_2 are 1 having changed their values
(α) simultaneously from (0,0) to (1,1), or
(β) successively from (0,0) via (0,1) to (1,1), or
(γ) successively from (0,0) via (1,0) to (1,1).

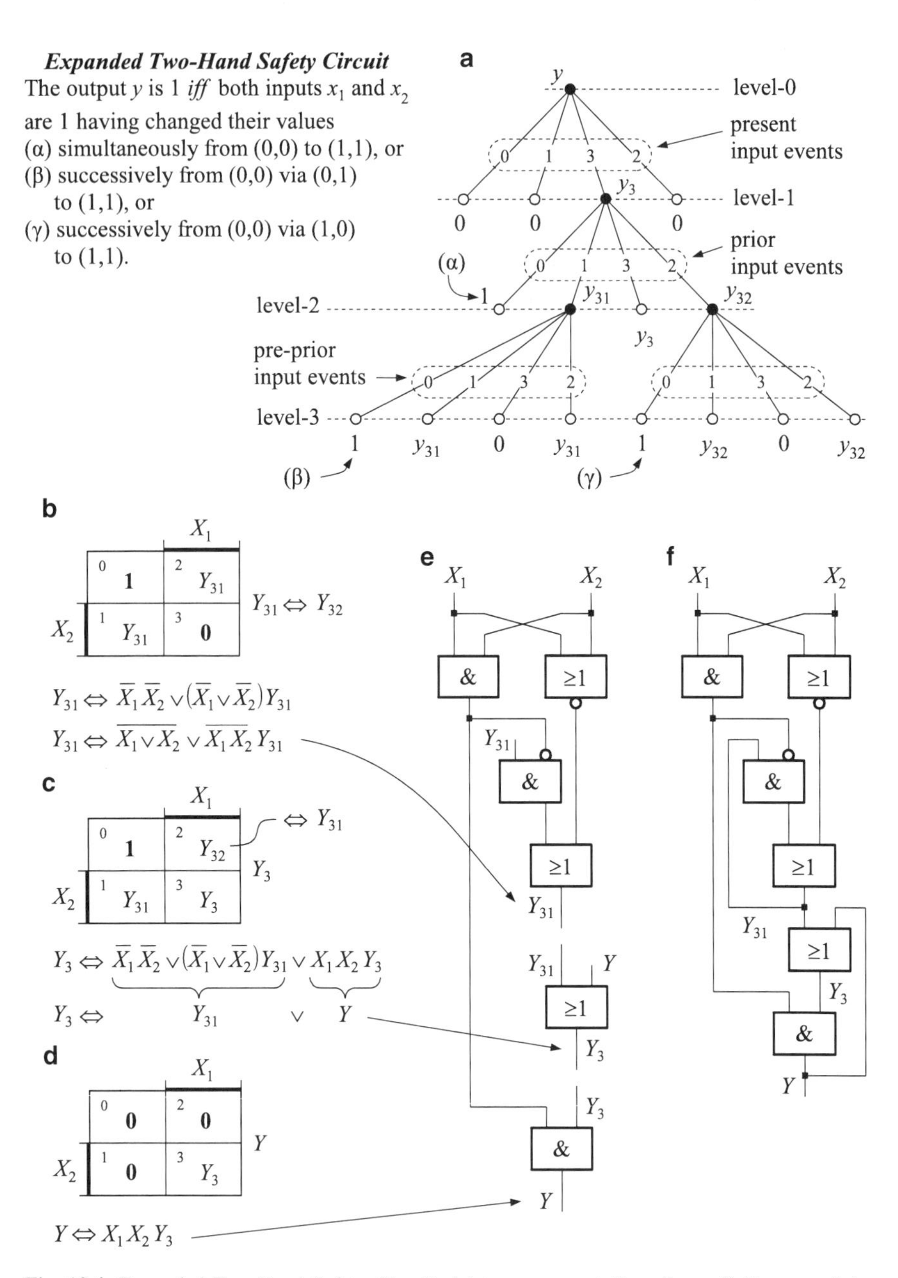

Fig. 13.4 Expanded Two-Hand Safety Circuit: (**a**) tree representation; (**b, c, d**) K-maps of the elementary sub-trees and their calculation; (**e**) interim logic circuits; (**f**) final logic circuit

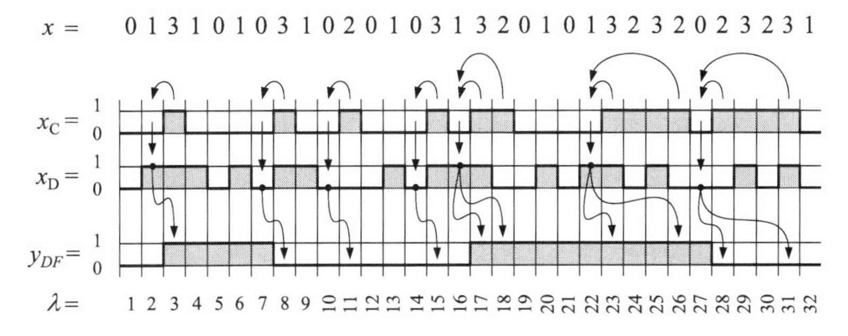

Fig. 13.5 Events graph of the D-flipflop: The value of the output y_{DF} remains unchanged as long as x_C is 0, and equals the value x_D had prior to when x_C became 1. The *arrows* emphasize the output values caused when and while the control input x_C is 1

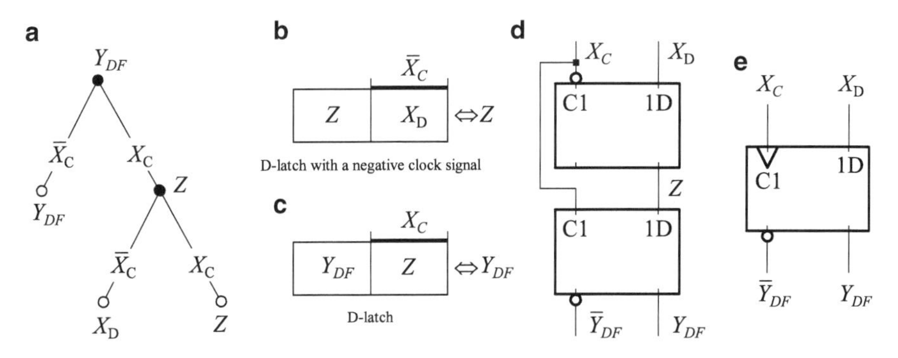

Fig. 13.6 D-flipflop: (**a**) tree representation; (**b, c**) K-maps of the elementary sub-trees (note that the K-maps are those of the D-latches); (**d**) logic circuit; (**e**) the symbol for the D-flipflop

Each elementary sub-tree of the word-recognition tree of sub-figure (a) is rewritten as a reduced K-map, sub-figures (b, c), these being the completely reduced K-maps of D-latches. D-latches are, by the way, the only latches whose K-maps can be completely reduced. Do note, that having obtained these K-maps, we need *no* further calculation to draw the circuit of sub-figure (d). The IEEE-symbol for this circuit is shown in sub-figure (e).

A circuit closely related to the D-flipflop is the toggle circuit, called a T-flipflop. In its basic form it is a circuit with a single input, and a single output. The input is said to toggle the output, meaning that the previous output value is memorised and output in inverted form.

Example 13.6 (T-flipflop). The value of y remains unchanged when the single input, the toggle input x_T, is 0, whereas y equals the complementary value it had prior to when x_T became 1.

The input–output behaviour is conveniently pictured in the events graph of Fig. 13.7a: Whenever the present input x_T is 1 (e.g., for $\lambda = 2, 4, 6, 8$), its prior

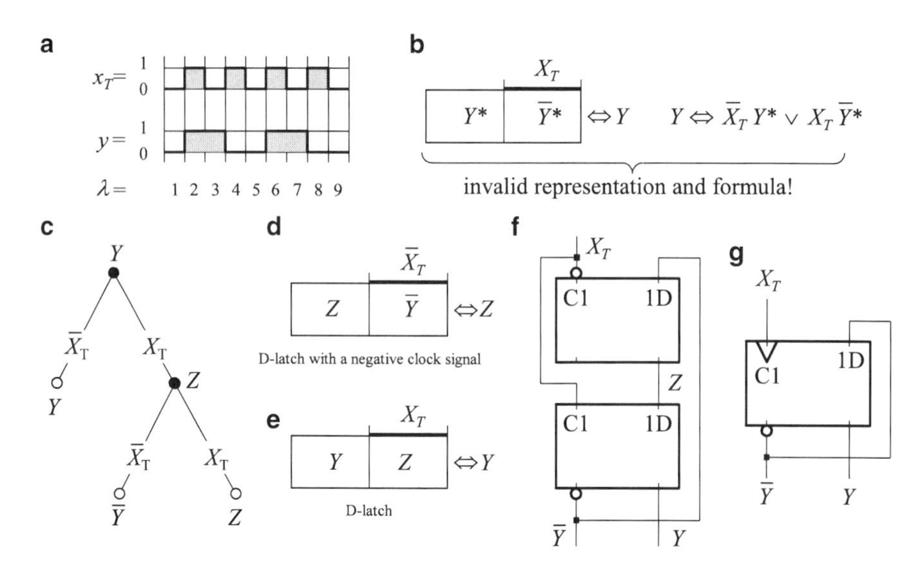

Fig. 13.7 T-flipflop: (**a**) its events graph; (**b**) an unsuccessful, because invalid, attempt at expressing the circuit's behaviour in a K-map; (**c**) tree representation; (**d, e**) rewriting the elementary sub-trees as K-maps (note that the K-maps are those of the D-latches); (**f**) logic circuit; (**g**) logic circuit using the symbol for the D-flipflop

value will, of course, have been 0; as shown in the events graph, an associated prior output value is, in these cases, always the negation of the present output value. One is strongly tempted to describe this behaviour as in the K-map of sub-figure (b) which, when evaluated, leads to the formula $Y \Leftrightarrow \overline{X}_T Y^* \vee X_T \overline{Y}^*$ which, in turn, reduces to $Y \Leftrightarrow \overline{Y}^*$ when X_T is **1**. The output would oscillate meaning that the circuit thus described would not be well behaved. If the time delay were made zero, we would even get the contradiction $Y \Leftrightarrow \overline{Y}$. We must thus refrain from using the simplistic and invalid representation of sub-figure (b).

The correct way to describe the T-flipflop formally is to use a word-recognition tree as shown in Fig. 13.7c stating that, when the input x_T is 1 (i.e., X_T is true) and its prior value 0 (i.e., $\overline{X}_T$ is true), the output is the negation, $\overline{Y}$, of the prior value of the output. The toggle circuit is the basis for binary counting circuits, and for certain shift registers. Its specification in a word-recognition tree, and the development of the logic circuit is demonstrated in Fig. 13.7.

13.4 JK-Flipflop

The **JK-flipflop** is an extension of the set-reset latches (SR-latches) of Fig. 11.20 such that when both inputs are 1, the *inverted* value of the output is memorised. By convention, the J-input is the setting, the K-input the resetting input.

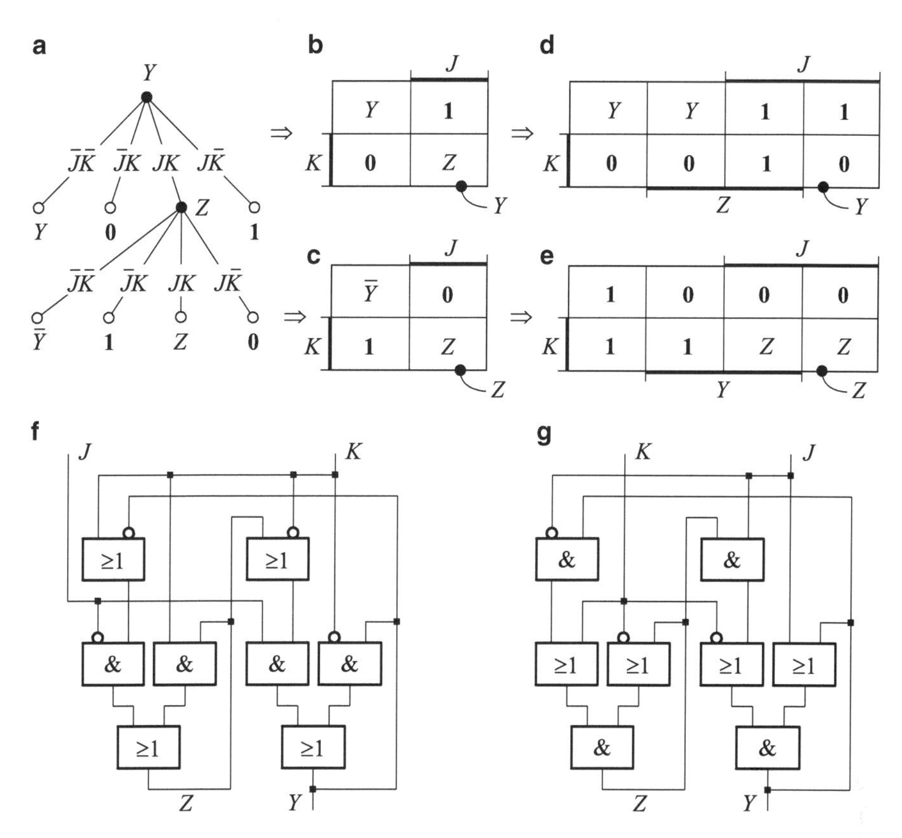

Fig. 13.8 JK-flipflop: (**a**) tree representation; (**b, c**) K-maps associated with the elementary sub-trees; (**d, e**) K-maps in standard form; (**f**) disjunctive logic circuit; (**g**) conjunctive logic circuit

This means that the word-recognition tree of the circuit is developed as follows—see Fig. 13.8a: the leaf of the $J\overline{K}$-edge is marked **1**, that of the $\overline{J}K$-edge **0**, and the end-node for the memorising input event $\overline{J}\ \overline{K}$ is marked Y. When the input event JK is present, the value of the output Y equals that of an internal variable, the variable arbitrarily called Z. The specification of the JK-flipflop requires Z to have complementary values to those of Y. Therefore, each leaf of the elementary Z-sub-tree carries the complementary values of the equivalent edges of the basic Y-tree.

The elementary sub-trees of Fig. 13.8a are rewritten as the reduced K-maps of sub-figures (b) and (c). These reduced K-maps are expanded so as to contain only the K-map's output variable as a map-entered variable—see sub-figures (d) and (e)—the thus obtained K-maps being standard K-maps for latches allowing us to apply the generalised feedback evaluation formulas. Using the disjunctive form $Y \Leftrightarrow U_1 \vee \big(U_{1S} \vee U_y\big)Y$, (10.4), we get:

$$Y \Leftrightarrow J(\overline{K} \vee Z) \vee \overline{K}Y, \qquad\qquad Z \Leftrightarrow \overline{J}(K \vee \overline{Y}) \vee KZ.$$

The conjunctive form $Y \Leftrightarrow V_0\big(V_{0S}V_y \vee Y\big)$, (10.4), leads to:

$$Y \Leftrightarrow (\overline{K} \vee JZ)(J \vee Y), \qquad\qquad Z \Leftrightarrow (K \vee \overline{J}Y)(\overline{K} \vee Z).$$

The circuits associated with these formulas are shown in Figs. 13.8f, g. It is interesting to note a subtle difference between these circuits. In the disjunctive circuit of sub-figure (f) there is a disparity between the number of times J is used as input to a gate (two times), and the number of times K is used (four times). Furthermore, the *inverted* value of Y is fed back to the sub-circuit of Z. On the other hand, in the conjunctive circuit, sub-figure (g), both inputs, J and K, are used in three instances as inputs to gates, and Y is used in positive form. Intuitively, I would prefer the latter circuit.

Chapter 14
Huffman's Flow Table

The word-recognition tree, put forth in the previous chapter, has two pronounced drawbacks: it does not readily lend itself to formulating the behaviour of more complicated sequential problems intuitively, and it is not helpful in analysing the behaviour of a given circuit. Both these problems are addressed by the so-called **flow table** conceived and introduced by Huffman (1954). This chapter presents the flow table as a **program** which governs the behaviour of an abstract automaton called a **sequential automaton** in this text. The sequential automaton, together with a flow table, produces an events graph: the sequence of input events of the events graph being the sequence of inputs to the automaton, the sequence of associated output values of the events graph being the successive and associated outputs of the automaton.

After discussing the principle of the sequential automaton, and explaining the **layout of a flow table**, we employ the latter to stringently specify a number of sequential problems. Writing flow tables that satisfy verbally stated problems is an exacting task that requires experience, insight, and intuition. If you are new to this field, I recommend you take your time in working through the examples of this chapter.

Only in a later chapter (after having mastered specification and synthesis) will we concern ourselves with the analysis of sequential circuits.

14.1 Moore-Type Sequential Automaton

In an events graph of a sequential circuit you will always find at least one input event which causes the output to be 0 in certain locations, and 1 in others. Nevertheless, there is a general principal that states what output value to expect whenever a new input event is entered. This principle is best explained by the behaviour of a simple tape-reading device we call a **(Moore-type) sequential automaton**. The input to a sequential automaton is an input event, the output is one of the numbers 0 or 1. The automaton reads a tape (of finite length) on which is written a program, this program

S.P. Vingron, *Logic Circuit Design*, DOI 10.1007/978-3-642-27657-6_14,

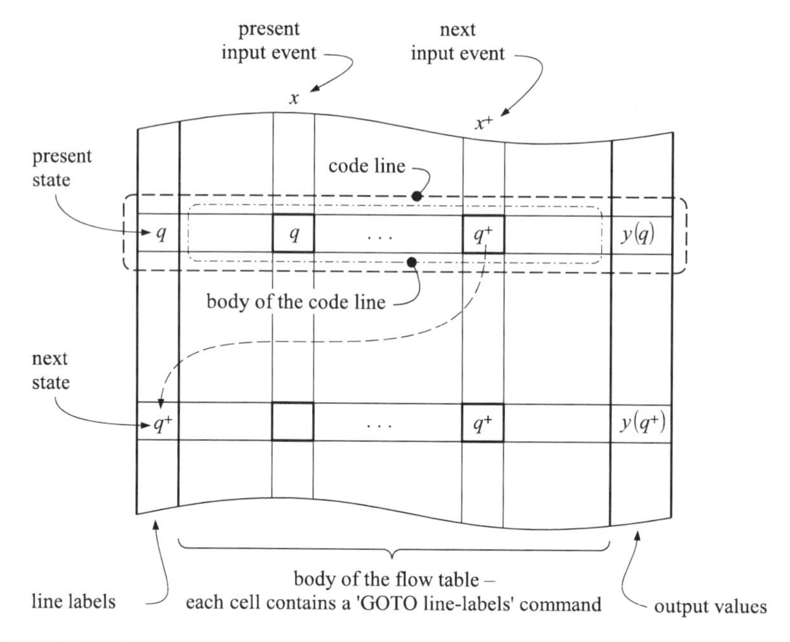

Fig. 14.1 Organisation of a **flow table**

defining unequivocally what new output value to expect whenever a new input event is presented as input to the automaton. The program written on the tape is called a *flow table*. We first look at the way the flow table is organised, and only then will we discuss how the sequential automaton reads the program and reacts to it.

As Fig. 14.1 shows, the program, or **flow table**, consists of labelled code-lines written one below the other. Each code line is referred to as an **(internal) state** of the automaton. The symbols used to label these code lines are chosen arbitrarily, but we usually use positive integers $\{1, 2, \ldots, s\}$, or letters of the alphabet $\{a, b, c, \ldots\}$. A calligraphic $\mathcal{Q}$ refers to the **set of states**, e.g., $\mathcal{Q} = \{1, 2, \ldots, s\}$, or $Q = \{a, b, c, \ldots\}$. A variable $q \in \mathcal{Q}$ that stands for any element of $\mathcal{Q}$ is called a **state variable**.

Each code line has the same tally of cells. The leftmost cell contains the line's label $q \in \mathcal{Q}$, i.e., the state variable q. The rightmost cell contains the output value $y(q)$—either a 0, or a 1—associated with the state q. The cells in between comprise the *body of the code line*, and each such cell contains a state variable q^+ which itself is some element of the set of states Q, the same set whose elements are used to label the lines of the flow table. There are 2^n cells in the body of the code line, each associated with one of the 2^n input events $x \in \mathcal{E}$, with $\mathcal{E} = \{0, 1, \ldots, 2^n - 1\}$. The cells in any given column are associated with the same input event x. This text refers to a cell (q, x) as the **condition** of the automaton.

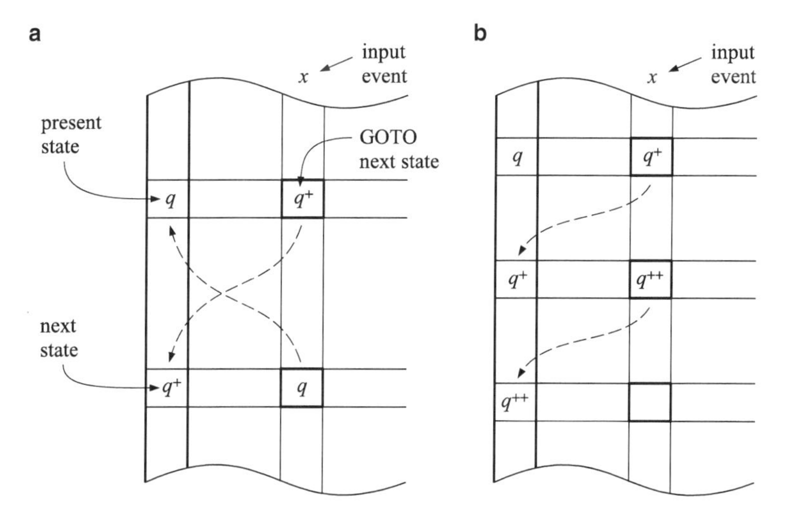

Fig. 14.2 Behaviour suppressed by the requirement of stability. (**a**) Buzzing, (**b**) autonomous changes of state

A **sequential automaton** is a device designed to read a flow table. To do so, it has *three* **reading heads** placed so that, together, they can read the information in a single code-line of the flow table. The leftmost reading head reads the line label, q, telling the automaton which line the other two heads are reading. The rightmost reading head reads the output value, $y(q)$, in the line labelled q, and sends this value to the automaton's output. Both these heads, the leftmost and the rightmost, are static. The middle reading head can read the content of any cell in the body of code line q—which cell is being read is determined by the input event x, present. This implies that the middle reading head can be moved to and from over the cells in the body of a code line as required.

The sequential automaton is fed a flow table it can read as described in the paragraph above. The automaton has the ability to move the tape, or flow table, upwards and downwards so that any line on the tape can be read. **The tape is always moved to the line** q^+ **specified as content to the present cell** (q, x) **being read; if** $q^+ = q$ **the tape remains static (i.e., is not moved)**.

Some terminology: If a cell (q, x), being read, contains the value q (i.e., the same label as the row label), then the automaton is said to be in a **stable condition**, the cell (q, x) itself also being called a **stable condition**; if the cell (q, x) contains a lable q^+ different to the row label q, the automaton is said to be in an **unstable condition**, the cell (q, x) itself also being called an **unstable condition**.

The sequential automaton is an **idealisation** determined by two things. **Firstly**, the tape is moved infinitely quickly from one line to another, meaning there is no time delay associated with this transition. This implies that, during the transition from one line to another, the input event x is invariant. **Secondly**, a flow table always abides by the **principle of stability**: When changing from a present condition (q, x) to a different next-condition (q, x^+) containing a next state q^+ such that $q^+ \neq q$, the condition (q^+, x^+) reached in the next state must be stable, i.e., it must have the value q^+ inscribed. This is visualised in Fig. 14.1. A consequence of stability is that **each *column* of a flow table must have at least one stable condition**. Furthermore, **if a column has only one stable condition, all cells in that column have the label of the stable state entered into them**, i.e., every cell 'points to' the (single) stable state—the **target state**.

The principle of stability effectively prohibits oscillating between states (so-called *buzzing*), visualised in Fig. 14.2a, and autonomous changes from one state to another, an example of this shown in Fig. 14.2b. Stability is the only restriction we place on the design of a flow table.

14.2 Primitive Flow-Table

The only restriction placed on a flow table is stability. But note that stability allows for *rows* with no stable conditions, a single stable condition, or even multiple stable conditions. When transforming a verbally specified sequential problem into a flow table, we will find it quite advantageous to restrict the number of stable conditions per row to one. A flow table that has no more than one stable condition per row is said to be a **primitive flow table**. As with the general flow table, the primitive flow table allows for rows containing no stable conditions. In this book we will always use a primitive flow table when transforming a verbally specified sequential problem into a flow table. Do not worry about creating redundant rows—these will automatically be removed during the synthesis procedure.

The primitive flow table is often pictured without line labels (see Fig. 14.3b), instead, emphasizing the only stable condition per line, as this condition contains the line label. Emphasizing a stable condition is usually done by drawing a circle around the label entered into the cell, which is convenient when working manually. In this book, though, a stable cell is characterised by a light grey background. The term 'stable state' q now refers equally to the row q, as it does to the stable condition or cell in that row. A phrase such as 'changing from state q to q^+' refers to changing from *the* stable condition in row q to *the* stable condition in row q^+.

In this book, the row labels of the primitive flow table are *not omitted*. As we will see in the next chapter, maintaining the row labels is necessary in developing an algorithm by which to eliminate all redundant rows and to develop a word-recognition tree. Furthermore, the stable conditions in *all* flow tables, whether primitive or not, are emphasized.

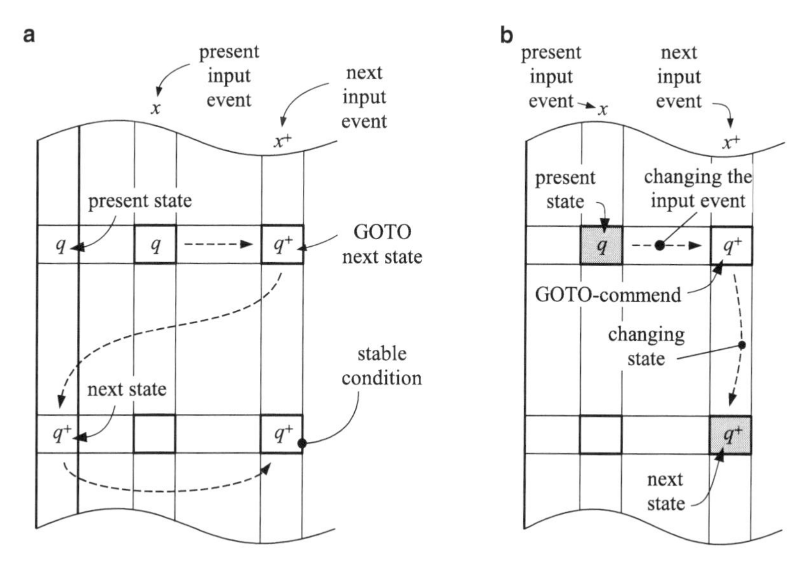

Fig. 14.3 Comparing the general flow table and its terminology, (**a**), with the primitive flow table and its terminology, (**b**)

14.3 Specifying Priority-AND Circuits

To compare a specification by word-recognition tree with that of a flow table, we follow at least some of the examples of the previous chapter. Developing the **flow table for the Priority-AND** is discussed in Fig. 14.4.

Sub-figure (a): We start by drawing a primitive flow table for the number of input events (if the problem has two input variables, the body of the flow table will consist of $2^2 = 4$ columns), without as yet deciding on the number of states (rows). According to the problem's specification (stated in the caption of Fig. 14.4) the output y can, but need not be 1 when both inputs are 1 (column 3), so that column 3 must at least contain two stable condition or cells—one to which we assign the output value 0, the other being assigned the output value 1. The problem's specification implicitly states that y is 0 when neither input is 1; therefore, any stable condition of one of the columns 0, 1, or 2 must be assigned an output value of 0.

Sub-figure (b): We now decide on what GOTO commands to enter into the still empty cells of column 3, sub-figure (a). This is done on a row by row basis. The choice, of course, is whether to enter the label c or d into a given cell. A GOTO-c command will cause a 0 output value, while a GOTO-d command will make the output become 1. When switching from the present input event (0,0) to (1,1) we decided in Sect. 13.1 that the output should be 0 so that we enter a GOTO-c command in cell $(a, 3)$. When switching from column 1 to column 3, x_2 is already 1 so that x_1 now changes its value after x_2 has become 1—in cell $(b, 3)$ we must thus enter a GOTO-c command. Finally, when switching from column 2 to column 3, x_1

a

q \ $x_1, x_2=$	0,0 ($x=0$)	0,1 ($x=1$)	1,1 ($x=3$)	1,0 ($x=2$)	y
a	a				0
b		b			0
c			c		0
d			d		1
e				e	0

b

q \ $x_1, x_2=$	0,0 ($x=0$)	0,1 ($x=1$)	1,1 ($x=3$)	1,0 ($x=2$)	y
a	a		c		0
b		b	c		0
c			c		0
d			d		1
e			d	e	0

c

q \ $x_1, x_2=$	0,0 ($x=0$)	0,1 ($x=1$)	1,1 ($x=3$)	1,0 ($x=2$)	y
a	a	b	c	e	0
b	a	b	c	e	0
c	a	b	c	e	0
d	a	b	d	e	1
e	a	b	d	e	0

Fig. 14.4 The **Priority-AND** problem of Example 13.1: 'The output y is 1 when both inputs x_1 and x_2 are 1, x_1 having become 1 prior to x_2'. (**a**) Estimating a minimal number of states; (**b**) concentrating on the GOTO commands of column 3; (**c**) choice of GOTO commands in those columns (0, 1, and 2) containing only a single stable condition

has become 1 prior to x_2 which only now changes its value from 0 to 1—to obtain a 1 output, we must enter a GOTO-d command into cell $(e, 3)$.

Sub-figure (c): The stable states a, b, and c have in common that they are, per verbal specification, associated with an output value of 0, *whatever the prior input event was*. Therefore, in the column containing stable state a, each of the other cells must contain a GOTO-a command. A like argument holds for columns b and c. As all cells in the flow table have GOTO commands entered, the **sequential problem is completely specified**, the development of the flow table accomplished.

We next discuss the development of the **flow table for the Expanded Priority-AND** of Example 13.2—see Fig. 14.5.

Sub-figure (a): According to the verbal specification of the Expanded Priority-AND, there is only one input word for which the output is to be 1, namely, the sequence (3,2,0) of input events, where the present input event is 3, its prior input event 2, and the pre-prior input event 0. First note that the output can, but need not be 1 when the present input event is 3. Therefore we enter two stable states into column 3, one (the stable state c) associated with the output 0, the other (the stable state d) with 1. When any of the input events 0, 1, or 2 is a present input event, the output must be 0. This means, any stable state in one of these columns must be assigned a 0 output value. But, how many stable states do we need in these columns? For column 2 we need two stable states: When the input event is 3, the output is only 1 if the input event 3 is preceded by the input-*word* (2,0); for any of the other preceding input-word (2,1), or (2,3) the output must be 0. This means that column 2 must have one stable state (f) which points to the stable state d, and another (e) which points to the stable c. For the moment there seems no need to assume the necessity of more than one stable state in each of the columns 0 and 1 so that we now continue with ...

Sub-figure (b): When switching from stable state a in column 0 to column 2, we must decide whether to enter a GOTO-e or a GOTO-f command into cell $(a, 2)$. To realise a 1 output for the input word (3,2,0) we must enter a GOTO-f command into cell $(a, 2)$. When switching from columns 1 or 3 to column 2 we must enter GOTO-e commands into the cells $(b, 2)$, $(c, 2)$, and $(d, 2)$ to ensure that going on

a

x= 0 1 3 2

q \ $x_1, x_2=$	0,0	0,1	1,1	1,0	y
a	a				0
b		b			0
c			c		0
d			d		1
e			c	e	0
f			d	f	0

b

x= 0 1 3 2

q \ $x_1, x_2=$	0,0	0,1	1,1	1,0	y
a	a		c	f	0
b		b	c	e	0
c			c	e	0
d			d	e	1
e			c	e	0
f			d	f	0

c

x= 0 1 3 2

q \ $x_1, x_2=$	0,0	0,1	1,1	1,0	y
a	a	b	c	f	0
b	a	b	c	e	0
c	a	b	c	e	0
d	a	b	d	e	1
e	a	b	c	e	0
f	a	b	d	f	0

Fig. 14.5 The **Expanded Priority-AND** problem of Example 13.2. (**a**) Estimating a minimal number of states and choosing the meaning of the stable states e and f; (**b**) completing the columns with two stable states; (**c**) completing columns 0 and 1 that have one stable state each

to column 3 produces a 0 output. When switching from either column 0 or 1 to column 3 we want the output to be 0 which we achieve by entering GOTO-c into the cells $(a, 3)$ and $(b, 3)$.

Sub-figure (c): The GOTO commands for the cells $(a, 2)$, $(a, 3)$, $(b, 2)$, $(b, 3)$,—i.e., for all transitions of the inputs from the stable states a and b to the columns 2 and 3—could be decided unambiguously making the introduction of (redundant) stable states in the columns 2 and 3 unnecessary. It is good practice to refrain from introducing redundant, and thus unnecessary, stable states. But no harm is done if they are introduced. As columns 0 and 1 each have only one stable state, all cells in these columns 'point to' the stable state of the respective column.

Having entered GOTO commands into all cells of our flow table, the specification of the sequential problems is completed. The latter problem shows quite clearly, that designing a flow table that satisfies a given sequential problem is all but trivial.

14.4 Sampling and Synchronising

In this section we develop *flow tables* for the Sampling Problem (originally formulated in Sect. 9.2 and pictured in the events graph of Fig. 9.2), and the Synchronisation Problem (first discussed in Sect. 13.3, its events graph shown in Fig. 13.5).

Example 14.1 (Sampling Circuit or D-Latch). A so-called data input x_D is **sampled**, i.e., its value is passed on to the output, y_{DL}, during usually brief time intervals which are defined by a so-called control or sampling input x_C being 1; between sampling, i.e., while x_C is 0, the output y_{DL} retains (memorises) the value of x_D last sampled (this is visualised in Fig. 9.2).

The development of the flow table for the Sampling Problem is given in Fig. 14.6. In sub-figure (a) we estimate the number of stable states: When the control input x_C is 1, which is the case in columns 2 and 3, the output value equals that of x_D

a

	$x=0$	1	3	2	
	$x_C, x_D=$				
q	0,0	0,1	1,1	1,0	y_{DL}
a			a		1
b				b	0
c	c				0
d	d				1
e		e			0
f		f			1

b

	$x=0$	1	3	2	
	$x_C, x_D=$				
q	0,0	0,1	1,1	1,0	y_{DL}
a			a	b	1
b			a	b	0
c	c		a	b	0
d	d		a	b	1
e		e	a	b	0
f		f	a	b	1

c

	$x=0$	1	3	2	
	$x_C, x_D=$				
q	0,0	0,1	1,1	1,0	y_{DL}
a	d	f	a	b	1
b	c	e	a	b	0
c	c	e	a	b	0
d	d	f	a	b	1
e	c	e	a	b	0
f	d	f	a	b	1

Fig. 14.6 The **Sampling Problem** of Example 14.1. (**a**) Estimating a minimal number of states and choosing output values assigned to the stable states c, d and e, f; (**b**) completing the columns with single stable states; (**c**) completing columns 0 and 1 that have two stable states

meaning that each column, as a first guess, has one stable state, the value assigned to it equalling the value of x_D in the respective column. When x_C is 0, which is the case in columns 0 and 1, the output retains its previous value, meaning it can be either 0 or 1, so that each of the columns must have two stable states, one for an output value of 0, the other for an output value of 1.

Sub-figure (b): If each of the columns 2 and 3 has *only* one stable state, then (as follows from stability) all cells in such a column have the label of the stable state of the column entered into them. And, of course, without a pressing reason no further stable states are introduced into columns 2 and 3.

Sub-figure (c): In columns 0 and 1, x_C being 0, the previous output values are stored. For instance, switching from stable e (to which the output value 0 is assigned) to column 0, we must enter c into cell $(e, 0)$ to go to that stable state in column 0 to which the output 0 is assigned. Like considerations for each unassigned cell of columns 0 and 1 let us complete the flow table as shown in sub-figure (c). While no finesse was needed in developing the flow table for the Sampling Problem, synchronisation has an interesting twist to it.

Example 14.2 (Synchronising Circuit or D-Flipflop). The output y_{DF} remains unchanged as long as the control or synchronisation input x_C is 0, and equals the value the data input x_D had prior to when x_C became 1 (this is visualised in Fig. 13.5).

The flow table for the Synchronisation Problem is developed in Fig. 14.7. It is, by now, established practice to first estimate the minimal number of states for the flow table; for our example, the result of this estimation is shown in sub-figure (a), and argued as follows. When the control input x_C is 0 (this being the case in columns 0 and 1) the output remains unchanged, meaning, in each column we need one stable state that is assign output value 0, and another that is assigned output value 1. This accounts for the necessity of creating stable states a, b, c, and d, and their associated outputs.

When x_C is 1 (columns 2 and 3), the output equals a previous value (either a 0, or a 1) of the data input x_D, this necessitating (at least) two stable states for each of

a

q \ $x = $ ($x_C, x_D =$)	0 (0,0)	1 (0,1)	3 (1,1)	2 (1,0)	y_{DF}
a	a				0
b	b				1
c		c			0
d		d			1
e			e		0
f			f		1
g				g	0
h				h	1

b

q \ $x = $ ($x_C, x_D =$)	0 (0,0)	1 (0,1)	3 (1,1)	2 (1,0)	y_{DF}
a	a	c	e	g	0
b	b	d	e	g	1
c	a	c	f	h	0
d	b	d	f	h	1
e	a	c	e	?	0
f	b	d	f		1
g	a	c		g	0
h	b	d		h	1

c

q \ $x = $ ($x_C, x_D =$)	0 (0,0)	1 (0,1)	3 (1,1)	2 (1,0)	y
a	a	c	e	g	0
b	b	d	e	g	1
c	a	c	f	h	0
d	b	d	f	h	1
e	a	c	e	0 3 2 g	0
f	b	d	f	1 3 2 h	1
g	a	c	0 2 3 e	g	0
h	b	d	1 2 3 f	h	1

Fig. 14.7 The **Synchronisation Problem** of Example 14.2. (**a**) Estimating a minimal number of states and choosing output values assigned to the stable states; (**b**) completing cells defined by **direct transitions** from one stable state to another; (**c**) completing cells defined my **multiple transitions** from one stable state to another

the columns 2 and 3—one to which the output value 0 is assigned, the other for the output value 1. Sub-figure (a) documents these considerations.

Sub-figure (b): Each empty cell in column 0 of sub-figure (a) can be specified by considering a **direct transition** from any prior stable-state of some other column to the appropriate present stable-state in column 0. The considerations are this: When switching from a prior stable-state q (not in column 0) with an associated output value $y(q)$, to column 0, we must enter into cell $(q, 0)$ that stable state q^+ whose associated output-value $y(q^+)$ equals $y(q)$. For instance, when switching from stable state d to column 0, we must enter b into cell $(d, 0)$, because stable state b causes the output to be 1, the same as the stable state d does. Like considerations hold for column 1.

Still considering sub-figure (b), we now concentrate on columns 2 and 3, columns in which x_C is 1. According to the specification, when x_C is 1, the output equals the value x_D had when x_C was 0, i.e., *prior* to when x_C became 1. Quite clearly, this only covers the transitions from the stable states of the columns 0 and 1 to either column 2, or column 3 allowing us to inscribe state variables into the top four rows of columns 2 and 3.

But, we certainly have a problem with the still unspecified cells $(e, 2)$, $(f, 2)$, $(g, 3)$, and $(h, 3)$ as the following discussion for cell $(e, 2)$ shows. When switching from the stable state e to column 2, we have the option of either entering a GOTO-g or a GOTO-h command into cell $(e, 2)$. But, and I beg you to ponder this argument, as x_C has remained 1 during this transition, we have no value of x_D to refer to for a prior 0-value of x_C. Not being able to decide whether to enter a GOTO-g or a GOTO-h command into cell $(e, 2)$ at this stage, I have, for the time being, entered a question mark.

Sub-figure (c): In deciding what to enter into the open cells $(e, 2)$, $(f, 2)$, $(g, 3)$, and $(h, 3)$ of sub-figure (b), we touch a sensitive topic, this being, that the flow table only enables us to document *direct transitions*, transitions from one stable state to another. For instance, consider the GOTO command to be entered into cell $(e, 2)$: Entering GOTO-g means we go from stable state e to stable state g (where the output is 0); entering GOTO-h means we go from stable state e to stable state h (where the output is 1). According to the specification of our example, we cannot decide on one of these GOTO commands on the basis of a *direct* transition because, switching *to* one of the stable states g or h *from* stable state e, the input x_C remains 1 meaning, there is no prior value of x_C to which we can associate x_D. We need to know when x_C was last 0 before stable state e was entered. The only way stable state e can be entered is to follow one of the GOTO-e commands in cell $(a, 3)$ or $(b, 3)$; these cells can only be reached from stable a or b (where both x_C and x_D are 0). Now, knowing that x_D was 0 prior to x_C being 1, we know that y must take on the value of x_D, i.e., of 0 when switching to column 2; this is only possible when entering GOTO-g into cell $(e, 2)$.

So, to decide on what command to enter into cell $(e, 2)$, GOTO-g or GOTO-h, we need to consider two successive direct transitions between stable states. These transitions are documented by writing not the successive *states*, but the successive *input events* of the stable states into the cell being discussed, and underlining the direct transition to the cell. In our example, we enter the sequence $0\ \underline{3\ 2}$ into cell $(e, 2)$. This sequence provides the argument, or reason, for entering the GOTO-g command into cell $(e, 2)$.

The notation developed above is used extensively in the next example, and is essential in all more complicated problems. So, please take your time in understanding how it is used by following the examples of multiple state transitions conscientiously. In sub-figure (c), do make sure you understand how the double transitions $1\ \underline{3\ 2}$, $0\ \underline{2\ 3}$, $1\ \underline{2\ 3}$ entered into the cells $(f, 2)$, $(g, 3)$, $(h, 3)$, respectively, explain the GOTO commands entered into these cells.

14.5 Passed-Sample Problem

The example discussed here is one of those problems that do not readily lend themselves to an intuitive and direct specification by word-recognition tree. It has a certain similarity with the Sampling Problem of Example 14.1, but goes further back to prior input events (hence its name).

Example 14.3 (Passed-Sample Problem). The output y is 0 whenever the sampling signal x_1 is 0, but when $x_1 = 1$ the value of y equals that value the data signal x_2 had the last time x_1 was 1.

To get a feeling for the problem (before developing the flow table), you might like to consider the events graph which visualises what is meant by a 'passed-sample'. To draw the events graph we *choose* (more or less arbitrarily) a sequence of input events. This provides us with the sequences for x, x_1 and x_2 shown in Fig. 14.8. According to the verbal specification, whenever $x_1 = 0$, the output y is also 0. This accounts for the zero output in the locations 2–5, then 7–9, 12, 13, 15, and so on. To see how to determine the output value y when $x_1 = 1$ let us start at location 6. The arrow marked a leads us back to location 1 which is 'the last time x_1 was 1'. In this location the value of x_2 is zero (follow arrow b). It is this value of x_2 we assign to the value of y in location 6 (follow arrow c). Corresponding arrows are shown for location 10. The remaining arrows (for the locations 11, 14, 16–18, 23, 27, 28, 31, and 33–35) are equivalent to arrow c. You might like to answer the question, why the initial output value is undefined for the starting input event chosen.

We now start developing the flow table. The various steps are shown in Figs. 14.9 and 14.10.

Figure 14.9a: We first try to estimate a minimal number of rows needed. Due to the output y being 0 when x_1 is 0, we assume that columns 0 and 1 will each have (at least) one stable state to which we must assign the output value 0. These stable states are arbitrarily chosen to be e and f. On the other hand the example specifies that *when* $x_1 = 1$ *the value of* y *equals that the data signal* x_2 *had the last time* x_1 *was 1*. For the time being the only thing we want to deduce from this formulation is that whenever $x_1 = 1$, i.e., in either of the columns 2 and 3, the output y will sometimes be 0, and sometimes be 1, necessitating each column to have (at least) two stable states—one to output 0, the other to output 1. The number of stable states so far specified is not necessarily enough to express the above example in a flow table as we discuss next.

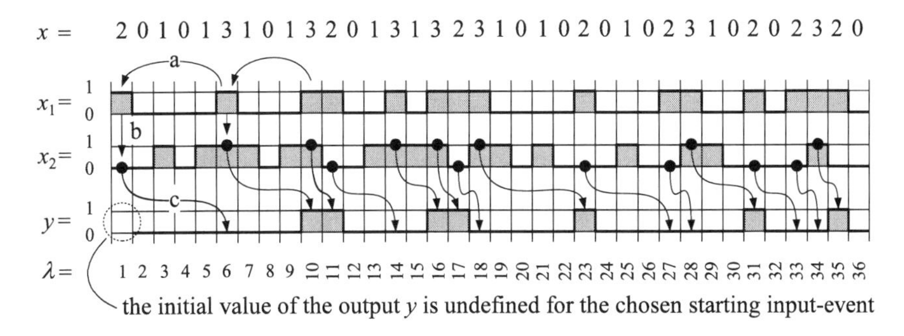

Fig. 14.8 An events graph for the **Passed-Sample Problem**

a

q \ $x_1, x_2 =$	0,0 ($x=0$)	0,1 ($x=1$)	1,1 ($x=3$)	1,0 ($x=2$)	y
a			a		0
b			b		1
c				c	0
d				d	1
e		e			0
f	f				0

b

q \ $x_1, x_2 =$	0,0 ($x=0$)	0,1 ($x=1$)	1,1 ($x=3$)	1,0 ($x=2$)	y
a			a		0
b			b		1
c				c	0
d				d	1
e		e	a	c	0
f	f		a	c	0
g		g	b	d	0
h	h		b	d	0

c

q \ $x_1, x_2 =$	0,0 ($x=0$)	0,1 ($x=1$)	1,1 ($x=3$)	1,0 ($x=2$)	y
a			a	d	0
b			b	d	1
c			a	c	0
d			a	d	1
e		e	a	c	0
f	f		a	c	0
g		g	b	d	0
h	h		b	d	0

Fig. 14.9 First steps in developing the flow table for the **Passed-Sample Problem**. (**a**) Preliminary estimate of the minimal number of rows. (**b**) Revised estimate. (**c**) All GOTO commands determined by direct transitions from one stable state to another

Figure 14.9b: Still considering Fig. 14.9a, for the moment, let us ask which of the GOTO commands, GOTO-a or GOTO-b, should we enter into cell $(e, 3)$? Whichever we enter, we exclude the possibility of going to the alternative stable condition in column 3. As shown in Fig. 14.9b, if we choose to enter GOTO-a into cell $(e, 3)$ we need to introduce a second stable condition into column 1, in our example the stable g, from which we can GOTO-b when switching to column 3—see cell $(g, 3)$. These considerations necessitate the introduction of the states or rows g and h together with the GOTO commands a to d in the bottom right quadrant of the flow table of Fig. 14.9b.

Figure 14.9c: In this figure we consider the GOTO commands for the top right quadrant of the flow table where $x_1 = 1$. For input transitions within this quadrant let us reformulate the original specification '*when $x_1 = 1$ the output y takes on the value x_2 had the last time x_1 was 1*' to be read as '*when x_2 changes its value from 0 to 1 (x_1 remaining 1), the new output value must be 0, this being the previous value of x_2*', and as '*when x_2 changes its value from 1 to 0 (with x_1 remaining 1), the new output value must be 1, this being the previous value of x_2*'. Thus, when switching from column 2 to column 3 we must GOTO-a, while when switching from column 3 to column 2 we must GOTO-d.

Figure 14.10a: We now concentrate on the GOTO commands to be entered into the top left quadrant of the flow table. When considering only the *direct* transition from column 2 to column 0 it is not possible to decide whether to enter a GOTO-f or a GOTO-h command into the cells $(c, 0)$ or $(d, 0)$ because both stable states, f and h, correctly let the output become 0. But, considering the **double transitions**, from column 2 to column 0, and back from column 0 to column 2 (or on to column 3), which we write as the *succession of events* 2 0 2 and 2 0 3, we see clearly that GOTO-f (as contrasted with GOTO-h) must be entered into both cells $(c, 0)$ and

a

q \ $x_1,x_2=$	0,0 ($x=0$)	0,1 ($x=1$)	1,1 ($x=3$)	1,0 ($x=2$)	y
a	3 0 2 3 0 3 h	3 1 2 3 1 3 g	a	d	0
b	3 0 2 3 0 3 h	3 1 2 3 1 3 g	b	d	1
c	2 0 2 2 0 3 f	2 1 2 2 1 3 e	a	c	0
d	2 0 2 2 0 3 f	2 1 2 2 1 3 e	a	d	1
e		e	a	c	0
f	f		a	c	0
g		g	b	d	0
h	h		b	d	0

b

q \ $x_1,x_2=$	0,0 ($x=0$)	0,1 ($x=1$)	1,1 ($x=3$)	1,0 ($x=2$)	y
a	h	g	a	d	0
b	h	g	b	d	1
c	f	e	a	c	0
d	f	e	a	d	1
e	2 1 0 2 2 3 0 3 f	e	a	c	0
f	f	2 0 1 2 2 0 1 3 e	a	c	0
g	3 1 0 2 3 1 0 3 h	g	b	d	0
h	h	3 0 1 2 3 0 1 3 g	b	d	0

c

q \ $x_1,x_2=$	0,0 ($x=0$)	0,1 ($x=1$)	1,1 ($x=3$)	1,0 ($x=2$)	y
a	h	g	a	d	0
b	h	g	b	d	1
c	f	e	a	c	0
d	f	e	a	d	1
e	f	e	a	c	0
f	f	e	a	c	0
g	h	g	b	d	0
h	h	g	b	d	0

Fig. 14.10 Continuing the development of the flow table for the **Passed-Sample Problem**. (**a**) Documenting the double transitions (between columns) that enable us to decide on GOTO commands. (**b**) Documenting those triple transitions (between columns) that enable us to decide on GOTO commands. (**c**) The final flow table

$(d, 0)$ to let the output become 0 when switching from column 0 to columns 2 or 3. The above considerations are documented by entering the double transition 2 0 2 into the cells $(c, 0)$ and $(d, 0)$ as 2 0 2. As in the previous example, **Underlining** marks the direct transition being considered within the multiple transition. The subsequent direct transition in our example would be denoted as 2 0 2, and would be entered into cell $(f, 2)$. But, transitions (whether direct or multiple transitions) are only entered into cells *not* containing a GOTO command, and only when the transition is used to argue or explain a GOTO entry. Allow me to ask you to ponder the GOTO commands for the remaining cells of the top left quadrant with the help of the double transitions written into these cells.

Figure 14.10b: The GOTO commands to be entered onto the bottom left quadrant cannot be defined by considering *direct* or even *double transitions*. For instance, if you try to specify the GOTO command for cell $(e, 0)$, you *must* start in column 2 to be able to GOTO-e—i.e., the stable e in cell $(e, 1)$—from where you can switch to cell $(e, 0)$. But whether we must enter a GOTO-f or a GOTO-h command into cell $(e, 0)$ cannot yet be decided. Considering as third transition—after $(2, 1)$ and

$(1,0)$—the transition $(0,2)$, we know from the verbal specification of our example that the output y must be 0, this being '*the value* x_2 *had the last time* x_1 *was* 1'. This third transition requires us to switch from column 0 to column 2. If we do so from stable h, the GOTO-d in cell $(h,2)$ sends us to row d where the output y is 1, whereas 0 as required. On the other hand, switching from the stable f of column 0 to column 2 sends us to row c where the output y is 0, as desired. We must therefore enter GOTO-f into cell $(e,0)$.

As cell $(e,0)$ defines the *direct* transition from column 1 to column 0, the triple transition 2 1 0 2, when entered into cell $(e,0)$, is denoted as $2\ \underline{1\ 0}\ 2$. The transitions $\underline{2\ 1}\ 0\ 2$ would be entered in the cells $(c,1)$ and $(d,1)$, while $2\ 1\ \underline{0\ 2}$ would be entered into cell $(f,2)$. Again, allow me to ask you to take the time to ponder the remaining entries in the bottom left quadrant.

Figure 14.10c: In the final flow table we of course enter only the GOTO commands, omitting all multiple transitions that led to the GOTO entries.

14.6 Expanded Two-Hand Safety Problem

The Expanded Two-Hand Safety Problem was introduced *en passant* in Sect. 13.2 as Example 13.4, its word-recognition tree and circuit developed in Fig. 13.4. For later use, we now develop a flow table for the problem restated here:

Example 14.4 (Expanded Two-Hand Safety Problem). The output y is 1 when both inputs x_1 and x_2 are 1 having changed their values

(α) simultaneously from (0,0) to (1,1), or
(β) successively from (0,0) via (0,1) to (1,1), or
(γ) successively from (0,0) via (1,0) to (1,1).

The various stages of the flow table's development are shown in the sub-figures of Fig. 14.11 which we now discuss.

Figure 14.11a: This sub-figure reflects the individual problems stated in steps (α), (β), and (γ). The direct transition, (α), from input event 0 to input event 3 is expressed by entering d into cell $(a,3)$.

The double transition, (β), from input event 0, via input event 1, on to input event 3 (written as input sequence 0,1,3) is split into successive, direct transitions from input event 0 to input event 1 (which is written as $\underline{0,1},3$), followed by the direct transition from input event 1 to input event 3 (which we write as $0,\overline{1,3}$). These direct transitions determine the entries b and d in the cells $(a,1)$ and $(b,3)$, respectively. Similarly, the successive, direct transitions $\underline{0,2},3$ and $0,\underline{2,3}$ tell us to enter the GOTO-commands c and d into the cells $(a,2)$ and $(c,3)$, respectively.

We next discuss the transitions between the stable states b and c repeating and emphasizing an argument put forth Sect. 13.2. These transitions require two

a

q \ $x_1, x_2 =$	0,0 ($x=0$)	0,1 ($x=1$)	1,1 ($x=3$)	1,0 ($x=2$)	y
a	a	b (0 1 3)	d	c (0 2 3)	0
b	a	b	d (0 1 3)	c	0
c	a	b	d (0 2 3)	c	0
d	a	•	d	•	1

do not GOTO-b (cell $(d,1)$) do not GOTO-c (cell $(d,2)$)

b

q \ $x_1, x_2 =$	0,0 ($x=0$)	0,1 ($x=1$)	1,1 ($x=3$)	1,0 ($x=2$)	y
a	a	b	d	c	0
b	a	b	d	c	0
c	a	b	d	c	0
d	a	f (3 1 3)	d	g (3 2 3)	1
e	a		e		0
f	a	f	e (3 1 3)		0
g	a		e (3 2 3)	g	0

c

q \ $x_1, x_2 =$	0,0 ($x=0$)	0,1 ($x=1$)	1,1 ($x=3$)	1,0 ($x=2$)	y
a	a	b	d	c	0
b	a	b	d	c	0
c	a	b	d	c	0
d	a	f	d	g	1
e	a	f (3 1 3)	e	g (3 2 3)	0
f	a	f	e	g	0
g	a	f	e	g	0

Fig. 14.11 Expanded Two-Hand Safety Problem

inputs to change their values simultaneously which, in common switching theory, is deplorable practice, and thus forbidden (although I wouldn't know how to enforce such a ban for input values a user is per definition allowed to change at will). The approach taken here is not to prohibit simultaneous change, rather to say how the circuit is expected to react *if and when* two or more input values change simultaneously.

So, for our example, we argue that the circuit, when in stable state d, need not know whether it reached this state via stable b or stable c, in effect, equating these two stable states. This is expressed by entering c into cell $(b, 2)$, and b into cell $(c, 1)$ allowing us to switch indiscriminately between the stable states b and c.

Figure 14.11b: In sub-figure **(a)** there are still two cells without GOTO-commands: $(d, 1)$ and $(d, 2)$. We may not enter b and c, respectively, into these cells because we would then be able to get a 1-output for an input sequence 1,3 or 2,3 which contradicts the problem's statement.

We need to append further rows, as shown in sub-figure **(b)**, to cope with the double transitions 3,1,3 and 3,2,3 when starting in stable state (d) where the output is 1. When coming back to column 3, the output must be 0 necessitating the introduction of the new stable state (e) which is reached from the newly introduced stable states (f) and (g).

Figure 14.11c: The double transitions 3,1,3 and 3,2,3 when starting in stable state (e) where the output is 0, must lead back to the stable state (e): Thus the entries $\underline{3,1},3$ for the GOTO-command f and $3,\overline{2,3}$ for the GOTO-command g in the respective cells $(e,1)$ and $(e,2)$.

14.7 From Flow Table to Events Graph

Although the flow table is a precise and unambiguous specification of a sequential switching problem, it does not have the vividly descriptive property of the events graph. But the latter is easily obtained from the former, as the example of the Priority-ANDs of Fig. 14.12 show. First, you choose a sequence of input events thought to be appropriate to the sequential problem. For this sequence you then proceed as the example in the insert states: When switching from a stable state in the flow table (e.g., from stable state a to the next input event (say, to input event or column 2), you follow the GOTO command to its stable state entering the associated output value of the flow table into the events graph. It is recommended to write the succession of stable states ($q_{p\&}$ and $q_{xp\&}$) into the events graph.

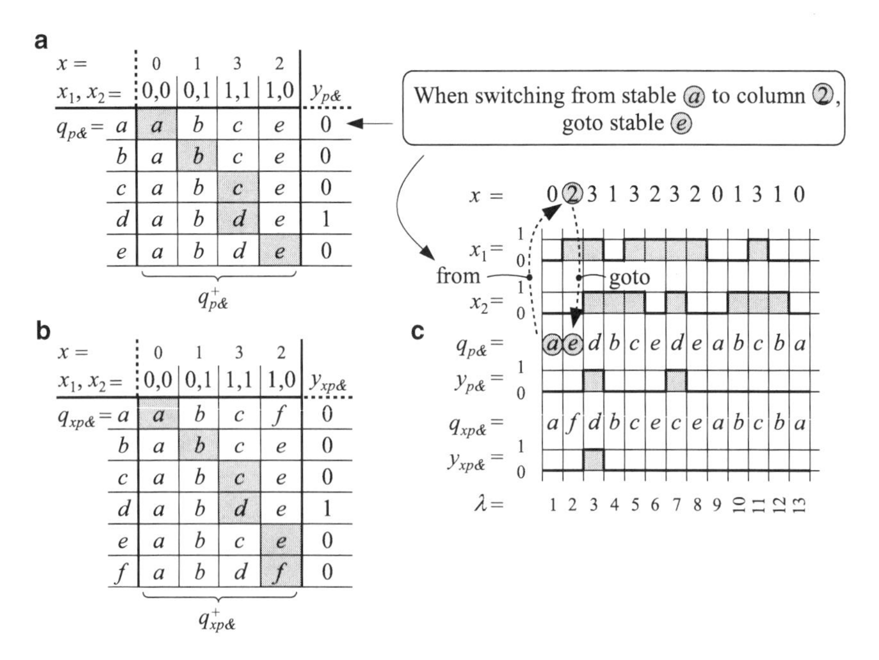

a

$x =$	0	1	3	2	
$x_1, x_2 =$	0,0	0,1	1,1	1,0	$y_{p\&}$
$q_{p\&} = a$	a	b	c	e	0
b	a	b	c	e	0
c	a	b	c	e	0
d	a	b	d	e	1
e	a	b	d	e	0
	$q^+_{p\&}$				

b

$x =$	0	1	3	2	
$x_1, x_2 =$	0,0	0,1	1,1	1,0	$y_{xp\&}$
$q_{xp\&} = a$	a	b	c	f	0
b	a	b	c	e	0
c	a	b	c	e	0
d	a	b	d	e	1
e	a	b	c	e	0
f	a	b	d	f	0
	$q^+_{xp\&}$				

Fig. 14.12 Developing and comparing events graphs for the Priority-AND Circuits. (**a**) Flow table for the **Priority-AND Problem**, taken from Fig. 14.4c. (**b**) Flow table for the **Expanded Priority-AND Problem**, taken from Fig. 14.5c. (**c**) Developing the events graphs for a chosen sequence of input events

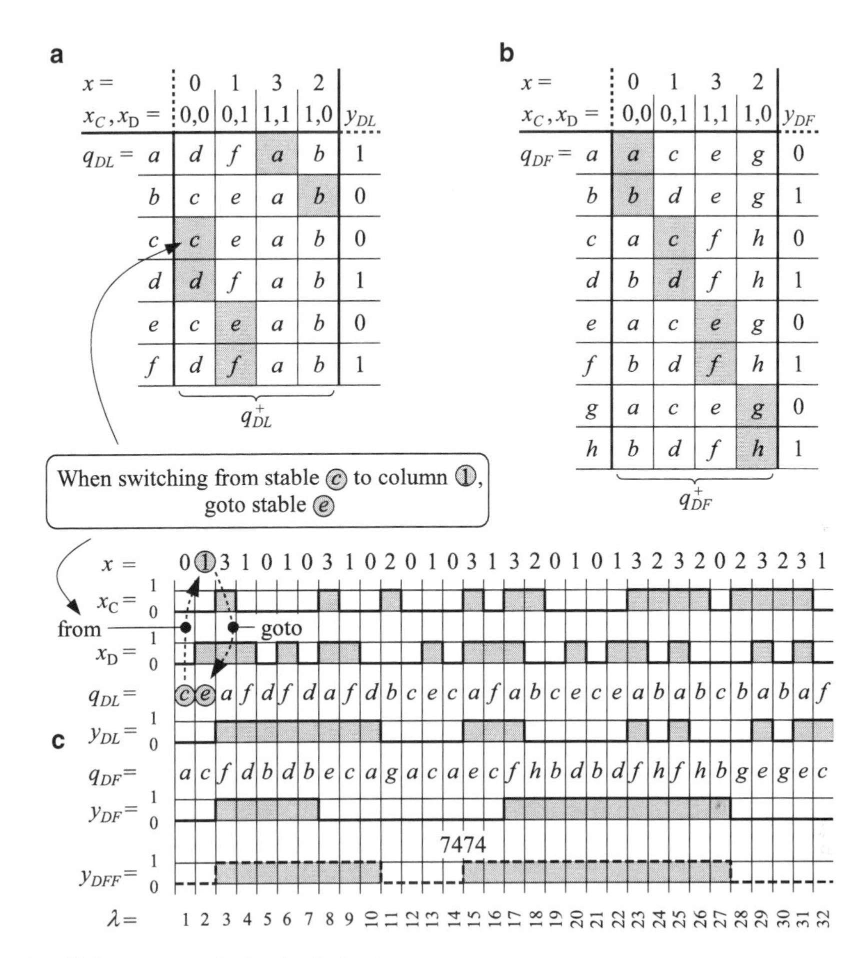

Fig. 14.13 Events graphs for the D-circuits

When comparing related circuits, such as Priority-AND Circuits, or D-latches and D-flipflops, it is good practice to use the same succession of input events. In this way, the events graph of Fig. 14.13 depicts the difference in the output behaviour of the D-latch (output y_{DL}), and that of the D-flipflop (output y_{DF}) quite clearly. As a matter of interest, the output (here referred to as y_{DFF}) of the industry design D-flipflop, the 7474 flipflop, is also added. This flipflop is considered later in quite some detail.

Chapter 15
State-Encoding by Iterative Catenation*

At first sight a flow table gives hardly a hint as how to design a binary circuit. But, as this chapter shows, there is a strait forward method to develop the associated word-recognition tree, this, subsequently, allowing us to calculate and design a circuit as discussed in Chap. 13. The word-recognition tree is obtained, firstly, by a *binary encoding of the states of the flow table*, and, secondly, by an *appropriate interpretation of the state encoding*.

While Huffman's encoding of states is an unsystematic affair, the approach taken here is algorithmic. Section 15.1 presents the first and basic step of state encoding (or state assignment, as it is mostly called) this being a simple **catenation** process. This step is repeated, iterated, as often as necessary to obtain the encoded flow table and, thereby, the word-recognition tree. The iteration process—called **iterative catenation**—is discussed in Section 15.2.

The subsequent sections contain examples for encoding flow tables, and show how to interpret the encoding to obtain a word-recognition tree. The flow tables used are those of the previous chapter. Where they refer to examples of Chap. 13, the results obtained in the present chapter can be checked against those of that chapter.

15.1 Catenation: From Moore to Mealy

Catenation was first used in Chap. 8 when developing combinational circuits by composition. Then, in Chap. 12 it was used in the development of latches, especially of robust, predominantly setting and resetting latches (PSR-latches). In this section, it is catenation that enables a solution to the foremost problem of Huffman's synthesis procedure—that of *systematically* encoding the states of the flow table.

The example of Fig. 15.1 will help us to visualise the general principle. A flow table—sub-figure (a)—actually consists of two distinct tables, one, called the **transition table** (sub-figure (b)), containing all GOTO-commands, the other, called

S.P. Vingron, *Logic Circuit Design*, DOI 10.1007/978-3-642-27657-6_15,

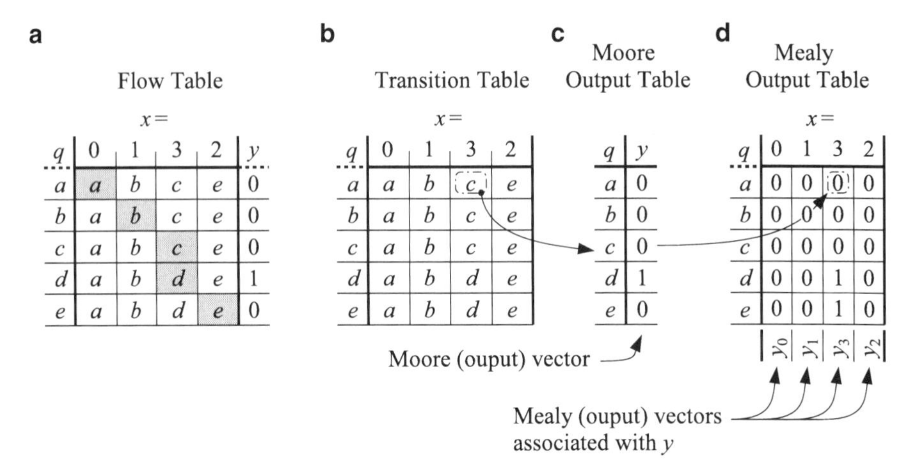

Fig. 15.1 Developing the primary Mealy output table (**d**) by catenating the transition table (**b**) and the Moore output table (**c**)

the **Moore output table** (sub-figure (c)), containing all output values. The transition *table* specifies a specific **transition function** $g : \mathcal{Q} \times \mathcal{E} \mapsto \mathcal{Q}$—where $\mathcal{Q}$ is the set of state labels q, and $\mathcal{E}$ is the set of input events x. The Moore output-*table* specifies a certain **Moore output function** $h : \mathcal{Q} \mapsto \{0, 1\}$. The functions g and h, together, are referred to as the **Moore model** of a sequential circuit.

The **catenation** of the transition function g and the Moore output function h, i.e., the expression $h\big(g(q, x)\big)$, is called a **Mealy output function** $f : \mathcal{Q} \times \mathcal{E} \mapsto \{0, 1\}$. Its output values, $f(q, x) = h\big(g(q, x)\big)$, are entered into a table called a Mealy output table—sub-figure (d).

The arrows from sub-figures (b) to (c), and on to sub-figure (d), visualise an example of this catenation for the argument, or cell, $(a, 3)$. The cell's content, c, is taken as argument for h which outputs 0, this, then, being the value entered into cell $(a, 3)$ of the Mealy output table: $h\big(g(a, 3)\big) = h(c) = f(a, 3)$. Like arrows from every cell of the transition table, via the Moore output-table, to the Mealy output-table, would visualise how the first two tables are catenated to obtain the last. Note that whether a cell in the transition table is a stable or non-stable condition is of no importance in the catenation process.

The output values of the Moore table are referred to as the Moore vector or variable. The output values of each column of the Mealy table are called a Mealy vector or variable. The Mealy variables are given the same names as those of the Moore variable, indexing each variable name by the column it describes. Thus, if the Moore variable is called y, the associated Mealy variables are y_0, y_1, y_3, and y_2, see sub-figures (c) and (d). As you can see in the next section, this already points to the development of a word-recognition tree.

15.2 Iterative Catenation

Here, in Fig. 15.2, we continue the state-encoding example of the previous section. The example is easily recognisable as the priority-AND, but, in this instance, not the sequential problem is of interest, rather, the *iterative catenation* procedure being developed.

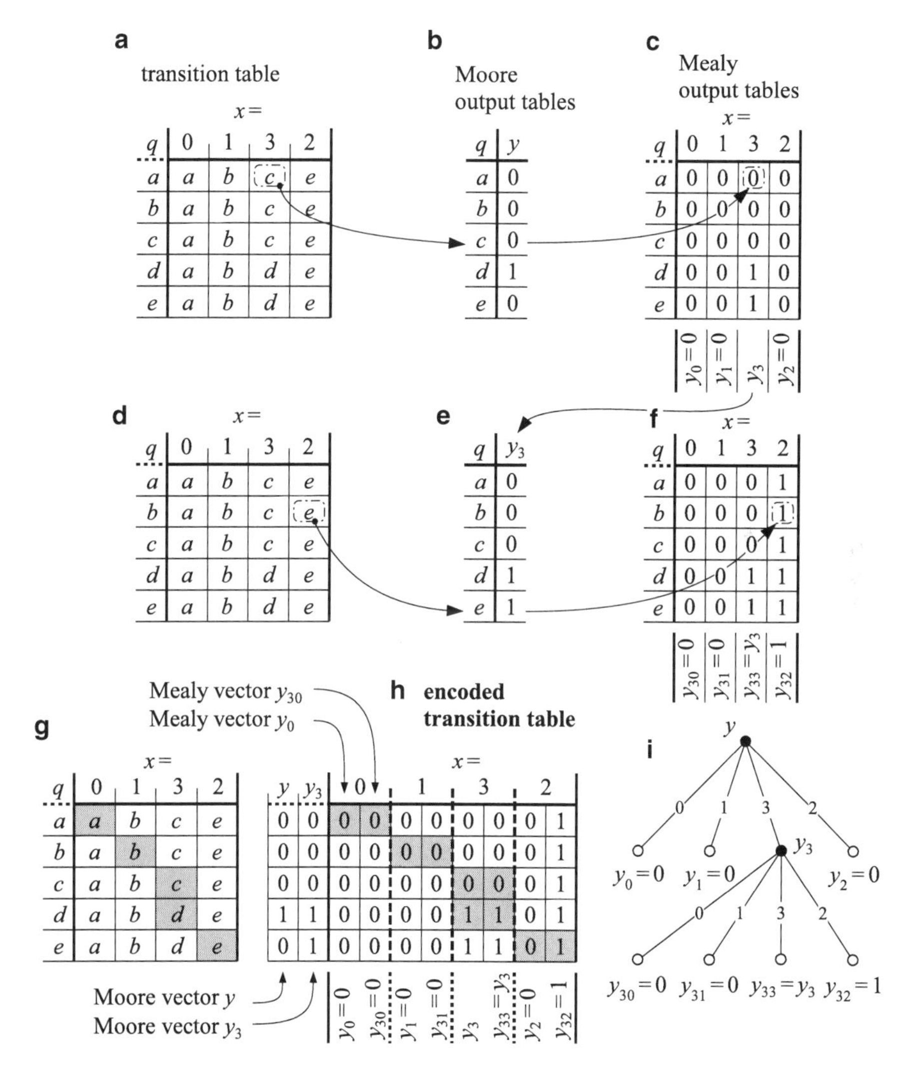

Fig. 15.2 Developing the encoded transition table and the associated word-recognition tree. (**a**)–(**c**): First iteration step. (**d**)–(**f**): Second iteration step. (**g**)–(**i**): **iterative catenation**

Obtaining the primary **Mealy table**, in the first iterative step, is repeated in sub-figures (a)–(c), the only difference to Fig. 15.1 being that, where possible, the Mealy variables have been assigned values. If all values in a certain column are the same, then its Mealy variable is said to be **determinate**, and the value entered into any cell of the column is assigned to the variable. Thus, we write $y_0 = 0$, $y_1 = 0$, and $y_2 = 0$.

If a column of a Mealy table contains both 0s *and* 1s, the column or the column's Mealy variable is a candidate for being **non-determinate** which in fact it is if the column's values are not row-wise identical, or row-wise inverted to the values of the Moore vector used for the catenation. The Mealy vector y_3, being neither identical nor inverse to the Moore vector y (compare sub-figures (b) and (c)), is therefore *non-determinate*.

The **objective of iterative catenation** is to find all non-determinate Mealy output vectors. This is done by repeating the catenation process for all these non-determinate Mealy vectors until no more non-determinate vectors are created. In our example, y_3 is non-determinate. Iterative catenation means, we use y_3 to define a new *Moore* output function $h_3 : \mathcal{Q} \mapsto \{0, 1\}$ specified by the table of sub-figure (e) where the output vector is y_3. The new Mealy function, resulting from the catenation $f_3(q, x) = h_3\big(g(q, x)\big)$, is pictured in the table of sub-figure (f). As easily recognisable, all its Mealy variables are determinate so that the iteration process terminates.

Now, take a look at—no, ponder—the so-called **encoded transition table** of sub-figure (h): How was it developed? It is a combination of the Moore tables (b) and (e), and the Mealy tables (c) and (f) such that the *Moore* vectors y and y_3 (in that order) determine the rows of the table, while every associated *Mealy* vector y_x and y_{3x} is entered into the input-events column x in the same order as the Moore vectors are used. This table is not merely *any* encoding of the transition table, it is *the* binary encoding of the transition table, this, inasmuch, as its columns can be directly interpreted as the nodes of a word-recognition tree (see sub-figure (i)).

The sub-figures (g)–(i) comprise a shorthand notation for developing a sequential circuit. It is this shorthand notation which is meant when speaking of **iterative catenation**, and it is this notation that is explained in the next section, and that we use from now on.

15.3 Expanded Priority-AND

When starting to develop an encoded transition table (without first developing Mealy output tables) one has no hint as to the number of iterations needed. It is therefore prudent to start with an empty transition table that provides enough horizontal space for a reasonable number of Moore, and associated Mealy vectors. In this book, I use a standardised transition table, like the one shown in Fig. 15.3, that allows for four Moore output vectors (when using two inputs). Restrictions of this kind are, of course, not present when working with a computer.

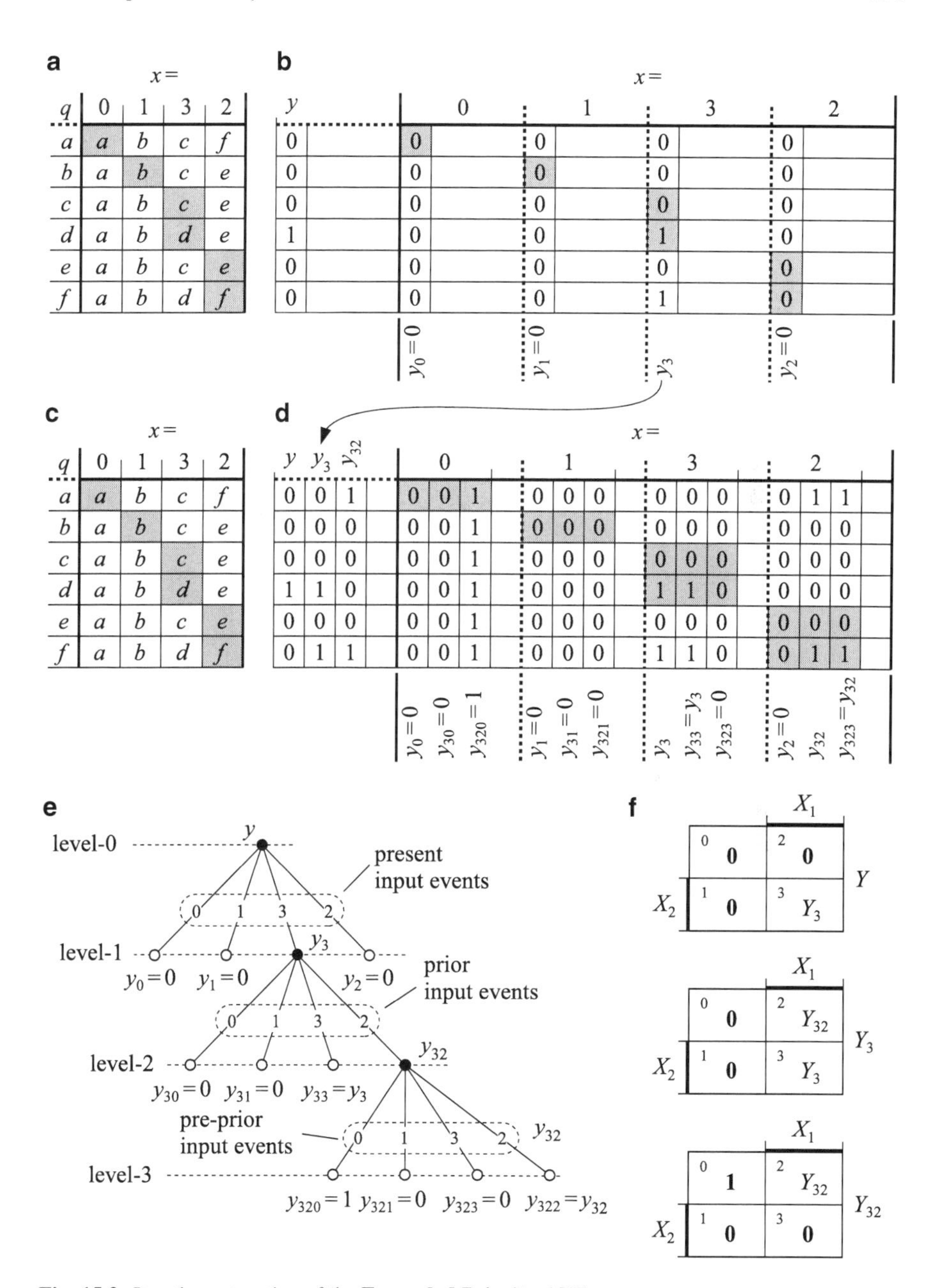

Fig. 15.3 Iterative catenation of the **Expanded Priority-AND**

In this section, the expanded priority-AND is used to show how the iterative catenation procedure evolves—Fig. 15.3. Sub-figure (a) points out that we start by drawing the transition table, i.e., the flow table without its output values. We then draw a wider, empty version of the transition table—sub-figure (b)—capable of incorporating a number of Moore, and associated Mealy vectors. In this table we enter the output values as the **primary Moore vector** y, creating the associated **primary Mealy vectors** y_0, y_1, y_3, y_2 by catenation—the result of this step is shown in sub-figure (b).

Search for the non-determinate Mealy vectors (in our example, the only one being y_3), and add them, as **secondary Moore vectors**, to the table. The only secondary Moore vector in our example is y_3 which is duly added—sub-figure (d). You repeat the catenation process of each secondary Moore vector with the transition table. For y_3, the results of this catenation are the secondary Mealy vectors $y_{3,0}, y_{3,1}, y_{3,3}, y_{3,2}$ shown in sub-figure (d).

These secondary Mealy vectors are, in turn, searched for those non-determinate, the only one in our example being $y_{3,2}$. In the next iteration step, $y_{3,2}$ is used as the **ternary Moore vector**, the result of the associated catenation leading to the **ternary Mealy vectors** $y_{3,2,0}, y_{3,2,1}, y_{3,2,3}, y_{3,2,2}$. Here, the iterative process terminates, as all ternary Mealy vectors are determinate. The result is the encoded transition table of sub-figure (d).

Developing the word-recognition tree from the encoded transition table is straight forward. Every Moore variable is connected to Mealy variables it initiated, sub-figure (e). Finally, the word-recognition tree is transcribed into K-maps, sub-figure (f), from which the circuit is calculated.

15.4 Two-Hand Safety Circuits

The iterative catenation processes shown in Figs. 15.4 and 15.5 are straight forward and presumably need no commenting. The arrows from the encoded transition tables to the word-recognition trees stress how their nodes correspond to the Mealy variables of the encoded transition tables.

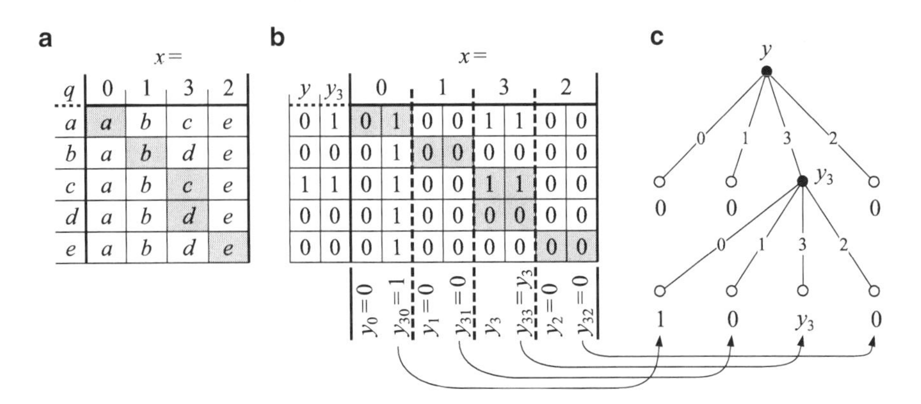

Fig. 15.4 Iterative catenation of the **Two-Hand Safety Circuit**

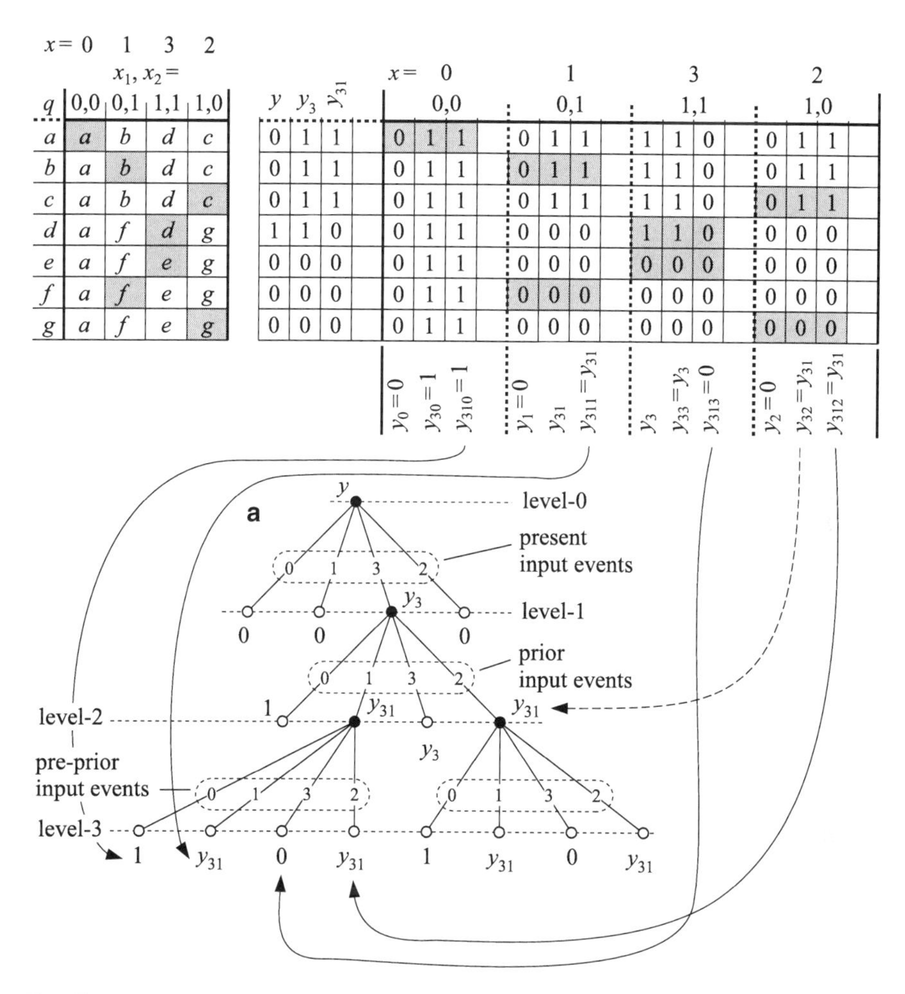

Fig. 15.5 Iterative catenation of the **Expanded Two-Hand Safety Circuit**

15.5 D-Latch and D-Flipflop

In Fig. 15.6 we take a closer look at the development of the encoded flow table of the D-latch (or sampling problem). The prime thing to note is that the Mealy vectors y_0 and y_1 are identical with the Moore vector y. They are thus determinate, and must be realised as feedback signals of y. The word-recognition tree (sub-figure (b)), and the associated K-map, thus clearly represent a latch.

The synchronisation problem, see Fig. 15.7, as we already know, is closely related to the sampling problem. Developing the encoded flow table should, by now, be routine so that explaining how to obtain the encoded transition table of sub-figure (a) again seems unnecessary. But it should be pointed out that Mealy vector y_2 is identical to Mealy vector y_3 so that node y_2 sprouts the same sub-tree as node y_3.

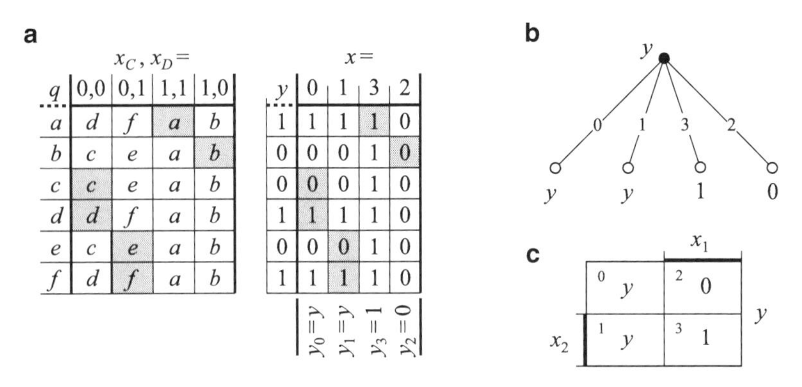

Fig. 15.6 Developing the encoded flow table, (**b**), for the **D-latch**

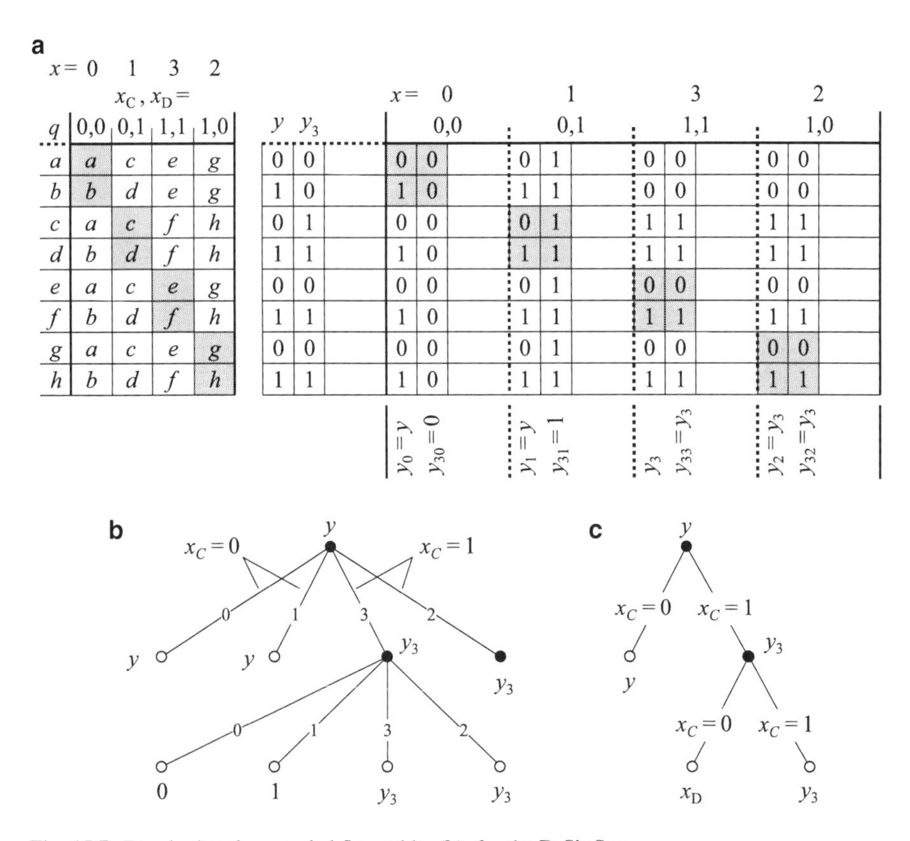

Fig. 15.7 Developing the encoded flow table, (**b**), for the **D-flipflop**

Further, note the way in which the 'full' word-recognition tree of sub-figure (b) may now be expressed as the 'reduced' tree of sub-figure (c). Do not take this reduction too lightly, or you won't be able to explain the x_D node-value in sub-figure (c).

15.6 Passed-Sample Circuit

The Passed-Sample Problem (Sect. 14.5 Example 14.3) is one of those problems which do not lend themselves well to a simple, intuitive formulation of its word-recognition tree. In fact, even developing its flow table is quite a meticulous process, as Sect. 14.5 shows. In this section, we want to use the flow table of Fig. 14.10c as a starting point from which to develop a circuit for the passed-sample problem.

The iterative catenation process by which to obtain the encoded transition table is unfolded step by step in Fig. 15.8, sub-figure (a) showing the *result of the first catenation step*. One will, of course, begin with an encodable transition table, empty, except for the output values y entered as the primary Moore vector (try to imagine this situation in sub-figure (a)). The result of catenating the original transition table with the primary Moore output (table) leads to values entered for the primary Mealy vectors y_0, y_1, y_3, and y_2. As you can see, y_0 and y_1 are determinate ($y_0 = 0$ and $y_1 = 0$) while y_3 and y_2 are not, so that these latter Mealy vectors are entered as secondary Moore vectors, giving us sub-figure (a) as shown.

Sub-figure (b) shows the result of the second catenation step, the catenation of the original transition table with the secondary Moore output (table) y_3. This catenation leads to the values of the secondary Mealy vectors $y_{3,0}$, $y_{3,1}$, $y_{3,3}$, and $y_{3,2}$ which we now check for being determinate or not. At this point we must state **determinateness** somewhat more precisely: **A Mealy vector whose values consisted of 0s and 1s is determinate if it equals an existing Moore vector, or equals the inverted form of an existing Moore vector**. We check the first of the secondary Mealy vectors, $y_{3,0}$, against the existing Moore vectors y, y_3, y_2 recognising that it is non-determinate: $y_{3,0}$ is therefore immediately added, as ternary Moore vector, to the list of existing Moore vectors. Now, the next secondary Mealy vector, $y_{3,1}$, is checked against the expanded list of Moore vectors, and proves to be determinate—$y_{3,1} = y_{3,0}$. Likewise, $y_{3,3}$, and $y_{3,2}$ are determinate— $y_{3,3} = y_3$ and $y_{3,2} = 0$. At this point, we have finished the development of sub-figure (b).

Sub-figure (c) shows the result of the next catenation step, the catenation of the original transition table with the secondary Moore output (table) y_2. The obtained Mealy vectors, $y_{2,0}$, $y_{2,1}$, $y_{2,3}$, and $y_{2,2}$, prove to be determinate—$y_{2,0} = y_{3,0}$, $y_{2,1} = y_{3,0}$, $y_{2,3} = 1$, and $y_{2,2} = y_2$. This is documented in sub-figure (c).

The encoded transition table is completed in Fig. 15.9a which contains the result of catenating the original transition table with the ternary Moore output (table) $y_{3,0}$. As you can see, all ternary Mealy vectors are determinate—$y_{3,0,0} = y_{3,0}$, $y_{3,0,1} = y_{3,0}$, $y_{3,0,3} = 1$, and $y_{3,0,2} = 0$. Depicting the encoding process step by step, as in Fig. 15.8, was only done for your convenience. In actual work, all these steps are drawn in one and the same encoded transition table starting as in Fig. 15.8a and ending as in Fig. 15.9a.

The unaccustomed look of the word-recognition tree is due to drawing the elementary sub-tree $y_{3,0}$ only once, and then only indicating to which nodes it should be appended. Possibly, you can recognise this sub-tree as that of the D-latch. This

a

q \ $x_1, x_2 =$	0,0	0,1	1,1	1,0
a	h	g	a	d
b	h	g	b	d
c	f	e	a	c
d	f	e	a	d
e	f	e	a	c
f	f	e	a	c
g	h	g	b	d
h	h	g	b	d

y	y_3	y_2		$x =$ 0 (0,0)	1 (0,1)	3 (1,1)	2 (1,0)
0	0	1		0	0	0	1
1	1	1		0	0	1	1
0	0	0		0	0	0	0
1	0	1		0	0	0	1
0	0	0		0	0	0	0
0	0	0		0	0	0	0
0	1	1		0	0	1	1
0	1	1		0	0	1	1
				$y_0 = 0$	$y_1 = 0$	y_3	y_2

b

q \ $x_1, x_2 =$	0,0	0,1	1,1	1,0
a	h	g	a	d
b	h	g	b	d
c	f	e	a	c
d	f	e	a	d
e	f	e	a	c
f	f	e	a	c
g	h	g	b	d
h	h	g	b	d

y	y_3	y_2	y_{30}	$x =$ 0 (0,0)		1 (0,1)		3 (1,1)		2 (1,0)	
0	0	1	1	0	1	0	1	0	0	1	0
1	1	1	1	0	1	0	1	1	1	1	0
0	0	0	0	0	0	0	0	0	0	0	0
1	0	1	0	0	0	0	0	0	0	1	0
0	0	0	0	0	0	0	0	0	0	0	0
0	0	0	0	0	0	0	0	0	0	0	0
0	1	1	1	0	1	0	1	1	1	1	0
0	1	1	1	0	1	0	1	1	1	1	0
				$y_0 = 0$	y_{30}	$y_1 = 0$	$y_{31} = y_{30}$	y_3	$y_{33} = y_3$	y_2	$y_{32} = 0$

c

q \ $x_1, x_2 =$	0,0	0,1	1,1	1,0
a	h	g	a	d
b	h	g	b	d
c	f	e	a	c
d	f	e	a	d
e	f	e	a	c
f	f	e	a	c
g	h	g	b	d
h	h	g	b	d

y	y_3	y_2	y_{30}	$x =$ 0 (0,0)			1 (0,1)			3 (1,1)			2 (1,0)		
0	0	1	1	0	1	1	0	1	1	0	0	1	1	0	1
1	1	1	1	0	1	1	0	1	1	1	1	1	1	0	1
0	0	0	0	0	0	0	0	0	0	0	0	1	0	0	0
1	0	1	0	0	0	0	0	0	0	0	0	1	1	0	1
0	0	0	0	0	0	0	0	0	0	0	0	1	0	0	0
0	0	0	0	0	0	0	0	0	0	0	0	1	0	0	0
0	1	1	1	0	1	1	0	1	1	1	1	1	1	0	1
0	1	1	1	0	1	1	0	1	1	1	1	1	1	0	1
				$y_0 = 0$	y_{30}	$y_{20} = y_{30}$	$y_1 = 0$	$y_{31} = y_{30}$	$y_{21} = y_{30}$	y_3	$y_{33} = y_3$	$y_{23} = 1$	y_2	$y_{32} = 0$	$y_{22} = y_2$

Fig. 15.8 Starting to develop the encoded transition table for the **Passed-Sample Problem**

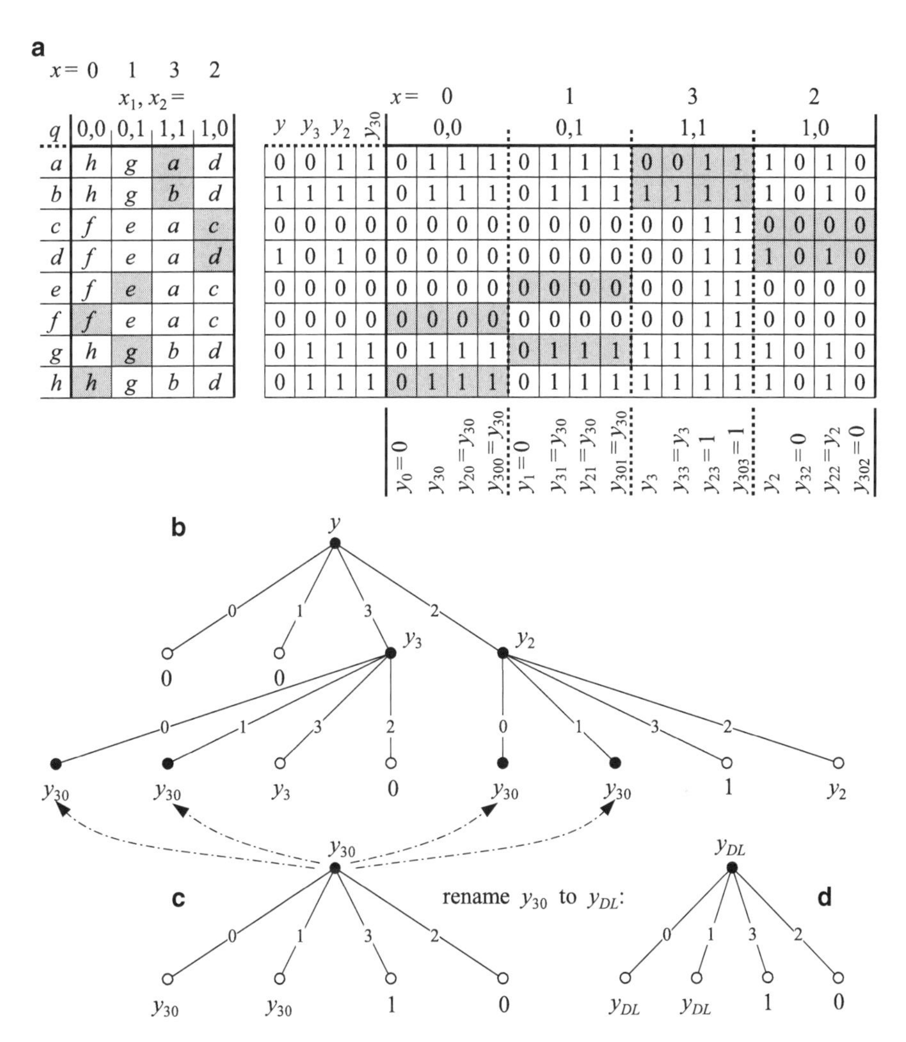

Fig. 15.9 Completing the development of the encoded transition table for the Passed-Sample Problem

is the reason for renaming $y_{3,0}$ to y_{DL}, and thereby documenting that this sub-tree is not specific to any single node.

Calculating the circuit is, in the main, a question of drawing the K-maps associated with the root and the *internal nodes* of the word-recognition tree, these K-maps having been developed in great detail in Fig. 15.10. As in the calculations of Chap. 13, we start with the internal node (or nodes) in the highest level, i.e., with y_{DL}. If we use y_{DL} as input to y_2, it is prudent to express y_{DL} in its *disjunctive* form, as a comparison of sub-figures (a) and (b) suggests. (The disjunctive formula, (10.4), for evaluating sub-figure (a) uses the contents of cell 0, 1 and 3. As these cells

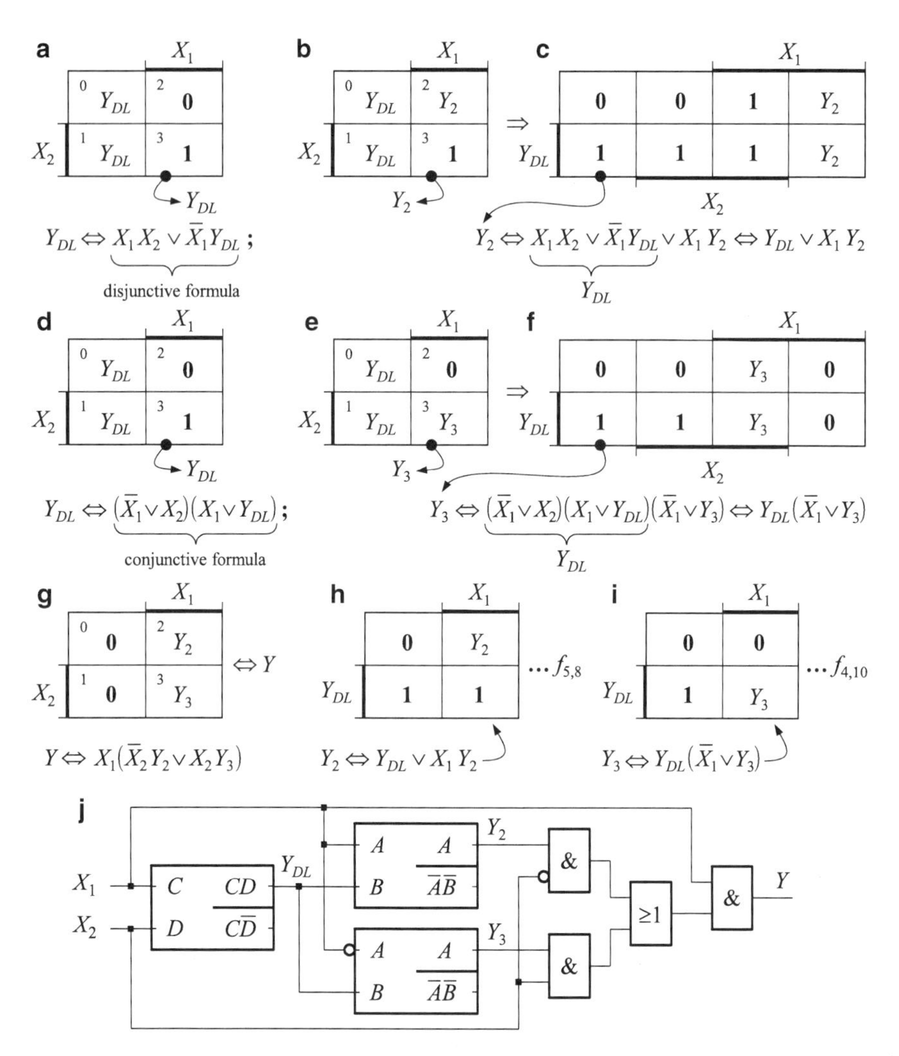

Fig. 15.10 Calculating the Passed-Sample Circuit

recur identically in the K-map for y_2 of sub-figure (b) we will be able to use y_{DL} to partially express y_2 thus simplifying its formula.) To evaluate the K-map for y_2 of sub-figure (b) we redraw it, extracting the map-entered variable y_{DL}, obtaining sub-figure (c). This K-map is then evaluated using (10.4)—see the calculation in sub-figure (c).

When calculating y_3—sub-figure (e)—we use y_{DL} as a map-entered variable. Here it is advisable to use the *conjunctive* formula (10.4) to evaluate y_{DL}—see sub-figure (d). The calculation of y_3 then leads to the result shown in sub-figure (f).

The formulas obtained for Y_2 and Y_3—sub-figures (c) and (f)—have the singular property of being dependant on only the two variables Y_{DL} and X_1, the other input, X_2, playing no part. These formulas can thus be expressed by the K-maps of sub-figures (h) and (i)—K-maps that represent the elementary latches $f_{5,8}$ and $f_{4,10}$ as you can see when looking back at Fig. 11.12. The elementary latches $f_{5,8}$ and $f_{4,10}$ are not independent of one another: Inverting the local A-input of the symbol of one latch transforms it into that of the other latch.

The final calculation to be done is that for the output Y, see sub-figure (g). This, together with the latches for Y_{DL}, Y_2, and Y_3, leads to the circuit of sub-figure (j).

15.7 Incompletely Specified Flow Tables

It is not uncommon for practical problems to be incompletely specified. A flow table is incompletely specified, or **hazy**, if at least one of its cells contains no GOTO-command, i.e., is empty. Such a cell is usually marked with a dash to emphasize that the cell's GOTO-command was not simply forgotten. The haziness of a flow table also carries over to its encoding. The encoding of a hazy flow table is discussed here using the following example problem. Incomplete specification is simulated here by requiring that input variables never change their values simultaneously allowing us to choose freely what should happen in such cases.

Example 15.1 (Expanded Priority-AND Variant). The output y is 1 when both inputs x_1 and x_2 are 1 having changed their values from $(0, 0)$ via $(0, 1)$ to $(1, 1)$. Having become 1, the output only becomes 0 when the inputs are again $(0, 0)$ having change their values from $(1, 1)$ via $(0, 1)$ to $(0, 0)$. The input variables never change their values simultaneously.

Figure 15.11a: The process that leads to the flow table is not detailed here, but please do check the flow table meticulously to see how it describes the Expanded Priority-AND Variant, concentrating especially on state labels e–j. Note that the output variable and its values have been detached from the flow table and written as Moore vector to the otherwise still empty, encoded transition table.

Figure 15.11b: Determining the primary Mealy vectors y_0, y_1, y_3, and y_2 is straight forward. y_0 is non-determinate (as by no choice of either 0 or 1 for the dashes in y_0 can y_0 be made row-wise identical or row-wise inverted to the Moore vector y), and is thus immediately used and written as a (secondary) Moore vector.

Now consider the Mealy vector y_1 comparing its symbols row-wise with those of the Moore vector y. As a dash can be replaced by either 0 or 1, we can enforce $y_1 = y$ (making it determinate) by replacing the dash in row d of y_1 by 0, and that of row g by 1.

Checking y_3 against Moore vectors y and y_0, you will see that y_3 is non-determinate so that it must be added as Moore vector. y_2 can be make determinate by enforcing $y_2 = y$.

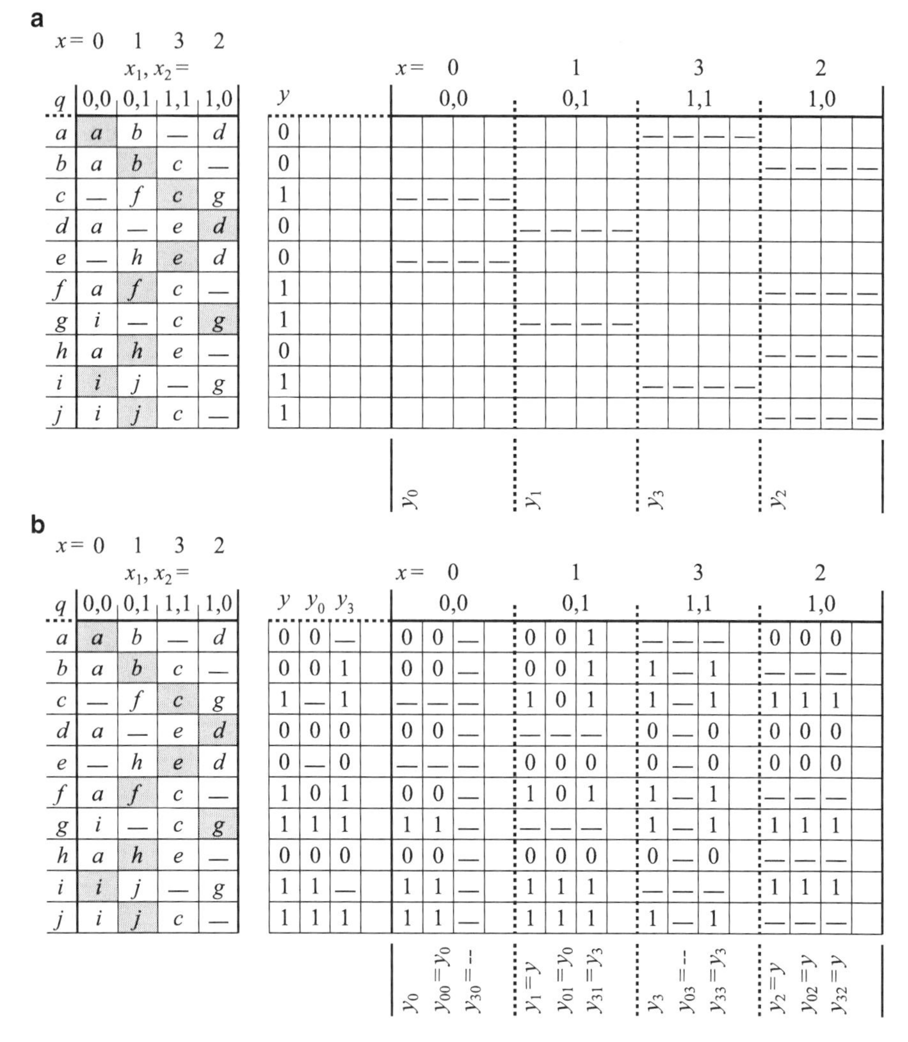

Fig. 15.11 Encoding the incompletely specified flow table of the **Expanded Priority-AND Variant**. (**a**) The unspecified GOTO-commands in the given flow table are transferred to equivalent cells in the still empty encoded transition table. (**b**) The encoded transition table

Applying iterative catenation for the Moore vector y_0 provides us with the Mealy vectors $y_{0,0}$, $y_{0,1}$, $y_{0,3}$, and $y_{0,2}$, each of which is (or can be made to be) determinate. Of these the Mealy vector $y_{0,3}$ catches the eye, the dashes allowing us to choose all of its components freely. The dashes in $y_{0,3}$ come about because stable state c of the transition table maps to the dash in row c of the Moore vector y_0, this dash being used for all cells in $y_{0,3}$ that correspond to the occurrences of c in column 3

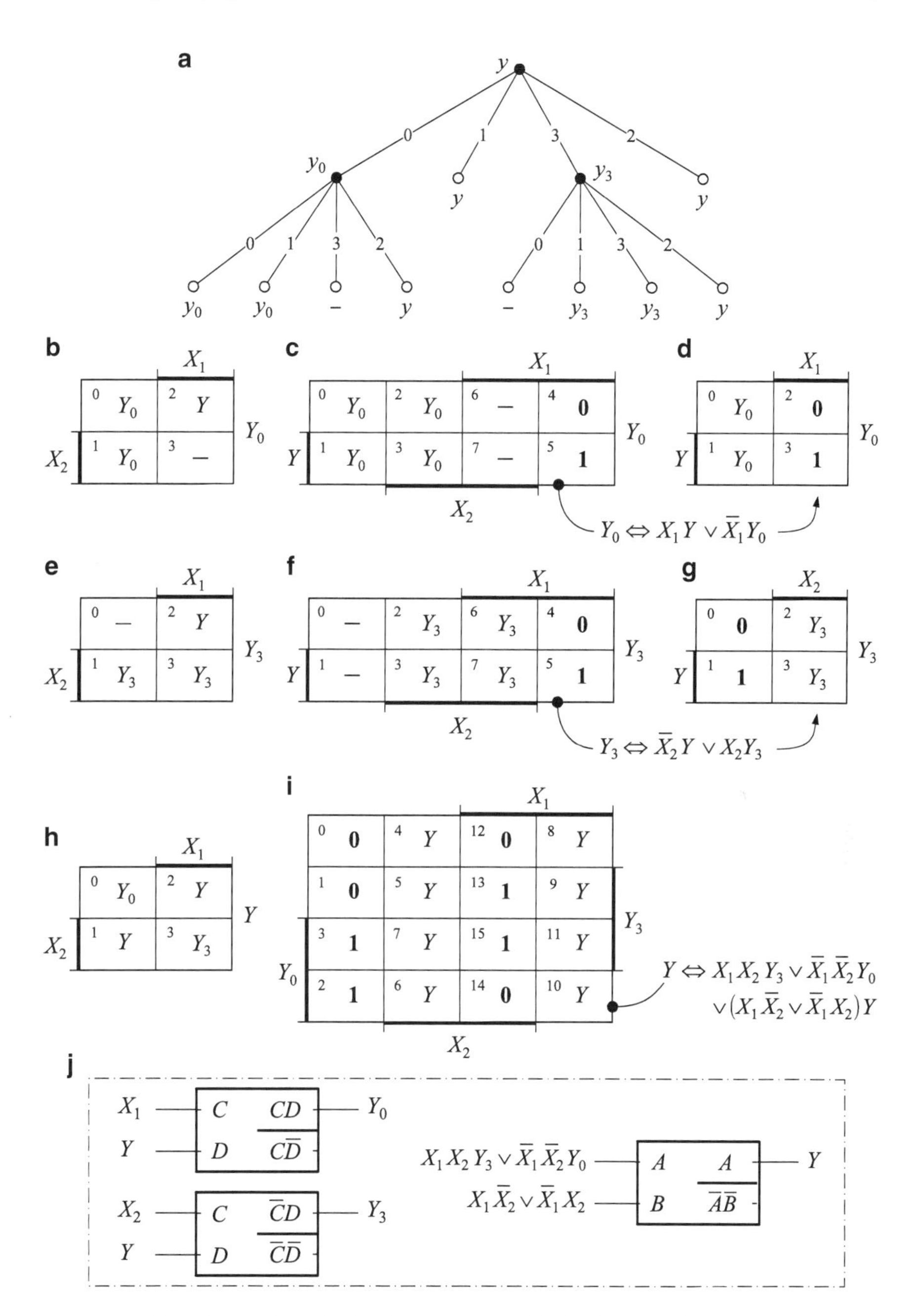

Fig. 15.12 Calculating a circuit for the **Expanded Priority-AND Variant**

of the transition table. The remaining dashes in $y_{0,3}$ are due to stable e mapping to a dash in y_0.

Iterative catenation is next applied to the Moore vector y_3. The obtained Mealy vectors, $y_{3,0}$, $y_{3,1}$, $y_{3,3}$, and $y_{3,2}$, are again determinate allowing us to terminate the iterative catenation process.

The encoded flow table of Fig. 15.11b allows us to draw the word-recognition tree of Fig. 15.12a. To design the circuit, we need the K-maps of sub-figures (b), (e), and (h), K-maps which are associated with the elementary sub-trees whose roots are y_0, y_3, and y of the word-recognition tree, respectively. To calculate the K-maps, we extract those map-entered variables which differ from the output variable, hereby obtaining the K-maps of sub-figures (c), (f), and (i). The formula for Y_0, calculated from the *three*-dimensional sub-figure (c), is independent of X_2 allowing us to express it by the *two*-dimensional K-map of sub-figure (d). This is the K-map of the D-latch where X_1 is used as clock signal C, and Y as data signal D (refer to Fig. 11.9). Similar considerations lead from the K-map of sub-figure (e) to that of sub-figure (g) this being the K-map of the D-latch with a *negative* clock signal C.

The K-map for the output Y, sub-figure (h), is redrawn with the map-entered variables Y_0 and Y_3 extracted, sub-figure (i). Evaluating the output variable Y as shown and substituting $X_1 X_2 Y_3 \vee \overline{X_1}\,\overline{X_2} Y_0$ by an auxiliary variable—say, Z_1—and $X_1 \overline{X_2} \vee \overline{X_1} X_2$ by Z_2 leads to the formula $Y \Leftrightarrow Z_1 \vee Z_2 Y$. This is the formula of latch $Y_{3,8}$ of Fig. 11.11. The principal layout of the circuit is shown in sub-figure (j). Replace the formulas $X_1 X_2 Y_3 \vee \overline{X_1}\,\overline{X_2} Y_0$ and $X_1 \overline{X_2} \vee \overline{X_1} X_2$ by their combinational circuits and connect all equally named variables.

Chapter 16
Circuit Analysis

In this chapter we look into the two main aspects of circuit analysis. **Firstly**, how can one find and describe the **external behaviour**, the so-called input–output behaviour, of a circuit? An answer is more often necessary than meets the eye, as it is still quite common to develop circuits intuitively, thus not being able to guarantee certain subtleties in their IO-behaviour. But you can also view this aspect as that of reverse engineering, or as a first step in circuit optimisation. Depending on which of the above points you want to emphasize, you might formulate your answer as a flow table, an events graph, or a word-recognition tree. **Secondly**, and this is the aspect of **internal behaviour**, we want to look into how the internal latches work together to avoid malfunctioning of the circuit. This malfunctioning can, as we will see, cause transient erroneous output signals in the event of **non-critical races** between internal latches, or permanently wrong output signals in the event of what we call **critical races** between internal latches.

16.1 Analysing a Circuit's External Behaviour

The circuit shown in Fig. 16.1 is one of two called D-flipflop, the other being the circuit that was developed in Fig. 13.6 and pictured in its sub-figure (d). To distinguish between them, the circuit of this section is referred to as the **(industry) 7474 D-flipflop**. It is our goal to describe its IO-behaviour by *flow table* and *word-recognition tree*, and to give an example of its behaviour in the form of an *events graph*.

The first step in analysis is to work toward a flow table for which purpose we develop the equations of the internal latches and the output. To this end, it is advisable to state the logic expressions of the outputs of *each* logic gate as shown in Fig. 16.1, this automatically leading to the formulas for the internal latches and the output:

$$Y_2 \Leftrightarrow \overline{C\,\overline{\overline{DY_3}\,Y_2}}, \qquad Y_3 \Leftrightarrow \overline{Y_2 C\,\overline{DY_3}}, \qquad Y \Leftrightarrow \overline{Y_2\,\overline{Y\,Y_3}}.$$

S.P. Vingron, *Logic Circuit Design*, DOI 10.1007/978-3-642-27657-6_16,

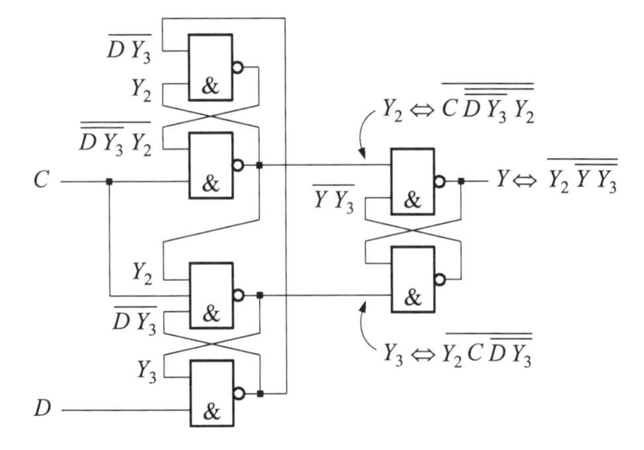

Fig. 16.1 The industry **7474 D-flipflop**

To conveniently enter these formulas into a flow table (preferably organised as a K-map) they are best transformed into AND-to-OR forms. The calculations for Y_2 and Y_3 are:

$$\begin{aligned} Y_2 &\Leftrightarrow \overline{C\overline{\overline{DY_3}Y_2}} \\ &\Leftrightarrow \overline{C} \vee \overline{DY_3}Y_2 \\ &\Leftrightarrow \overline{C} \vee (\overline{D} \vee \overline{Y_3})Y_2 \\ &\Leftrightarrow \overline{C} \vee \overline{D}Y_2 \vee \overline{Y_3}Y_2 \end{aligned} \qquad \begin{aligned} Y_3 &\Leftrightarrow \overline{Y_2C\overline{DY_3}} \\ &\Leftrightarrow \overline{Y_2C} \vee DY_3 \\ &\Leftrightarrow \overline{Y_2} \vee \overline{C} \vee DY_3 \end{aligned}$$

When transforming the output Y, we must insert the formulas obtained for Y_2 and Y_3:

$$\begin{aligned} Y &\Leftrightarrow \overline{Y_2\overline{YY_3}} \\ &\Leftrightarrow \overline{Y_2} \vee YY_3 \\ &\Leftrightarrow \overline{\overline{C} \vee \overline{D}Y_2 \vee \overline{Y_3}Y_2} \vee Y(\overline{Y_2} \vee \overline{C} \vee DY_3) \\ &\Leftrightarrow C(D \vee \overline{Y_2})(Y_3 \vee \overline{Y_2}) \vee Y\overline{Y_2} \vee Y\overline{C} \vee YDY_3 \\ &\Leftrightarrow (CD \vee C\overline{Y_2})(Y_3 \vee \overline{Y_2}) \vee Y\overline{Y_2} \vee Y\overline{C} \vee YDY_3 \\ &\Leftrightarrow CDY_3 \vee C\overline{Y_2}Y_3 \vee \underbrace{CD\overline{Y_2} \vee C\overline{Y_2}}_{C\overline{Y_2}} \vee Y\overline{Y_2} \vee Y\overline{C} \vee YDY_3 \\ &\Leftrightarrow CDY_3 \vee \underbrace{C\overline{Y_2}Y_3 \vee C\overline{Y_2}}_{C\overline{Y_2}} \vee Y\overline{Y_2} \vee Y\overline{C} \vee YDY_3 \\ &\Leftrightarrow CDY_3 \vee C\overline{Y_2} \vee Y\overline{Y_2} \vee Y\overline{C} \vee YDY_3 \end{aligned}$$

calculation continued:

$$
\begin{aligned}
Y &\Leftrightarrow CDY_3 \vee C\overline{Y_2} \vee Y\overline{Y_2}(C \vee \overline{C}) \vee Y\overline{C} \vee YDY_3 \\
&\Leftrightarrow CDY_3 \vee \underbrace{C\overline{Y_2} \vee Y\overline{Y_2}C}_{C\overline{Y_2}} \vee \underbrace{Y\overline{Y_2}\,\overline{C} \vee Y\overline{C}}_{Y\overline{C}} \vee YDY_3 \\
&\Leftrightarrow Y_3CD \vee \overline{Y_2}C \vee Y\overline{C} \vee YY_3D
\end{aligned}
$$

Each of the AND-to-OR formulas obtained for Y, Y_3, and Y_2 are used to develop the K-maps of Fig. 16.2a–c the rows of which are defined by the internal latches. Combining the individual K-maps for y, y_3, and y_2 leads to the composite K-map of sub-figure (d). This K-map is the encoded transition table. For further use, we rewrite it in non-encoded form shown in sub-figure (e). To achieve our first goal, the flow table, we append the output vector y to the transition table—sub-figure (f).

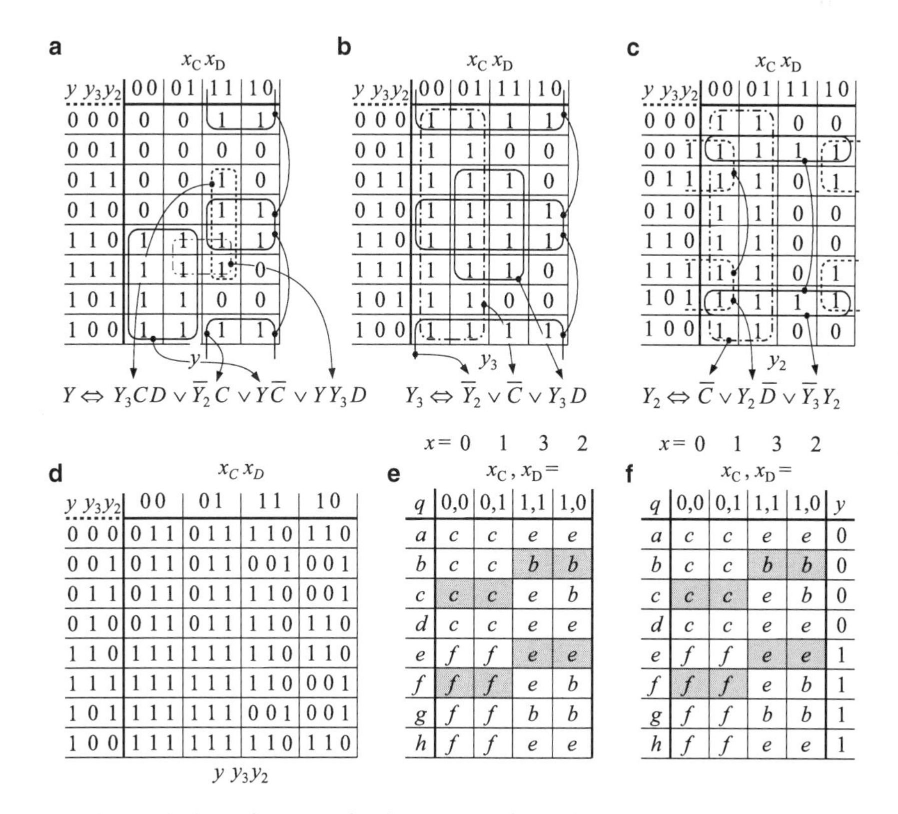

(a) $x_C x_D$

$y\ y_3 y_2$	00	01	11	10
0 0 0	0	0	1	1
0 0 1	0	0	0	0
0 1 1	0	0	1	0
0 1 0	0	0	1	1
1 1 0	1	1	1	1
1 1 1	1	1	1	0
1 0 1	1	1	0	0
1 0 0	1	1	1	1

(b) $x_C x_D$

$y\ y_3 y_2$	00	01	11	10
0 0 0	1	1	1	1
0 0 1	1	1	0	0
0 1 1	1	1	1	0
0 1 0	1	1	1	1
1 1 0	1	1	1	1
1 1 1	1	1	1	0
1 0 1	1	1	0	0
1 0 0	1	1	1	1

(c) $x_C x_D$

$y\ y_3 y_2$	00	01	11	10
0 0 0	1	1	0	0
0 0 1	1	1	1	1
0 1 1	1	1	0	1
0 1 0	1	1	0	0
1 1 0	1	1	0	0
1 1 1	1	1	0	1
1 0 1	1	1	1	1
1 0 0	1	1	0	0

(d) $x_C x_D$

$y\ y_3 y_2$	00	01	11	10
0 0 0	0 1 1	0 1 1	1 1 0	1 1 0
0 0 1	0 1 1	0 1 1	0 0 1	0 0 1
0 1 1	0 1 1	0 1 1	1 1 0	0 0 1
0 1 0	0 1 1	0 1 1	1 1 0	1 1 0
1 1 0	1 1 1	1 1 1	1 1 0	1 1 0
1 1 1	1 1 1	1 1 1	1 1 0	0 0 1
1 0 1	1 1 1	1 1 1	0 0 1	0 0 1
1 0 0	1 1 1	1 1 1	1 1 0	1 1 0

$y\ y_3 y_2$

(e) $x =$ 0 1 3 2; $x_C, x_D =$

q	0,0	0,1	1,1	1,0
a	c	c	e	e
b	c	c	b	b
c	c	c	e	b
d	c	c	e	e
e	f	f	e	e
f	f	f	e	b
g	f	f	b	b
h	f	f	e	e

(f) $x =$ 0 1 3 2; $x_C, x_D =$

q	0,0	0,1	1,1	1,0	y
a	c	c	e	e	0
b	c	c	b	b	0
c	c	c	e	b	0
d	c	c	e	e	0
e	f	f	e	e	1
f	f	f	e	b	1
g	f	f	b	b	1
h	f	f	e	e	1

Fig. 16.2 Developing a flow table for the industry **7474 D-flipflop**. (**a**) The individual K-map describing the output Y; (**b**) the individual K-map describing the internal latch Y_3; (**c**) the individual K-map describing the internal latch Y_2; (**d**) the composite K-map: likewise, this is the encoded transition table; (**e**) rewriting the transition table in a non-encoded form; (**f**) appending the output y to the transition table to obtain the **flow table**

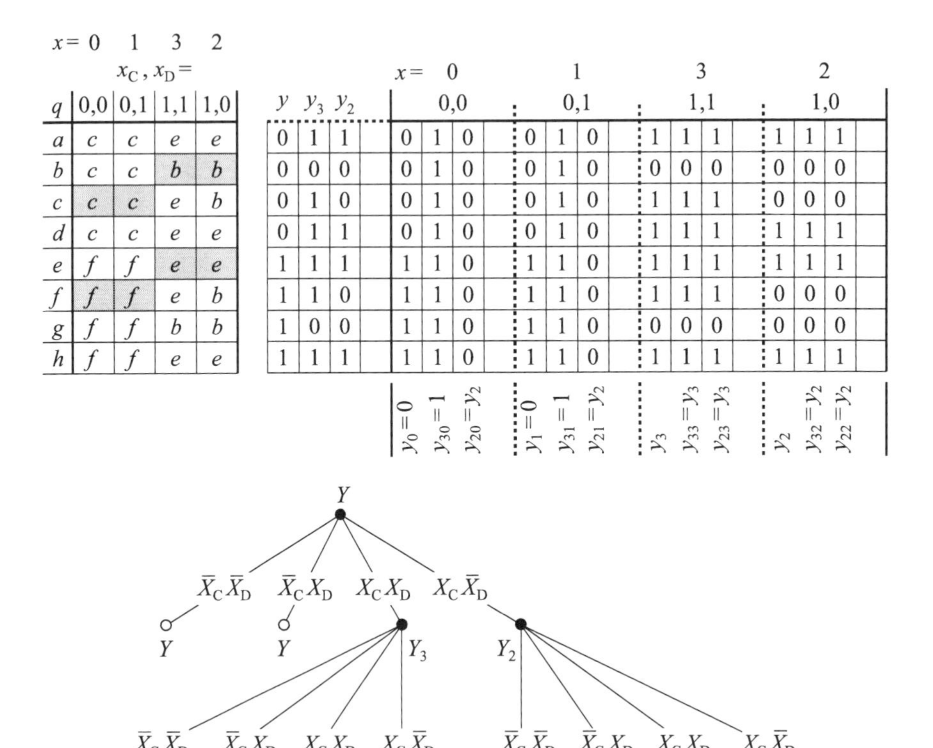

q	$x=0$ $x_C, x_D=$ 0,0	1 0,1	3 1,1	2 1,0
a	c	c	e	e
b	c	c	b	b
c	c	c	e	b
d	c	c	e	e
e	f	f	e	e
f	f	f	e	b
g	f	f	b	b
h	f	f	e	e

y	y_3	y_2	$x=0$ 0,0			1 0,1			3 1,1			2 1,0		
0	1	1	0	1	0	0	1	0	1	1	1	1	1	1
0	0	0	0	1	0	0	1	0	0	0	0	0	0	0
0	1	0	0	1	0	0	1	0	1	1	1	0	0	0
0	1	1	0	1	0	0	1	0	1	1	1	1	1	1
1	1	1	1	1	0	1	1	0	1	1	1	1	1	1
1	1	0	1	1	0	1	1	0	1	1	1	0	0	0
1	0	0	1	1	0	1	1	0	0	0	0	0	0	0
1	1	1	1	1	0	1	1	0	1	1	1	1	1	1
			$y_0=0$	$y_{30}=1$	$y_{20}=y_2$	$y_1=0$	$y_{31}=1$	$y_{21}=y_2$	y_3	$y_{33}=y_3$	$y_{23}=y_3$	y_2	$y_{32}=y_2$	$y_{22}=y_2$

Fig. 16.3 Developing the word-recognition tree for the industry **7474 D-flipflop**

The flow table of Fig. 16.2f enables us to develop an events graph after choosing a sequence of input events. In fact, this was already done in Fig. 14.13c and need not be repeated here.

The final task, developing the word-recognition tree from a given flow table, is by now hopefully routine. The process is summarized in Fig. 16.3. It is hardly possible to show the similarities and the differences between the normal D-flipflop (see Fig. 13.6) and the 7474 D-flipflop more clearly than by comparing their word-recognition trees.

16.2 Formalistic Analysis of State Transitions

The prime concern, when considering a circuit's **internal behaviour**, is to find out how the internal latches work together. To see how this analysis is carried out, we consider a specific circuit—say, the Passed-Sample Circuit of Fig. 15.10, a figure

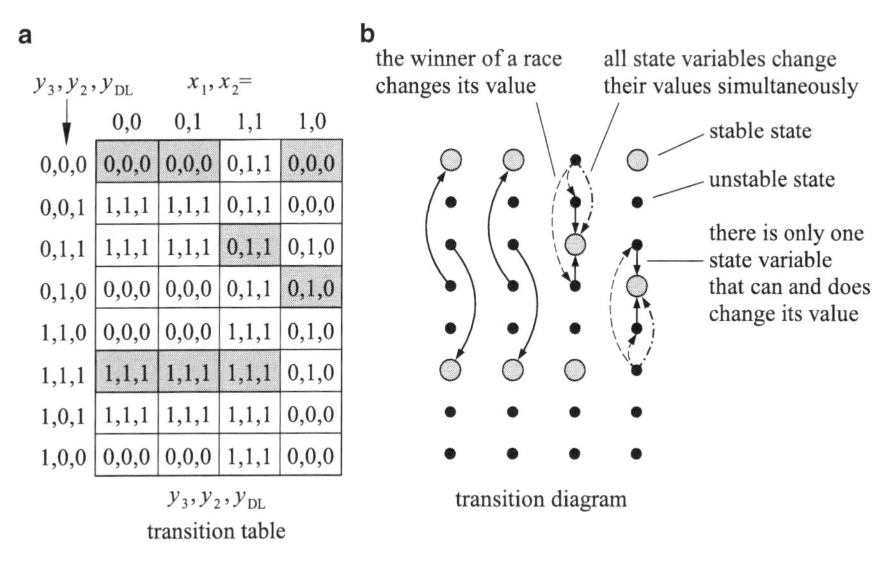

y_3,y_2,y_{DL} \ $x_1,x_2=$	0,0	0,1	1,1	1,0
0,0,0	0,0,0	0,0,0	0,1,1	0,0,0
0,0,1	1,1,1	1,1,1	0,1,1	0,0,0
0,1,1	1,1,1	1,1,1	0,1,1	0,1,0
0,1,0	0,0,0	0,0,0	0,1,1	0,1,0
1,1,0	0,0,0	0,0,0	1,1,1	0,1,0
1,1,1	1,1,1	1,1,1	1,1,1	0,1,0
1,0,1	1,1,1	1,1,1	1,1,1	0,0,0
1,0,0	0,0,0	0,0,0	1,1,1	0,0,0

Fig. 16.4 Developing the **transition diagram** for the **Passed-Sample Circuit**

we now refer to. The first thing to be done is to develop the logic formulas for each internal latch. Here, we can fall back on the formulas already stated in Fig. 15.10 writing them, for the same reason as in the previous section, in their AND-to-OR forms:

$$Y_3 \Leftrightarrow Y_{DL}\overline{X_1} \vee Y_{DL}Y_3, \qquad Y_2 \Leftrightarrow Y_{DL} \vee X_1Y_2, \qquad Y_{DL} \Leftrightarrow X_1X_2 \vee \overline{X_1}Y_{DL}.$$

These AND-to-OR forms allow us to develop the K-map of Fig. 16.4a which is of course the **transition table** of our Passed-Sample Circuit. The transition table is rewritten as a **transition diagram**—Fig. 16.4b—the large, shaded circles representing stable states, the small black circles the unstable states.

A single arrow with a **full shaft** from an unstable state (black circle) to a stable state (shaded circle) indicates that changing a *single* state variable leads to the indicated transition from the unstable to the stable state. These are well behaved transitions. A single arrow, drawn with a **dot-dashed shaft**, leading from an unstable state to a stable state indicates that all state variables that will change their values do so simultaneously. Simultaneous state-variable changes are an ideal case, but unlikely to occur. If you think of the state variables which will change their values as **racing** to do so, the likelihood is that one of them will win the race—the state reached is indicated by an arrow with a **dashed shaft**. For each possible outcome of a race, you draw an arrow with a dashed shaft.

Illustrating the above, we concentrate on the four top cells and circles of column $(x_1, x_2) = (1, 1)$ in the transition table and diagram of Fig. 16.4. When entering the top cell of column $(1, 1)$, from some other cell of the top row, two internal variables—y_3 and y_{DL}—are required to change their values. If they change their

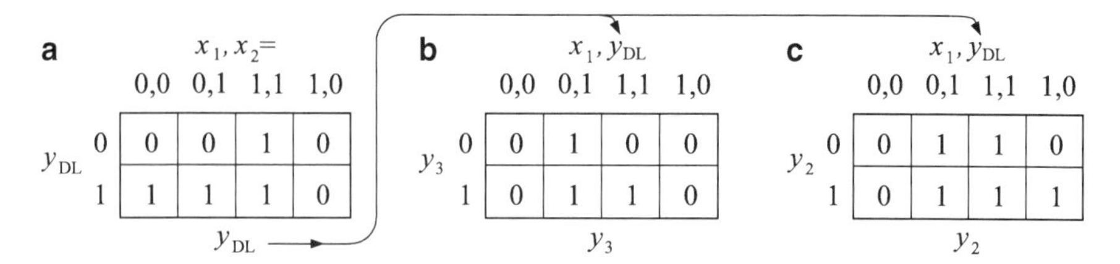

Fig. 16.5 Passed-Sample Circuit: depicting the individual transition tables for the actual signal flow in the internal latches

values simultaneously, we are sent to the *stable* state in row $(0, 1, 1)$ which is clearly intended by the layout of the transition table. If y_2 changes its value first, we are sent to the unstable state in row $(0, 0, 1)$, and from there to the final and desired destination in row $(0, 1, 1)$. On the other hand, if y_{DL} changes its value first, we are sent to $(0, 1, 0)$ and from there to the desired stable state in row $(0, 1, 1)$. Whichever internal variable wins the race, y_2 or y_{DL}, we finally reach the desired stable state. The **race** between these variables is said to be **non-critical**.

All races that do occur in Fig. 16.4 are non-critical, so let us construct an example of a **critical race**. Suppose the GOTO-command entered in unstable cell $(0, 0, 1; 1, 1)$ were $(1, 0, 1)$. From state $(0, 0, 1; 1, 1)$ we would be sent to $(1, 0, 1; 1, 1)$ and from there to the final and incorrect destination $(1, 1, 1; 1, 1)$. Thus, in the presence of critical race conditions, the circuit may malfunction.

The standard analysis, just discussed, is quite persuasive. But, alas, it has little to do with the actual signal flow in the circuit of Fig. 15.10 being analysed, and is therefore quite beside the point. The kind of analysis practised in the transition table of Fig. 16.4a implies the internal variables y_{DL}, y_3 and y_2 to be mutually dependent. But, neither do y_3 or y_2 influence each other nor do they influence y_{DL}. As it stands, only y_3 and y_2 are dependent on y_{DL} making y_{DL} a normal input variable to the y_3-latch and to the y_2-latch. This one-way signal flow is accentuated in the way the transition tables of Fig. 16.5 are organised; note that these transition tables a merely a rearranged restatement of the K-maps of Fig. 15.10a, h, i. These transition tables correctly give no indication of races, let alone critical races.

16.3 Realistic Analysis and Essential Hazards

The analysis method demonstrated in Fig. 16.4 is an important and necessary tool when used in those cases where the internal state variables are *mutually dependent*. But even then its results can only be helpful, but are not binding. An almost standard example is the **T-flipflop** which we now look into in some detail. The T-flipflop was developed in Fig. 13.7. Its design incorporates two D-latches. In Sect. 12.2 we took quite some pain to develop a robust D-latch on the basis of the Eccles-Jordan latch, the result leading to the circuit of Fig. 12.7b. Using this excellent design we can detail the T-flipflop as shown in Fig. 16.6b. To analyse this circuit we developing the

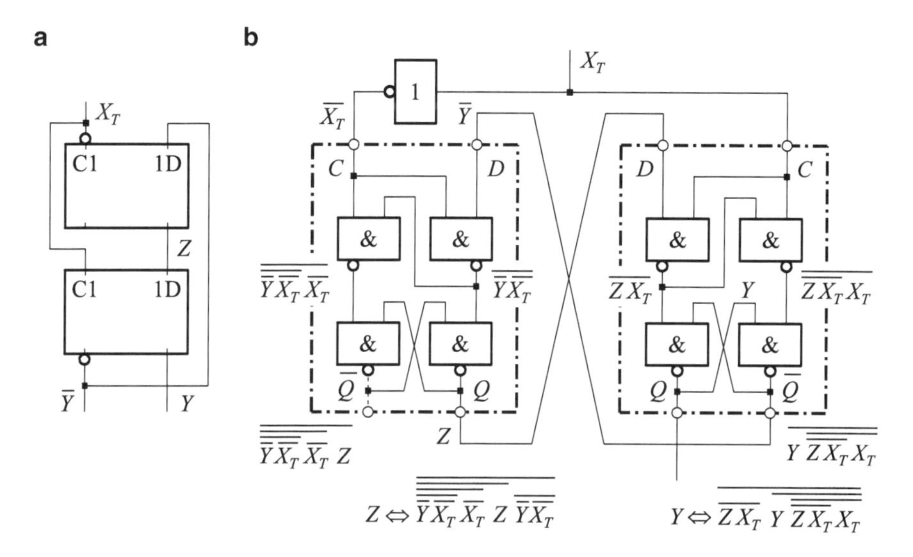

Fig. 16.6 Developing the algebraic description of a given **T-flipflop** implementation

algebraic description for each of its gates as shown in the figure. Please do take the trouble to verify these algebraic expressions.

To develop the transition table we rewrite the formulas for Y and Z as AND-to-OR expressions:

$$Y \Leftrightarrow \overline{\overline{ZX_T}\;\overline{Y\,\overline{\overline{ZX_T}\;X_T}}} \qquad Z \Leftrightarrow \overline{\overline{\overline{\overline{\overline{Y}}\;\overline{X_T}}\;\overline{X_T}\;Z}\;\overline{\overline{Y}\;\overline{X_T}}}$$

$$Y \Leftrightarrow ZX_T \vee Y\,\overline{\overline{ZX_T}\;X_T} \qquad Z \Leftrightarrow \overline{\overline{\overline{Y}}\;\overline{\overline{X_T}}\;\overline{X_T}}\;Z \vee \overline{Y}\;\overline{X_T}$$

$$Y \Leftrightarrow ZX_T \vee Y(ZX_T \vee \overline{X_T}) \qquad Z \Leftrightarrow (\overline{Y}\;\overline{X_T} \vee X_T)Z \vee \overline{Y}\;\overline{X_T}$$

$$Y \Leftrightarrow ZX_T \vee Y(Z \vee \overline{X_T}) \qquad Z \Leftrightarrow (\overline{Y} \vee X_T)Z \vee \overline{Y}\;\overline{X_T}$$

$$Y \Leftrightarrow ZX_T \vee YZ \vee Y\overline{X_T} \qquad Z \Leftrightarrow \overline{Y}Z \vee X_TZ \vee \overline{Y}\;\overline{X_T}$$

Nothing substantial is gained by using the right side of Z as a replacement for Z in the above formula for Y. The result is simply $Y \Leftrightarrow X_TZ \vee \overline{X_T}Y$. On the other hand, replacing $\overline{Y}$ in the formula for Z by the inverted expression for Y only leads to $Z \Leftrightarrow X_TZ \vee \overline{X_T}\;\overline{Y}$. I do encourage you to do these calculations although they are carried out below.

$$Y \Leftrightarrow ZX_T \vee YZ \vee Y\overline{X_T}$$

$$Y \Leftrightarrow (\overline{Y}Z \vee X_TZ \vee \overline{Y}\;\overline{X_T})X_T \vee (\overline{Y}Z \vee X_TZ \vee \overline{Y}\;\overline{X_T})Y \vee Y\overline{X_T}$$

$$Y \Leftrightarrow \underbrace{\overline{Y} X_T Z \vee X_T Z}_{X_T Z} \vee Y X_T Z \vee Y \overline{X_T}$$

$$Y \Leftrightarrow \underbrace{X_T Z \vee Y X_T Z}_{X_T Z} \vee Y \overline{X_T}$$

$$Y \Leftrightarrow X_T Z \vee \overline{X_T} Y$$

Here we have the abbreviated formula for Y. We next invert this formula and write it in its AND-to-OR form:

$$\overline{Y} \Leftrightarrow (\overline{X_T} \vee \overline{Z})(X_T \vee \overline{Y}) \Leftrightarrow X_T \overline{Z} \vee \overline{X_T}\,\overline{Y} \vee \overline{Y}\,\overline{Z}$$

We now use the AND-to-OR form of $\overline{Y}$ as a replacement in the formula for Z:

$$Z \Leftrightarrow \overline{Y} Z \vee X_T Z \vee \overline{Y}\,\overline{X_T}$$

$$Z \Leftrightarrow (X_T \overline{Z} \vee \overline{X_T}\,\overline{Y} \vee \overline{Y}\,\overline{Z}) Z \vee X_T Z \vee (X_T \overline{Z} \vee \overline{X_T}\,\overline{Y} \vee \overline{Y}\,\overline{Z}) \overline{X_T}$$

$$Z \Leftrightarrow \overline{X_T}\,\overline{Y}\,Z \vee X_T Z \vee \overline{X_T}\,\overline{Y} \vee \overline{X_T}\,\overline{Y}\,\overline{Z}$$

$$Z \Leftrightarrow X_T Z \vee \overline{X_T}\,\overline{Y}$$

The above formulas for Y and Z, whether abbreviated or not, lead to the same K-maps, Fig. 16.7a, b. Their composite form is the transition table of Fig. 16.7c. There is nothing wrong with this table: No races occur, let alone critical ones. In fact, we could not expect a better result—in theory at least.

Contrary to the ideal behaviour as predicted in the transition table, Fig. 16.7c, a *real-world T-flipflop can malfunction seriously*. For this to happen, it is necessary for the inverter of the circuit to delay the propagation of the inverted input signal, $\overline{X}_T$, say, by Δt. Even then, the malfunctioning will only occur when the input signal switches from **0** to **1**–it will not occur when X_T switches from **1** to **0**.

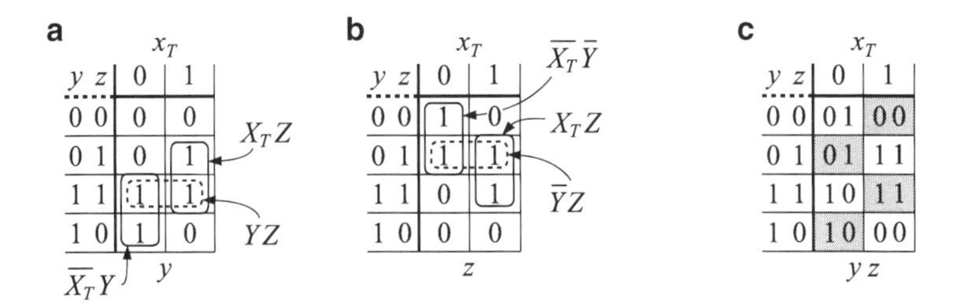

Fig. 16.7 Analysing the **T-flipflop**. (**a**) K-map for y; (**b**) K-map for z; (**c**) transition tables

We first take a detailed look at the signal flow when X_T switches from **1** to **0** showing that the circuit works flawlessly despite the delayed inversion of the input signal X_T. At this point, I must ask you to study the signal flow in Fig. 16.8 carefully, and the hopefully not too sparse explanations. Let it go without saying that starting at $(y, z; x_T) = (0, 0; 1)$ leads to an equally flawless end-result of $(y, z; x_T) = (0, 1; 0)$, but you are encouraged to verify this on your own.

Now concentrate on Fig. 16.9 where we start at $(y, z; x_T) = (0, 1; 0)$. In sub-figure (a) assume X_T to have been **0** long enough for the input to the negation also to be **0**. Make sure you understand the signal flow in this sub-figure (see the text to its right). Next, in sub-figure (b), we switch X_T to **1**, and consider the situation when this **1**-signal has not yet appeared at the output of the delay Δt. The volatile signal flow incurring is explained (as best I can) in the text next to the sub-figure. The difficulty one comes up against in visualising the signal flow lies in the permanently changing signals due to their oscillating. An oscillating circuit is of course completely useless (except as a door–bell buzzer, or a clock signal for an asynchronous circuit). The oscillation can easily be suppressed by **introducing the delay** Δt_y *greater than* Δt. With the help of the explanation next to sub-figure (c), you should quite easily be able to follow the (stable) signal flow in this sub-figure. We fair no better with respect to oscillation when starting at $(y, z; x_T) = (1, 0; 0)$ which you are encouraged to show in detail.

Unger (1969) called the serious circuit-malfunctioning of the kind discussed above for a real-world T-flipflop an **essential hazard**. But Unger (and much less I) was able to say what an essential hazard is. Unger's way out of this dilemma was to transfer the erroneous circuit behaviour to a design characteristic of the circuit's flow table by formulating: 'Definition—Essential Hazard: A sequential function *contains*[1] an essential hazard if there exists a state S_0 and an input variable x such that, starting with the system in S_0, three consecutive changes in x bring the system to a state other than the one arrived at after the first change in x'. Unger subsequently puts forth and proves the following 'Theorem on Essential Hazards and Delays: If a function contains an essential hazard, then any proper circuit realisation of it *must contain* at least one delay element'.

But, correcting the erroneous behaviour of a sequential circuit by inserting a delay (as was done in the T-flipflop example above) never changes the design of the circuit's flow table. By Unger's definition of an essential hazard, we are left with the impression that our circuit still malfunctions seriously. Thus, the definition not only does not say *what* an essential hazard is, it is furthermore not an adequate criterion to decide whether or not the circuit malfunction has been eliminated. The criticism is deeper yet, for in the next section we show that, in some cases at least, essential malfunctioning of a sequential circuit can be avoided without employing delays.

[1]emphasis by S.P. Vingron. Note: The definition does not say 'A sequential function *is* this or that...'.

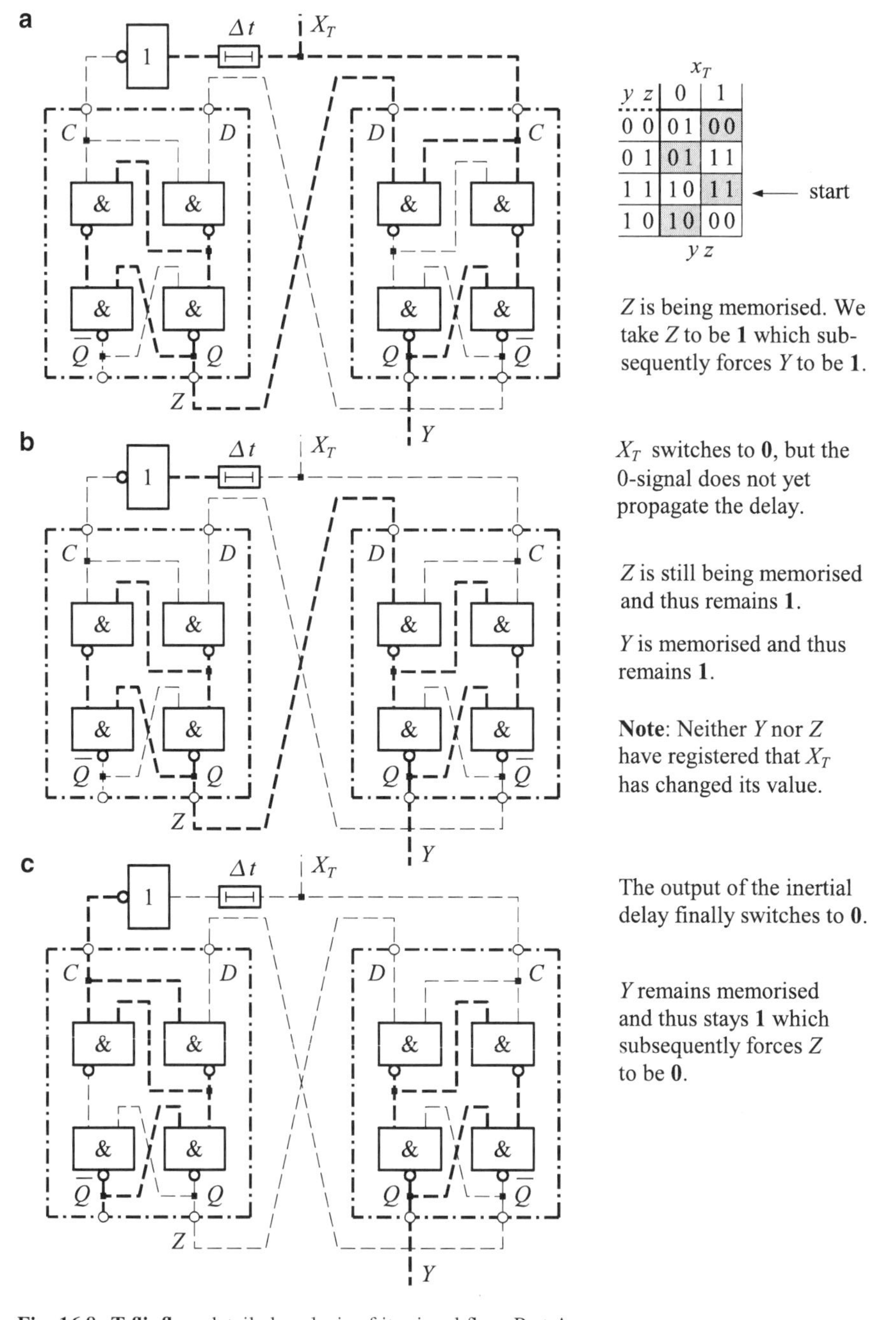

Fig. 16.8 T-flipflop: detailed analysis of its signal flow. Part A

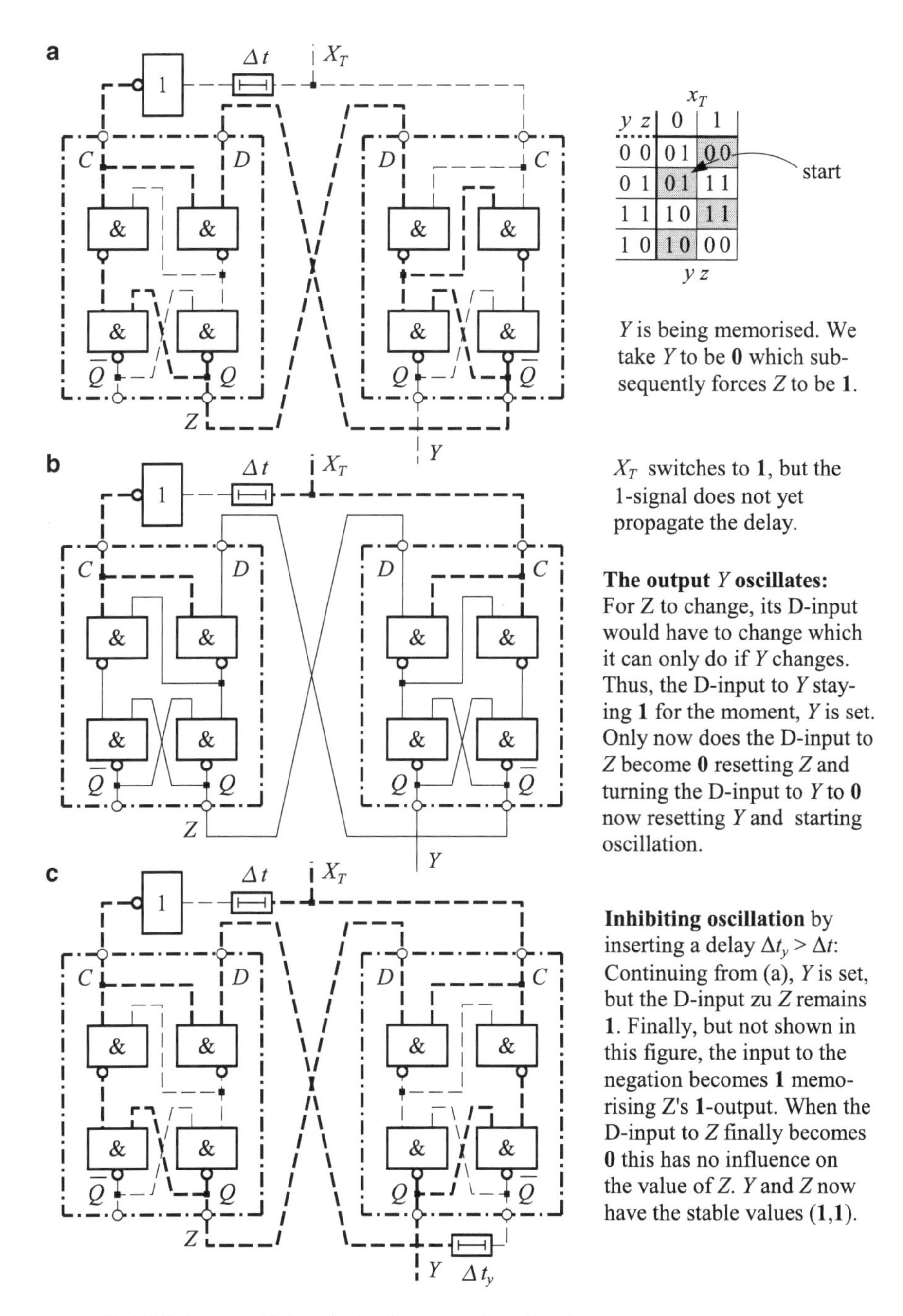

Fig. 16.9 T-flipflop: detailed analysis of its signal flow. Part B

16.4 Avoiding Essential Hazards

When introducing the concept of essential hazards, Unger showed that it was always possible to avoid them by inserting a carefully positioned delay of appropriate magnitude into the circuit—claiming, in fact, that this was the only way to avoid essential hazards. A counter example was presented (in the 1960s, before Unger (1969) paper was published) by the Swiss physicist H.H. Glättli with his remarkable, and intuitively developed version of the Expanded Two-Hand Safety Circuit designed especially to avoid the insertion of delays. The existence of such counter examples to Unger's theory on essential hazards does throw doubt on the theory.

We start by analysing the circuit for the Expanded Two-Hand Safety Problem, calculated in Fig. 13.4, as shown in Fig. 16.10. If you have worked through Sect. 16.1, there is no need to explain the formalism of Fig. 16.10, and we can go

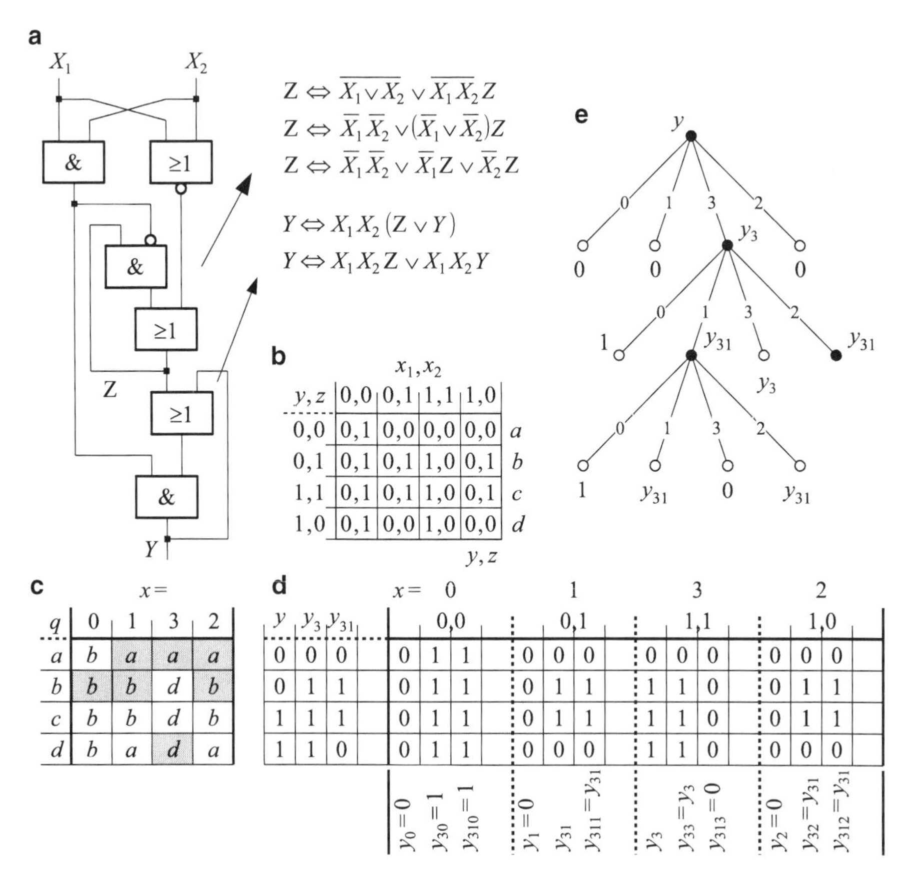

Fig. 16.10 Analysing the expanded two-hand safety circuit developed in Fig. 13.4

on to taking a detailed look at its sub-figure (c). This flow table has two essential hazards—one in $(b, 1)$, the other in $(b, 2)$. Starting, for instance, in $(b, 1)$, and letting x_1 change its value three times, you end up in $(a, 3)$, and not in the original condition $(b, 1)$, making $(b, 1)$ an essential hazard.

To visualise the effect of the essential hazard, you might like to follow the train of thought layed out here. While Z is **1** it prepares the output latch to be set, once X_1 and X_2 both become **1**. But as soon as both input signals, X_1 and X_2, become **1**, in an attempt to set the output latch Y, the signal Z, needed to set the output latch, is switched off. The question is what happens first: If the feedback of the output latch is established before Z is switched off, the circuit will work properly; *vice versa*, if Z is switched off before the feedback of the output latch is established, the circuit will malfunction. This isn't quite the behaviour to expect of a safety device.

We next look at Glättli's proposal depicted in Fig. 16.11a. Notice that the lead into the inverted input of the AND is not any more taken from the conjunction

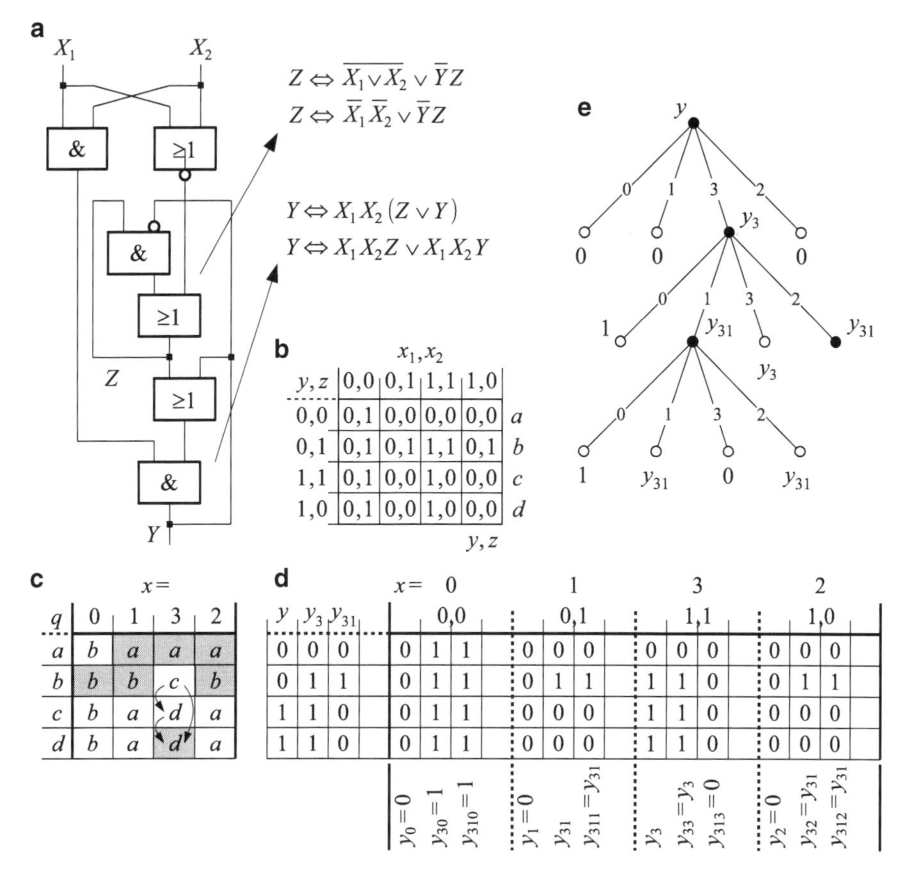

Fig. 16.11 Analysing Glättli's intuitive solution to his expanded two-hand safety problem specified in Example 13.4

$X_1 \wedge X_2$, but rather, is taken from the output Y implying that the internal latch cannot be reset before the output is **1**. Next, please conscientiously follow the formal analysis laid out in Fig. 16.11 because we shall soon run into difficulty. The first step is to develop the AND-to-OR formulas which we use to inscribe the values for (y, z) in the body of the **transition table** of sub-figure (b). Substituting each ordered pair of ((0,0), (0,1), [1,1), (1,0)) by the equally positioned state label of the quadruple $(a, b, c, d,)$, we come by the **flow table** shown in sub-figure (c). This flow table has an abnormality: state $(b, 3)$ is unstable because its inscribed next-state, c, itself leads to an unstable condition, the condition $(c, 3)$. In this case *we enforce the* **principle of stability** by entering the final state, d, as next-state into cell $(b, 3)$, which of course has no decremental influence on the input-output behaviour or the circuit. Although this is mandatory, it is (for space reasons) not shown in Fig. 16.11 but assumed for the development of iterative catenation in sub-figure (d). The word-recognition tree now obtained, and shown on sub-figure (e), is identical with that of Fig. 16.10e meaning that the circuits of Figs. 16.10a and 16.11a have an identical, *static*, input–output behaviour. Their dynamic behaviour differs: the **1**-signal of the internal latch Z, needed to set the output latch Y, is always retained until Y has become **1**. This effectively counteracts the essential hazard. Note that this contradicts Unger's theorem stating that an essential hazard can only be eliminated by introducing a delay.

To be honest, this section must end with a confession: Try as I did, I was not able to develop Glättli's circuit *systematically*. You are invited to better me, and then please show me how it is done.

Chapter 17
State Reduction*

State reduction refers to finding some or all mutually *equivalent states of a flow table*, and dropping (i.e., **merging**) from each set of equivalent states all but one state. **Equivalent states of a flow table** are those rows in the flow table's transition table that are mutually identical. All merged variants of a flow table have one and the same word-recognition tree which is why we call the merged transition tables **equivalent**. The rows that can be dropped are referred to as **redundant**. This concept of state reduction is as simple as it is novel, and is presented in place of the rather unwieldy, standard, recursive procedure.

In standard switching theory heavy emphasis is placed on designing circuits with a minimal number of latches in an attempt to reduce the cost and increase the reliability of a sequential circuit. Given p latches we can encode 2^p rows of a flow table. Vice versa, the smallest number of latches needed to encode a flow table with s rows is the smallest integer greater than or equal to the binary logarithm of s, formally, $\lceil \log_2 s$. In this context, merging a transition table is done in the hope of reducing the smallest number of latches needed.

In the first two sections we concentrate on merging **completely specified** flow tables considering two different merging procedures. In Sect. 17.1 we merge a primitive transition table into a non-primitive transition table with an associated **Moore** output table. (A primitive transition table has, at most, one stable state per row; the non-primitive transition table may contain multiple stable states per row.) In Sect. 17.2 we merge a primitive transition table into a different non-primitive transition table which, in turn, is associated with a **Mealy** output table. (Moore and Mealy output tables were introduced in Sect. 15.1 and illustrated in Fig. 15.1). The merged flow tables obtained in both sections are **equivalent** with the respective original flow tables meaning that the merged flow tables lead to the development of the same word-recognition trees as the respective original flow tables.

In Sect. 17.3 we take a look at the merging process for **incompletely specified** flow tables. By its very nature an incompletely specified flow table cannot be assigned a *unique* word-recognition tree making it obvious that any merged flow table will lead to a **non-equivalent** word-recognition tree. The best we can hope for is that both trees have prime behavioural features in common.

S.P. Vingron, *Logic Circuit Design*, DOI 10.1007/978-3-642-27657-6_17,

Merging leads directly to a second kind sequential automaton, the Mealy-type sequential automaton, an automaton capable of reading merged flow tables. This topic is briefly looked into in the **last section**.

17.1 Merging Toward a Moore Flow Table

In this section we discuss the simplest merging procedure where we are content in minimising the number of rows of the transition table such that the output table remains a Moore output table. We refer to this procedure as **Moore merging**. To illustrate it we use Fig. 17.1. This merging process transforms a *primitive* flow table (sub-figure (a)) into a *non-primitive* flow table of a minimum number of rows (sub-figure (d)), a table that still retains the property of assigning a single output value to

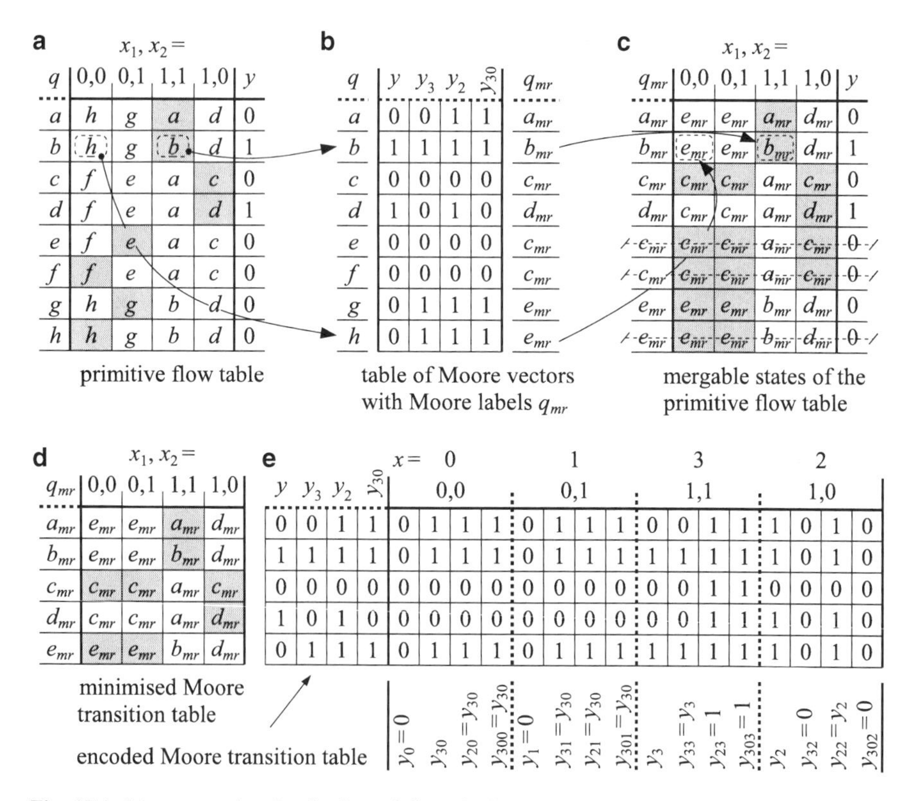

Fig. 17.1 Moore merging for the **Passed-Sample Problem**. (**a**) The given primitive flow table (see Fig. 14.10c); (**b**) Repeating the table of Moore vectors y, y_3, y_2, $y_{3,0}$ (taken from Fig. 15.9a) assigning the original encoding q to the *left*, and the Moore encoding q_{mr} to the *right*; (**c**) substituting all original row labels, q, and all GOTO-commands of the primitive flow table by the Moore labels, q_{mr} taken from sub-figure (b); (**d**) Merging sub-figure (c); (**e**) encoding sub-figure (d)

each row (in our example the output variable y has been transferred to the encoded Moore table). Most importantly, the original, primitive flow table and the minimised Moore flow table are **equivalent** meaning that both lead to the development of identical word-recognition trees. Moore merging can be described as follows, the steps being illustrated by the example in Fig. 17.1:

Step 1. Start with a given primitive flow table (such as the one shown in sub-figure (a)).

Step 2. For the given primitive flow table develop the binary encoded transition table (for our primitive flow table this was done in Fig. 15.9a).

Step 3. Extract (or concentrate on) that part of the binary encoded transition table containing the Moore vectors (the table of Moore vectors for our example—y, y_3, y_2, $y_{3,0}$,—is shown in sub-figure (b)). Give all *identical* rows of the table of Moore vectors equal labels, so-called **Moore labels** q_{mr} (in our example, a_{mr}, b_{mr}, c_{mr}, d_{mr}, e_{mr}) writing them to the right of the table of Moore vectors. To the left of the table of Moore vectors repeat the row vector q with its row labels (in our example (see sub-figure (b)) the row labels are $a, b, \dots, h$).

Step 4. Write a like table to the given, primitive flow table (see sub-figure (c) in our example) substituting each original row label, and each GOTO-command, q, by the Moore label, q_{mr}, contained in the same position of the vector if Moore labels as the position of the row label or the GOTO-command in vector q. (For our example this step is pictured by the arrows from sub-figures (a) to (b), and then on to (c)).

Step 5. Finally, merge the flow table developed in the previous step obtaining the minimised Moore flow table (see sub-figure (d) for our example).

The result of merging can easily be checked. Encoding the minimised Moore flow table must lead to Mealy vectors which, when used to develop a word-recognition tree, will provide us with the same tree as that for the original flow table. For our example, the binary encoding is done in Fig. 17.1e. I encourage you to draw its word-recognition tree and compare it with that of Fig. 15.9. A further aspect needs to be pointed out. The encoded Moore flow table of Fig. 17.1d need not be developed from the minimised Moore flow table: Merging the latter with the help of the Moore labels must lead to the former.

It is sometimes possible to reduce the number of rows of a transition table still further if we accept a Mealy instead of a Moore output table. Looking at our example, Fig. 17.1, it is obvious that reducing the number of rows from eight to five does not reduce the number of internal latches—we still need three. But, if it were possible to reduce the number of rows further, even only by one, two internal latches would then suffice to encode the four-row transition table. In the next section we thus discuss merging when using Mealy output tables.

17.2 Merging Toward a Mealy Flow Table

When trees were introduced in Sect. 13.1 we defined **internal nodes** as nodes that have output edges *and* an input edge. This excludes the root as an internal node, as the root has no input edge. The internal nodes stand for latches of the sequential circuit. As one and the same latch can sometimes represent more than one internal node, the number of latches needed is always less than or equal to the number of internal nodes. Let us refer to the Moore vectors associated with the internal nodes as **internal Moore vectors**. To encode the latches of a sequential circuit, it must suffice to employ the *internal* Moore vectors, i.e., without considering the output vector y, as was done in the previous section. The output y is then defined separately in a Mealy output table (see Fig. 15.1d).

Reconsidering the Passed Sample Problem, Fig. 17.2, note that only the Moore vectors y_3, y_2, $y_{3.0}$ of the internal nodes are used to encode the rows of the transition table. We call the symbols q_{ml}, which characterise identical rows, **Mealy labels**. They, together with the outputs y, are written as ordered pairs to the right of the table of the Moore vectors y_3, y_2, $y_{3.0}$. To carry out a **Mealy merging**, such that of Fig. 17.2, systematically, follow the steps put forth here:

Step 1. Start with a given primitive flow table (such as the one shown in **sub-figure (a)**).

Step 2. For the given primitive flow table develop the binary encoded transition table (for our primitive flow table this was done in Fig. 15.9a).

Step 3. Extract that part of the binary encoded transition table containing the internal Moore vectors (the table of inernal Moore vectors for our example—y_3, y_2, $y_{3.0}$,—is shown in sub-figure (b)). Give all *identical* rows of the table of internal Moore vectors equal labels, so-called **Mealy labels** q_{ml} (in our example, a_{ml}, b_{ml}, c_{ml}, and d_{mr}) writing them to the right of the table of internal Moore vectors together with the output values y (see the column of pairs q_{ml}, y). To the left of the table of Moore vectors repeat the row vector q with its row labels (in our example (see sub-figure (b)) the row labels are $a, b, \ldots, h$).

Step 4. Write a like table to the given, primitive flow table (see **sub-figure (c)** in our example) substituting each original *row label* q by the equally positioned Mealy label q_{ml} of sub-figure (b), and substituting each GOTO-command, q, of the primitive flow table by the equally positioned *pair* (q_{ml}, y) taken from sub-figure (b). (For our example this procedure is pictured by the arrows from sub-figures (a) to (b), and then on to (c)).

Step 5. Finally, merge the transition table developed in the previous step obtaining the minimised Mealy transition table (see sub-figure (d) for our example).

Note that the Mealy table of sub-figure (d) actually consists of *two* distinct tables: a transition table, and a Mealy output table as shown in Fig. 17.3. The transition table (after encoding it) allows us to calculate the latches of the circuit, while the Mealy output table allows us to calculate a combinational circuit for the output

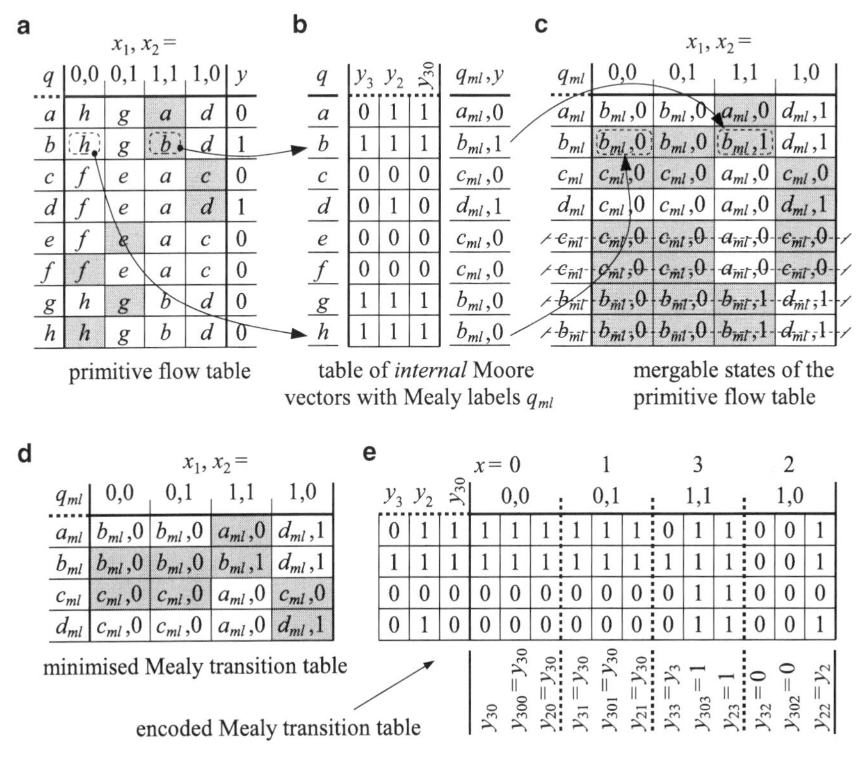

a primitive flow table

q	$x_1, x_2 =$ 0,0	0,1	1,1	1,0	y
a	h	g	a	d	0
b	h	g	b	d	1
c	f	e	a	c	0
d	f	e	a	d	1
e	f	e	a	c	0
f	f	e	a	c	0
g	h	g	b	d	0
h	h	g	b	d	0

b table of *internal* Moore vectors with Mealy labels q_{ml}

q	y_3	y_2	y_{30}	q_{ml}, y
a	0	1	1	a_{ml},0
b	1	1	1	b_{ml},1
c	0	0	0	c_{ml},0
d	0	1	0	d_{ml},1
e	0	0	0	c_{ml},0
f	0	0	0	c_{ml},0
g	1	1	1	b_{ml},0
h	1	1	1	b_{ml},0

c mergable states of the primitive flow table

q_{ml}	$x_1, x_2 =$ 0,0	0,1	1,1	1,0
a_{ml}	b_{ml},0	b_{ml},0	a_{ml},0	d_{ml},1
b_{ml}	b_{ml},0	b_{ml},0	b_{ml},1	d_{ml},1
c_{ml}	c_{ml},0	c_{ml},0	a_{ml},0	c_{ml},0
d_{ml}	c_{ml},0	c_{ml},0	a_{ml},0	d_{ml},1
~~c_{ml}~~	~~c_{ml},0~~	~~c_{ml},0~~	~~a_{ml},0~~	~~c_{ml},0~~
~~c_{ml}~~	~~c_{ml},0~~	~~c_{ml},0~~	~~a_{ml},0~~	~~c_{ml},0~~
~~b_{ml}~~	~~b_{ml},0~~	~~b_{ml},0~~	~~b_{ml},1~~	~~d_{ml},1~~
~~b_{ml}~~	~~b_{ml},0~~	~~b_{ml},0~~	~~b_{ml},1~~	~~d_{ml},1~~

d minimised Mealy transition table

q_{ml}	$x_1, x_2 =$ 0,0	0,1	1,1	1,0
a_{ml}	b_{ml},0	b_{ml},0	a_{ml},0	d_{ml},1
b_{ml}	b_{ml},0	b_{ml},0	b_{ml},1	d_{ml},1
c_{ml}	c_{ml},0	c_{ml},0	a_{ml},0	c_{ml},0
d_{ml}	c_{ml},0	c_{ml},0	a_{ml},0	d_{ml},1

e encoded Mealy transition table

y_3	y_2	y_{30}	$x=0$ 0,0			1 0,1			3 1,1			2 1,0		
			y_{30}	$y_{300}=y_{30}$	$y_{20}=y_{30}$	$y_{31}=y_{30}$	$y_{301}=y_{30}$	$y_{21}=y_{30}$	$y_{33}=y_3$	$y_{303}=1$	$y_{23}=1$	$y_{32}=0$	$y_{302}=0$	$y_{22}=y_2$
0	1	1	1	1	1	1	1	1	0	1	1	0	0	1
1	1	1	1	1	1	1	1	1	1	1	1	0	0	1
0	0	0	0	0	0	0	0	0	0	1	1	0	0	0
0	1	0	0	0	0	0	0	0	0	1	1	0	0	1

Fig. 17.2 Mealy merging for the **Passed-Sample Problem**. (**a**) The given primitive flow table (see Fig. 14.10c); (**b**) Repeating the table of internal Moore vectors y_3, y_2, $y_{3,0}$ (taken from Fig. 15.9a) assigning the original encoding q to the *left*, and the Mealy encoding, q_{ml}, with the output y, to the *right*; (**c**) substituting all original row labels, q, by equally positioned Mealy labels q_{ml}, and substituting all GOTO-commands of the primitive flow table by the equally positioned pair (q_{mr}, y) taken from sub-figure (b); (**d**) merging sub-figure (c); (**e**) encoding sub-figure (d)

q_{ml}, y

q_{ml}	$x_1, x_2 =$ 0,0	0,1	1,1	1,0
a_{ml}	b_{ml},0	b_{ml},0	a_{ml},0	d_{ml},1
b_{ml}	b_{ml},0	b_{ml},0	b_{ml},1	d_{ml},1
c_{ml}	c_{ml},0	c_{ml},0	a_{ml},0	c_{ml},0
d_{ml}	c_{ml},0	c_{ml},0	a_{ml},0	d_{ml},1

$\Rightarrow$

q_{ml}

q_{ml}	$x_1, x_2 =$ 0,0	0,1	1,1	1,0
a_{ml}	b_{ml}	b_{ml}	a_{ml}	d_{ml}
b_{ml}	b_{ml}	b_{ml}	b_{ml}	d_{ml}
c_{ml}	c_{ml}	c_{ml}	a_{ml}	c_{ml}
d_{ml}	c_{ml}	c_{ml}	a_{ml}	d_{ml}

y

q_{ml}	$x_1, x_2 =$ 0,0	0,1	1,1	1,0
a_{ml}	0	0	0	1
b_{ml}	0	0	1	1
c_{ml}	0	0	0	0
d_{ml}	0	0	0	1

Fig. 17.3 Splitting the minimised Mealy transition table of Fig. 17.2d into one for the states q_{mr} and one for the output y

variable y (of course using the same encoding as for the transition table). For the example shown here, we were successful in further reducing the number of rows of the transition table from five (Moore merging) to four showing that a Passed Sample Circuit can be realised using two latches. On the other hand, you can see that the to realise the output function y we will need a markedly more complicated logic function than in the case of Moore merging.

It is important to note that the word-recognition tree of the minimised Mealy table of a *completely specified* sequential problem is always identical with the word-recognition tree of the problem's primitive flow table. Any circuits developed from these (reduced and unreduced) tables are thus equivalent, i.e., are indistinguishable in their reaction to identical input words. As in the previous section, you are encouraged to check this claim by comparing the word-recognition tree that you can develop from the encoded Mealy transition table of Fig. 17.2e with the word-recognition tree of Fig. 15.9—as expected, they are identical.

17.3 Merging Incompletely Specified Tables

More often than not practical sequential problems are incompletely specified. The sequential problem used to discuss and illustrate the merging of *incompletely* specified flow tables is the Expanded Priority-AND Variant considered in-depth (but without merging) in Sect. 15.7. We use this example to emphasize the following points: **Firstly**, merging flow or transition tables leads to quite a number of alternative circuits, **secondly**, these circuits are mutually non-equivalent, and, **thirdly**, only one of these circuits is equivalent to the circuit developed from the non-merged flow table and whose word-recognition tree is shown in Fig. 15.12a.

The merging process for an incompletely specified problem assumes we have developed the binary encoding of the incompletely specified transition table. For the running example of this section, Example 15.1, this was done in Fig. 15.11b. The transition table and the table of Moore vectors are repeated in Fig. 17.4a, b. Note that, in the analysis to come, we do *not* attempt to choose GOTO-commands for empty cells in the transition table of sub-figure (a)—we choose values (0 or 1) for the empty (or don't care) cells in the table of Moore vectors of sub-figure (b) because identical rows of this table govern the merging process.

We start our analysis of sub-figure (b) by noting that cell (i, y_3) must have a 1 inscribed if we want row i to be identical with any other row, and thus mergable. Binary values for the other empty cells—(a, y_3), (c, y_0), and (e, y_0)—can be chosen freely. The eight possibilities are listed in sub-figures (c)–(j), the first four having four distinct rows (and therefore needing only two latches to be encoded), the second four having five distinct rows (which require three latches for their encoding).

Here we briefly interrupt the merging procedure to see which of the tables of Moore vectors, (c)–(j), leads to the word-recognition tree of Fig. 15.12a. The following considerations in this paragraph refer to Fig. 15.11b. It is necessary to point out a consequence of enforcing determinateness of Mealy vectors by equating them with certain Moore vectors. Specifically, choosing $y_{0,1}$ to equal y_0 by replacing the dashes in rows d and g of $y_{0,1}$ by 0 and 1, respectively, forces us, vice versa, to replace both dashes in rows c and e of the Moore vector y_0 by the value 0 equivalently placed in $y_{0,1}$. A like consideration, taking $y_{3,1} = y_3$ into account,

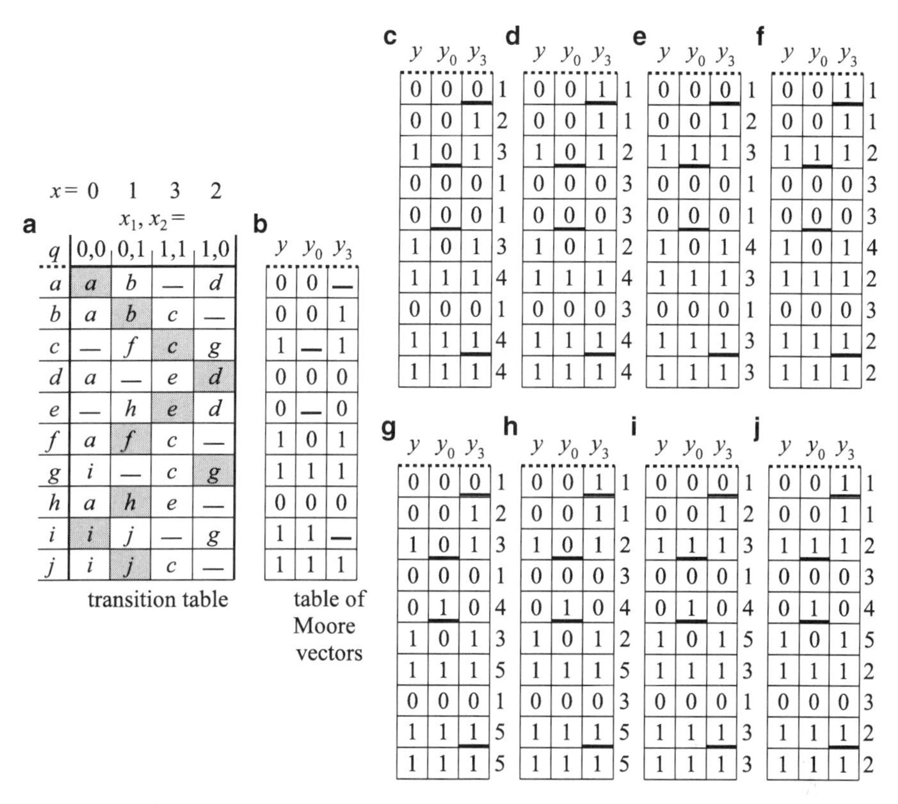

a transition table

q \ $x_1, x_2=$	0,0 ($x=0$)	0,1 ($x=1$)	1,1 ($x=3$)	1,0 ($x=2$)
a	a	b	—	d
b	a	b	c	—
c	—	f	c	g
d	a	—	e	d
e	—	h	e	d
f	a	f	c	—
g	i	—	c	g
h	a	h	e	—
i	i	j	—	g
j	i	j	c	—

b table of Moore vectors

y	y_0	y_3
0	0	—
0	0	1
1	—	1
0	0	0
0	—	0
1	0	1
1	1	1
0	0	0
1	1	—
1	1	1

c

y	y_0	y_3	
0	0	0	1
0	0	1	2
1	0	1	3
0	0	0	1
0	0	0	1
1	0	1	3
1	1	1	4
0	0	0	1
1	1	1	4
1	1	1	4

d

y	y_0	y_3	
0	0	1	1
0	0	1	1
1	0	1	2
0	0	0	3
0	0	0	3
1	0	1	2
1	1	1	4
0	0	0	3
1	1	1	4
1	1	1	4

e

y	y_0	y_3	
0	0	0	1
0	0	1	2
1	1	1	3
0	0	0	1
0	0	0	1
1	0	1	4
1	1	1	3
0	0	0	1
1	1	1	3
1	1	1	3

f

y	y_0	y_3	
0	0	1	1
0	0	1	1
1	1	1	2
0	0	0	3
0	0	0	3
1	0	1	4
1	1	1	2
0	0	0	3
1	1	1	2
1	1	1	2

g

y	y_0	y_3	
0	0	0	1
0	0	1	2
1	0	1	3
0	0	0	1
0	1	0	4
1	0	1	3
1	1	1	5
0	0	0	1
1	1	1	5
1	1	1	5

h

y	y_0	y_3	
0	0	1	1
0	0	1	1
1	0	1	2
0	0	0	3
0	1	0	4
1	0	1	2
1	1	1	5
0	0	0	3
1	1	1	5
1	1	1	5

i

y	y_0	y_3	
0	0	0	1
0	0	1	2
1	1	1	3
0	0	0	1
0	1	0	4
1	0	1	5
1	1	1	3
0	0	0	1
1	1	1	3
1	1	1	3

j

y	y_0	y_3	
0	0	1	1
0	0	1	1
1	1	1	2
0	0	0	3
0	1	0	4
1	0	1	5
1	1	1	2
0	0	0	3
1	1	1	2
1	1	1	2

Fig. 17.4 Listing all completely specified tables of Moore vectors (**c**)–(**j**) for the incompletely specified **Expanded Priority-AND Variant** repeated in (**a**) and (**b**) (See Example 15.1 and Fig. 15.11b)

forces us to replace the dashes in y_3 by 1. The result is the table of Moore vectors shown in Fig. 17.4d. It is only this table of Moore vectors that provides us with the same word-recognition tree as shown in Fig. 15.12a.

Returning to the merging process, we note that there is no pre-cognitive way to decide which of the tables of Moore vectors in Fig. 17.4, sub-figures (c)–(j), will provide us with circuits of desired properties (other than the number of latches needed). If the circuit is important enough to you, you will have to design a circuit or circuits for each individual table of Moore vectors evaluating each design by criteria you believe to be appropriate.

In our example, where only merging per se is of interest, we arbitrarily choose to use the table of Moore vectors of Fig. 17.4c. This choice and the completer merging process is documented in Fig. 17.5. A detailed explanation of the merging process would follow the lines laid down Sect. 17.1. Do compare the word-recognition trees of Figs. 17.5e and 15.12a. Obviously the behaviour of the circuits developed from these trees will differ slightly

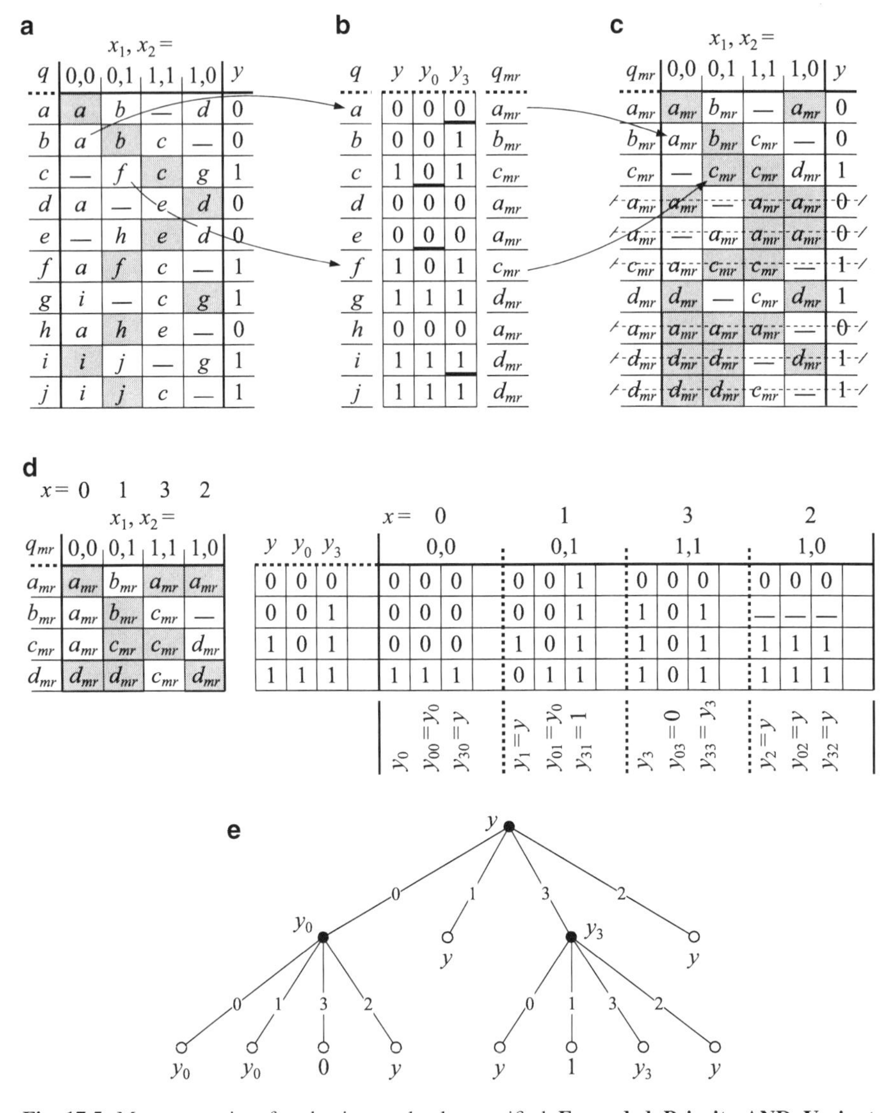

Fig. 17.5 Moore merging for the incompletely specified **Expanded Priority-AND Variant**. (**a**) The given primitive flow table; (**b**) choosing a completely specified table of Moore vectors, our choice being the table from Fig. 17.4c; (**c**)–(**e**) illustrating the normal merging process as for any completely specified problem

17.4 Mealy-Type Sequential Automaton

A Mealy-type flow table differs from a Moore-type flow table in that each cell of the former contains not only a GOTO row-label command, q, but also an output value, $y(q,x)$. The sequential automaton that can read and react to a Mealy-type flow table

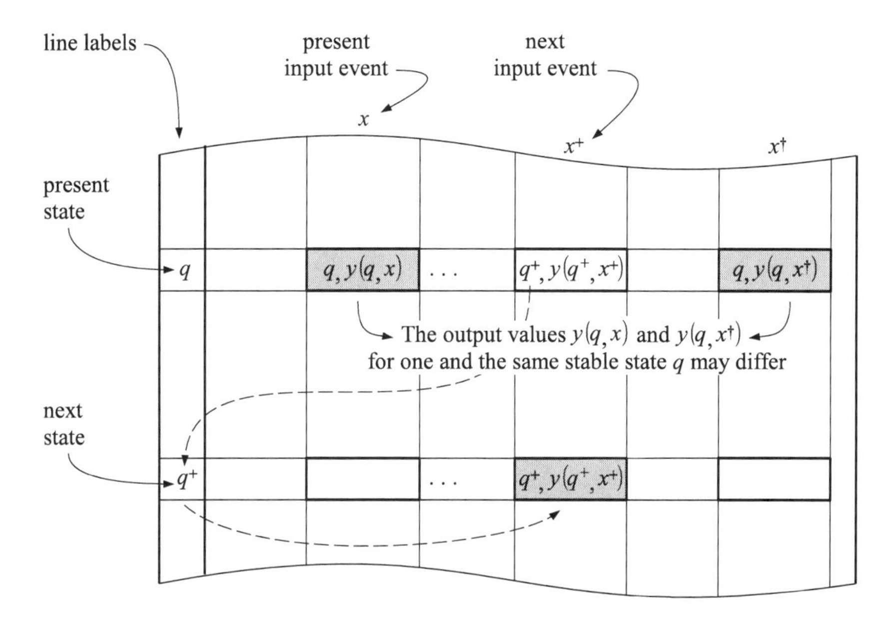

Fig. 17.6 Organisation of a Mealy flow table. The *shaded* cells represent stable states

differs in one significant point from the sequential automaton discussed in Sect. 14.1 and used to read Moore-type flow tables. The information contained in a cell of the flow table is simultaneously read by two heads. One head reads the GOTO row-label information, q, while the other reads the output value, $y(q, x)$. The row label read is used to move the tape to that row, whereas the output value read is sent to the output of the automaton. These two heads can read the information $q, y(q, x)$ in any given row; which cell in a given row is being read is determined by the value of the input event, x.

Figure 17.6 shows the principle organisation of a Mealy-type flow table. Consider the row labelled q: Any cell containing the same row label, q, represents a stable condition; these cells are emphasized by a light-grey background. When reading one of them, the tape remains static. An uncertainty is sometimes felt in specifying an output value for an *unstable* cell, for instance, for cell (q, x^+). Do we enter the output value of the next state, or of a previous state? The principle is, to enter the output value of the next state because the previous output, in our example, either $y(q, x)$ or $y(q, x^+)$, could differ. In those cases when the previous output values are all equal, this output value can be used as output value for $y(q^+, x^+)$. Sometimes, a choice of this kind can help to design a simpler output circuit.

Sequential problems are quite often specified using Mealy flow tables instead of Moore flows table or primitive (Moore) flow tables. There can be no objection to this practice. Personally, I prefer the primitive flow table for systematic reasons. It is the basic form of a flow table from which all other variants are derived by

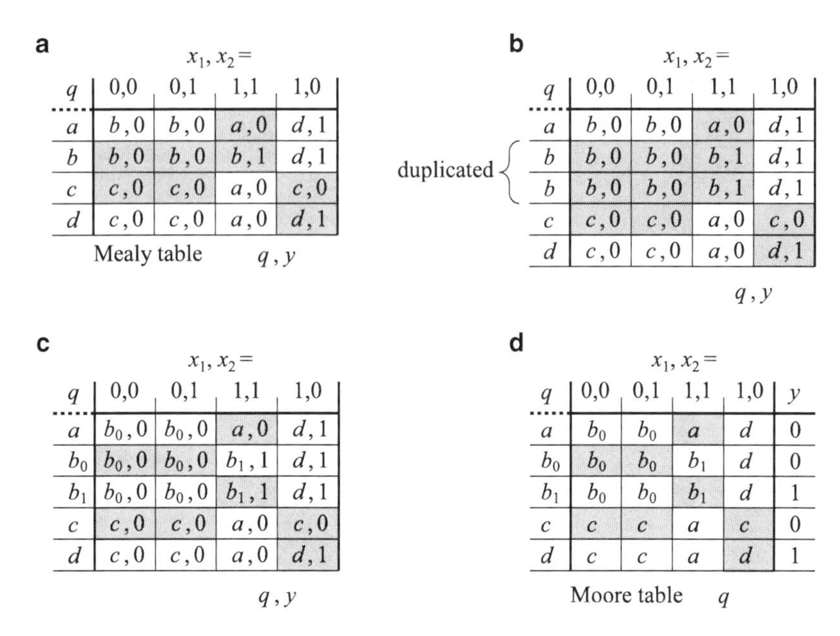

a Mealy table ($x_1, x_2 =$; entries q, y)

q	0,0	0,1	1,1	1,0
a	$b,0$	$b,0$	$a,0$	$d,1$
b	$b,0$	$b,0$	$b,1$	$d,1$
c	$c,0$	$c,0$	$a,0$	$c,0$
d	$c,0$	$c,0$	$a,0$	$d,1$

b ($x_1, x_2 =$; entries q, y; row b duplicated)

q	0,0	0,1	1,1	1,0
a	$b,0$	$b,0$	$a,0$	$d,1$
b	$b,0$	$b,0$	$b,1$	$d,1$
b	$b,0$	$b,0$	$b,1$	$d,1$
c	$c,0$	$c,0$	$a,0$	$c,0$
d	$c,0$	$c,0$	$a,0$	$d,1$

c ($x_1, x_2 =$; entries q, y)

q	0,0	0,1	1,1	1,0
a	$b_0,0$	$b_0,0$	$a,0$	$d,1$
b_0	$b_0,0$	$b_0,0$	$b_1,1$	$d,1$
b_1	$b_0,0$	$b_0,0$	$b_1,1$	$d,1$
c	$c,0$	$c,0$	$a,0$	$c,0$
d	$c,0$	$c,0$	$a,0$	$d,1$

d Moore table ($x_1, x_2 =$; entries q)

q	0,0	0,1	1,1	1,0	y
a	b_0	b_0	a	d	0
b_0	b_0	b_0	b_1	d	0
b_1	b_0	b_0	b_1	d	1
c	c	c	a	c	0
d	c	c	a	d	1

Fig. 17.7 From Mealy table to Moore table

merging. Merging a primitive flow table provides us with a (non-primitive) Moore table which merged leads to a Mealy flow table.

It now remains to be seen, how to obtain a Moore table from a given Mealy flow table. You can follow the procedure in Fig. 17.7. Assume we are presented with the Mealy table of sub-figure (a)—taken from Fig. 17.2d. Typical for a Mealy table is that it has at least one row (in our example, row b) with two or more stable condition, or cells, whose outputs differ. In cells $(b, 0)$ and $(b, 1)$ the output value is 0, while in cell $(b, 3)$ the output is 1. We start by duplicating each such row—see sub-figure (b).

Each of these rows are then given surrogate names that do not yet occur as state names—the method of choice being to distinguish the identical rows by giving one of the row names the index 0, and the other the index 1. At present, these row names (b_0 and b_1) exist only in the left-most column, and not in the body of the table. The surrogate name with the 0-index is now used as surrogate for all non-indexed occurrences of the state name in those cells where the output is 0. In our example, all occurrences of $(b, 0)$ are replaced by $(b_0, 0)$.The surrogate name with the 1-index is then used as surrogate for all non-indexed occurrences of the state name in those cells where the output is 1. In our example, all occurrences of $(b, 1)$ are replaced by $(b_1, 0)$. You can follow this in sub-figure (c).

The above procedure ensures that all stable cells in any row are associated with the same output value. This allows us to extract these output values, i.e., to concentrate them a single output column y—see sub-figure (d). The table obtained is of course normally not a primitive flow table, but this is of minor importance.

Chapter 18
Verifying a Logic Design*

The problem considered in this chapter is this: Having designed an asynchronous circuit according to a given word-recognition tree, or flow table, we next want to **verify** our logic design by developing a *minimal length* events graph with which to *completely test* the circuit's input–output behaviour. In the literature, this is usually referred to as **functional testing**. In general, the problem is seen as unsolvable. Mead and Conway (1980) put it this way: 'Complete functional testing of complex systems with internal sequencing is not possible in general, and most integrated system chips manufactured, even at 1978 levels of complexity, are not economically testable for even a small fraction of their possible internal states.'

Nevertheless, testing actual technical realisations is sometimes a necessity because even faultlessly developed sequential circuits can come up with unpleasant surprises. The solution to the verification problem is only possible due to the word-recognition tree, although we still need to take a closer look at its attributes. Periodically recurring sub-trees of the word-recognition tree are, as we will see, the key to finding finite length (in fact, minimal length) test input-sequences of input events. Furthermore, with the help of the flow table, it can be shown that these minimal length input-sequences suffice to test the asynchronous circuit completely.

18.1 End-Nodes and Their Event Graphs

As unexpected as it seems at first sight, a word-recognition tree can be transformed into a succession of input events, each input event causing, and thus being assigned, a well defined output value, either 0 or 1. In short, a word-recognition tree can be transformed into an events graph. To do this, we first need a basic understanding of how each end-node, or leaf, can be expressed by a partial events graph.

In Fig. 18.1 we use the example of the Passed-Sample Problem to show the connection between an end-node and a partial events graph. The input event 0, sub-figure (a), causes the output value y to be 0; we write this as $y_0 = 0$, meaning, whenever the input event is 0, the output is 0 whatever the previous input event,

S.P. Vingron, *Logic Circuit Design*, DOI 10.1007/978-3-642-27657-6_18,

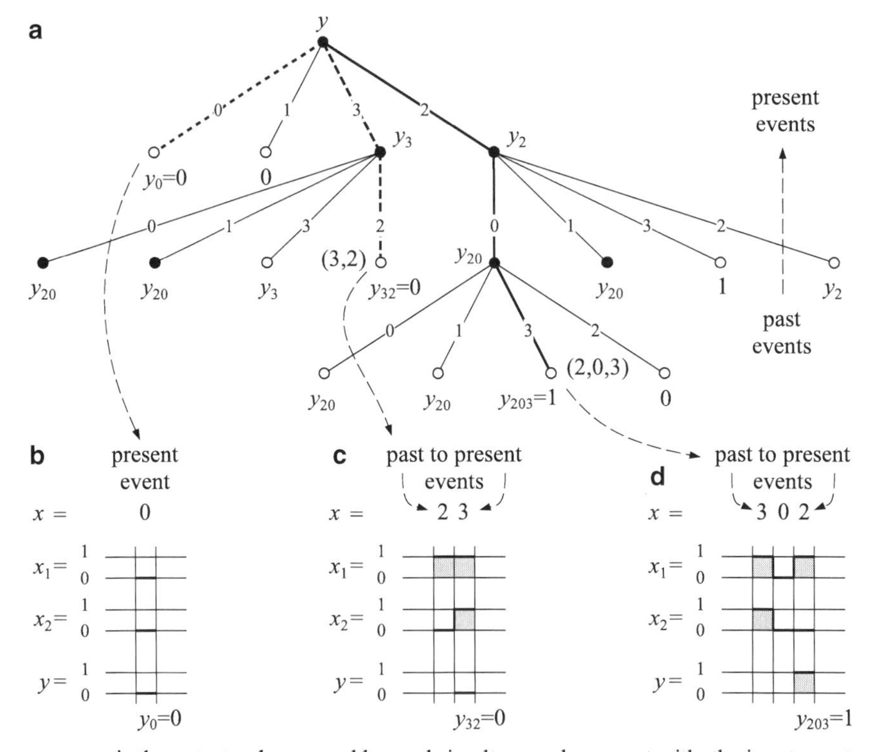

y_0 is the output value caused by, and simultaneously present with, the input event 0 (whatever any previous input event, or events).

y_{32} is the output value caused as long as the present input event is 3, but only when this was preceded by the input event 2.

y_{203} is the output value caused as long as the present input event is 2 when this was preceded by the input event 0 which, in turn, is preceded by the input event 3.

Fig. 18.1 Passed-Sample Problem. Expressing output values of end-nodes as partial events graphs

or events. This is the intended interpretation of the partial events graph of sub-figure (b).

Going one level higher, consider input word $(3, 2)$, marked by the leaf $(3, 2)$, and the output $y_{3,2} = 0$ it implies. It is important to realise that the output $y_{3,2}$ is only present when (you could say, while) the present input event is 3, but only if this input event was preceded by the input event 2. Expressing this in a partial events graph, sub-figure (c), note that we can only enter the output value in the location marked by input event 3—the output value in the location marked by input event 2 *remains undefined*. One more thing to note when drawing an events graph is that the succession of input events *from passed to present*, and their implied outputs, are entered *from left to right*. Any input event to the left is prior to the input event to its right.

As a final example take a look at leaf $(2, 0, 3)$ and its assigned output $y_{2,0,3} = 1$. Here again, the output value, 1, prevails simultaneously with the present input event, 2, but only when it was preceded by input event 0, and this by input event 3. The output values caused by the prior and pre-prior input events, 0 and 3, remain undefined by the specification of $y_{2,0,3}$. This is pictured in sub-figure (d).

In the next section we start redesigning the word-recognition tree to fit the requirements of verification.

18.2 Verification Tree and Verification Graph

To test a circuit, we supply its inputs with a special sequence of input events and measure the successive and associated output values; these must coincide with predefined output values for the test to be successful. What are the properties this 'special sequence of input events' must have?

(a) The input sequence must be of finite length.
(b) To enable economical testing, the sequence must be as short as possible.
(c) Two successive input events must differ in the value of one and only one input variable.
(d) The input sequence must enable a complete test of the circuit.

A **verification graph** is an events graph whose input sequence has the properties specified above. The reason for allowing only one input variable at a time to change its value, point (c), is that it is not reproducible for two or more specific input variables to change their values simultaneously. The problem of a complete test is non-trivial, its discussion deferred to the next section.

A *verification graph* is developed from a *verification tree*, which itself is developed from a *word-recognition tree*. Thus, we now turn to the **verification tree** and its development from a word-recognition tree. We demonstrate and discuss this development using the example laid out in Fig. 18.2. Sub-figure (a) repeats the word-recognition tree of the Passed-Sample Problem. We first abide by the property that 'Two successive input events must differ in the value of one and only one input variable'. Concentrating on the *elementary* sub-trees y_3 and y_2 of sub-figure (a)—'elementary' meaning disregarding all sub-trees of higher levels—you see that we retain only those edges contained in the sub-trees y_3 and y_2 of sub-figure (b). Let me elaborate for sub-tree y_3 of sub-figure (a). Its output is present when the current input event is 3; therefore, the prior input event may neither be 0 nor 3 meaning these edges must be dropped as pictured in sub-figure (b).

Dropping output edges of elementary sub-trees, as described above, is called **pruning** (the elementary sub-tree). Specifically: Pruning refers to *retaining* exactly those output edges whose Hamming weight differ from the Hamming weight of the current tree's input edge by 1. Put the other way around, pruning refers, (**a**), to dropping *the* output edge identical to the input edge, and, (**b**), those output edges whose Hamming weights differ from that of the tree's input edge by 2 or more.

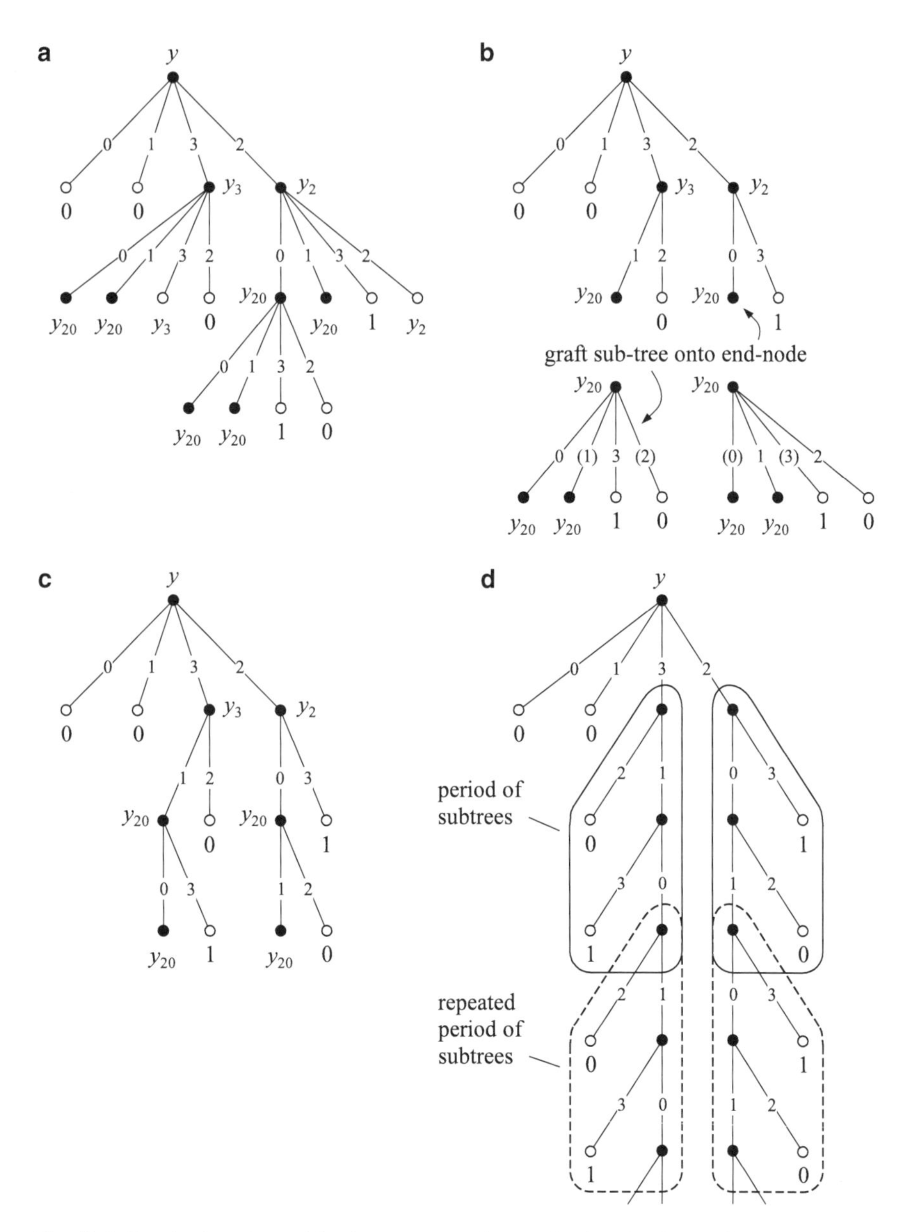

Fig. 18.2 Developing the **verification tree**, sub-figure (**d**), for the **Passed-Sample Problem** specified in sub-figure (**a**)

As the sub-trees y_3 and y_2 of sub-figure (b) show, the effect of pruning is always a drastic reduction in the number of output edges. This is combined with an equal reduction in the number of higher level sub-tree. In our example, the number of second level sub-trees, y_{20}, is reduced from four, in sub-figure (a), to two, in sub-figure (b).

You will notice that, in sub-figure (b), the second level sub-trees, y_{20}, are drawn slightly detached from the leaves of the level-one sub-trees, y_3 and y_2, just pruned. This is only done to emphasize that the pruning process starts at level one, and that the higher level sub-trees are, for the moment, disregarded. Having pruned the level-one sub-trees, we now attach, or reattach, the level-two sub-trees.

Attaching, or reattaching, a higher order sub-tree to the tree is referred to as **grafting** the sub-tree, as if it were a scion, to the tree. Specifically: Grafting refers to attaching an elementary sub-tree to an appropriate leaf of the same or another sub-tree, whereby the grafted sub-tree *is pruned* after having been attached. In sub-figure (b), the pruning to be done after attaching sub-trees y_{20} is indicated by writing the input events of the effected edges in parenthesis.

The preliminary verification tree we so far have obtained by grafting and pruning is shown in sub-figure (c). From here on, wherever possible, we continue the grafting process. To make sure you understand how this is done, please graft two further levels to the preliminary verification tree of sub-figure (c). The result should be the tree of sub-figure (d).

This process of grafting could go on ad infinitum. The question, of course, is **when to stop grafting**? The simplest case is when all end-nodes are assigned constant values (either 0 or 1). An example of this is the JK-flipflop discussed subsequently. The less obvious case for pruning a branch of the verification tree is after a group of sub-trees (or a single sub-tree) recurs. We call such a recurring group of sub-trees a period (of sub-trees). A period is recognised as a period by its repetition. The periods are carefully emphasized in sub-figure (d). The reason for pruning after the second period is that hence no new information is to be obtained. The result pictured in sub-figure (d) is the verification tree for the passed-sample problem.

The next step is to develop the verification graph from the verification tree. The principle for doing this is to ensure that the verification graph incorporates *all* partial verification graphs that can be read from the verification tree. Let us follow this principle with the help of Fig. 18.3 which shows a few of the possible partial verification graphs for our example.

We first note that each leaf assigned a constant output value is expressed as a partial verification graphs; see for instance sub-figures (b)–(e). Note that, in their elementary form, the partial verification graphs contain an output value only in their *present* (or most recent) location. The output values for the input events of the prior locations are obtained from lower-level leaves. An example for this is shown and commented in sub-figures (f) and (g), these almost completing the partial verification graphs of sub-figures (d) and (e). Sub-figures (f) and (g) also document that, in this example, the output values for the first occurrence of the input events 2 and 3 cannot be specified, as these input events represent *internal nodes* in level 1 of the verification tree. The result is that either output sequence in these sub-figures could start with a 0, or with a 1. Both cases must be taken into account.

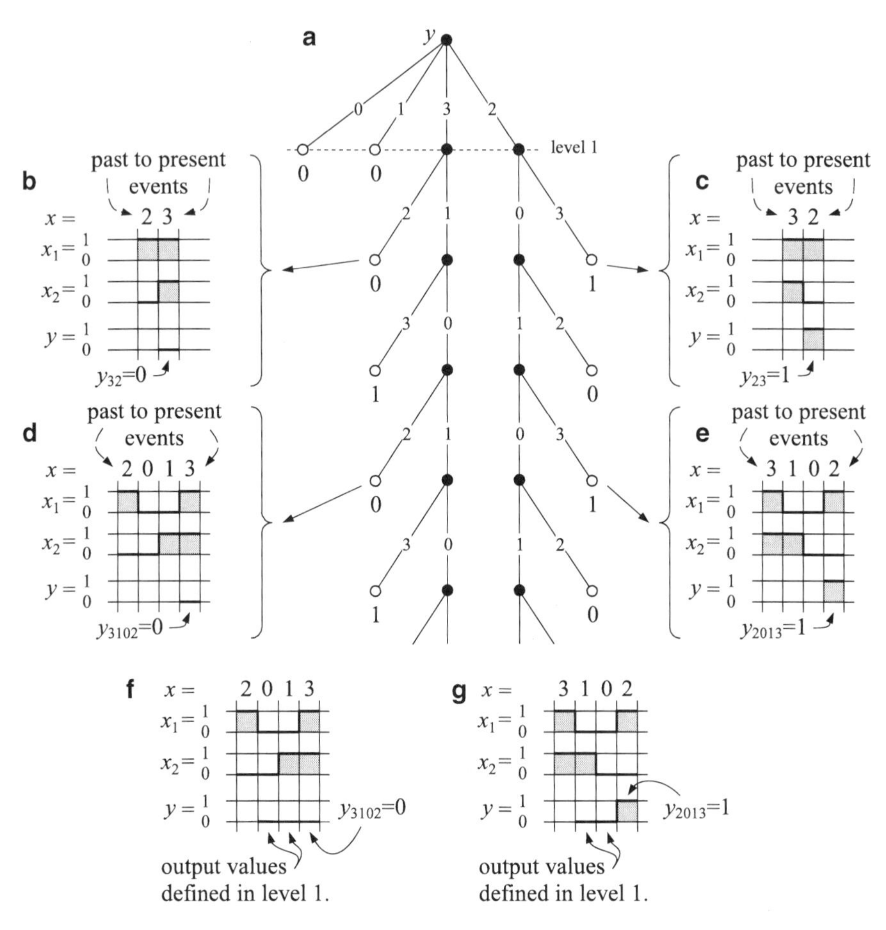

Fig. 18.3 Passed-Sample Problem: interpreting the end-nodes of the verification tree as partial verification graphs

Summarising, we are now able to write *all* possible partial verification graphs that can be read from verification tree in the form of input–output sequences (the top row consisting of a sequence of one or more input events, the bottom row consisting of the associated sequence of output values):

$$\begin{pmatrix} 0 \\ 0 \end{pmatrix} \qquad \begin{pmatrix} 1 \\ 0 \end{pmatrix}$$

$$\begin{pmatrix} 3\ 1\ 0\ 1\ 3 \\ 0\ 0\ 0\ 0\ 1 \end{pmatrix} \cdots \text{(a)} \qquad \begin{pmatrix} 3\ 1\ 0\ 1\ 3 \\ 1\ 0\ 0\ 0\ 1 \end{pmatrix} \cdots \text{(b)}$$

$$\begin{pmatrix}2&0&1&3\\0&0&0&0\end{pmatrix}\ldots(c)\qquad\begin{pmatrix}2&0&1&3\\1&0&0&0\end{pmatrix}\ldots(d)$$

$$\begin{pmatrix}3&1&3\\0&0&1\end{pmatrix}\ldots(e)\qquad\begin{pmatrix}3&1&3\\1&0&1\end{pmatrix}\ldots(f)$$

$$\begin{pmatrix}2&3\\0&0\end{pmatrix}\ldots(g)\qquad\begin{pmatrix}2&3\\1&0\end{pmatrix}\ldots(h)$$

$$\begin{pmatrix}2&0&1&0&2\\0&0&0&0&0\end{pmatrix}\ldots(i)\qquad\begin{pmatrix}2&0&1&0&2\\1&0&0&0&0\end{pmatrix}\ldots(j)$$

$$\begin{pmatrix}3&1&0&2\\0&0&0&1\end{pmatrix}\ldots(k)\qquad\begin{pmatrix}2&0&1&3\\1&0&0&1\end{pmatrix}\ldots(l)$$

$$\begin{pmatrix}2&0&2\\0&0&0\end{pmatrix}\ldots(m)\qquad\begin{pmatrix}2&0&2\\1&0&0\end{pmatrix}\ldots(n)$$

$$\begin{pmatrix}3&2\\0&1\end{pmatrix}\ldots(o)\qquad\begin{pmatrix}3&2\\1&1\end{pmatrix}\ldots(p)$$

In this example, as already stated, the output value associated with the first input event of an IO-sequence which has two or more input events is not defined, and is thus given both possible values, 0 and 1. For one and the same sequence of input events, we therefore get two IO-sequences—one starting with an output value of 0 (left column of the IO-sequences written above), the other with an output value of 1 (right column).

The final step is to chain all the IO-sequences together to obtain the **verification graph**. There are many ways in which the above IO-sequences (a)–(p) can be chained together. We now refer to Fig. 18.4. In doing so, each IO-sequence should, if possible, be used only once (not to obtain an unnecessarily long events graph. Also note that the ending input event and associated output event of an IO-sequence must equal the beginning input event and associated output event of the following IO-sequence. This is emphasized by the shaded areas of the chained IO-sequences. Furthermore, these shaded input events and associated output values are only entered once in the events graph. A word on the **start-up sequence** in chaining the IO-sequences. Whichever IO-sequence, (a)–(p), you want to start with, the *first* output value will be undefined while its last output value is always defined. The first IO-sequence is therefore only used to define a starting output value for the second IO-sequence.

The end-nodes or leaves of a verification tree are marked by constants (either a 0, or a 1). Following the edges from a leaf to the root provides a sequence of input events from past to present, an input sequence, which is used to draw the associated

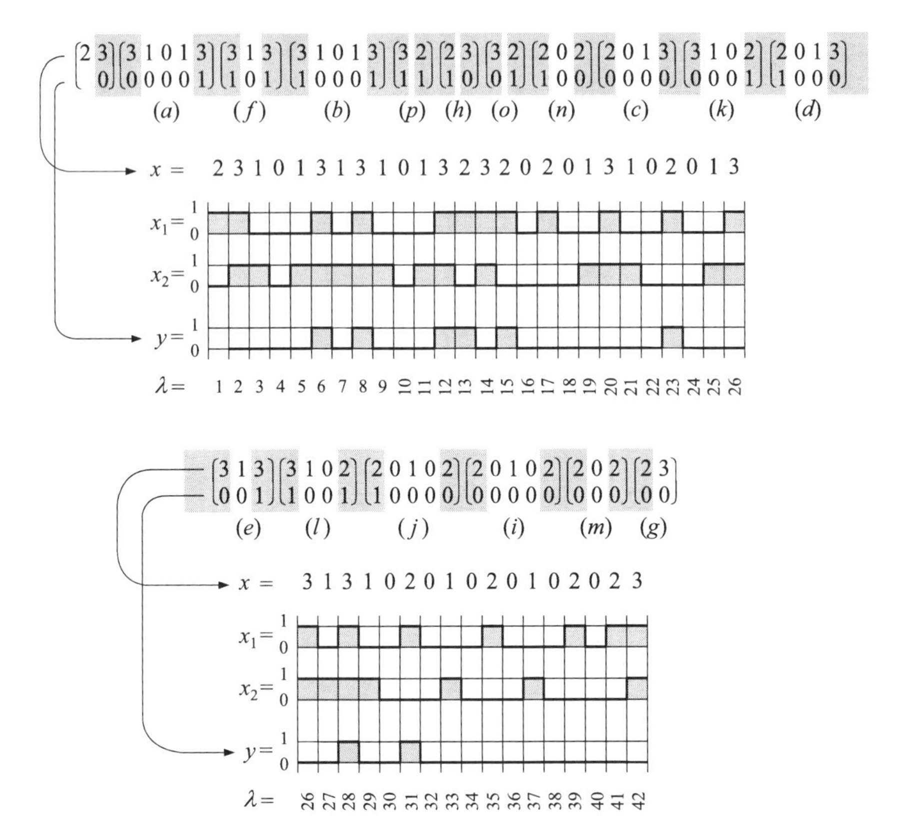

Fig. 18.4 Passed-Sample Problem: developing a **verification graph** from the partial IO-graphs, (a)–(p), read from the verification tree of Fig. 18.3

partial events graph with the more recent events always drawn to the right. All leaves must be represented in the final verification graph.

18.3 Verification Table

In the previous section we used the periodicity of sub-trees to prune the verification tree thereby limiting the length of the IO-sequences for which the circuit must be tested. In this section we show that this principle suffices meaning that no IO-sequences of greater length are needed than those which follow from the highest levels if the pruned verification tree.

The basic idea is that all IO-sequences together must allow us to test, or reach, each GOTO-command in the flow table. We document such a test in a **verification table**: Its **rows** contain those cells (q, x) of the flow table in which *unstable states*

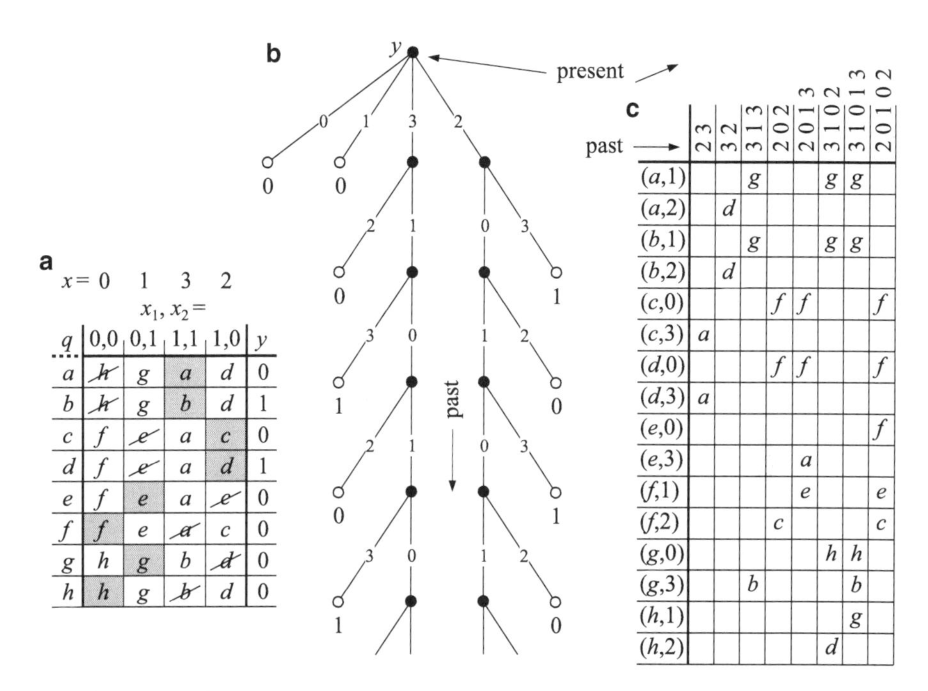

Fig. 18.5 **Passed-Sample Problem**: developing the **verification table**, (**c**), from the flow table, (**a**), and the verification tree, (**b**)

are inscribed; the **columns** mark the IO-sequences (of input events) read from the verification tree, and listed by increasing length; the **GOTO-commands** in any given column of the verification table document all possible *direct* transitions in the input sequence of that column.

We use the passed-sample problem to see how its verification table, shown in Fig. 18.5c, is developed. To determine the rows, we refer to the problem's flow table, repeated in sub-figure (a), from which we copy the coordinates (q, x) of the cells containing unstable GOTO-commands. The sequence in which we assign these coordinates to the rows of the verification table is irrelevant. To determine the columns, we refer to the problem's verification tree, repeated in sub-figure (b), copying all IO-sequences, from past to present (or from leaf to root), to the columns of the verification table.

Having defined the verification table by specifying its rows and columns, we now concentrate on the GOTO-commands to enter into its cells. The GOTO-commands are entered for one IO-sequence after another, i.e., for one column after another. The first column contains the direct transitions when switching from input event 2 to 3. Consulting the flow table, we see that the transitions from 2 to 3 go via the cells $(c, 3)$ and $(d, 3)$, and that both contain the unstable GOTO-a command. This GOTO-a command is therefore entered into rows $(c, 3)$ and $(d, 3)$ of column (2 3) of the verification table.

Jumping ahead, let us now see what to enter into the fifth column, the column marked by the input sequence (2 0 1 3). This input sequence consists of three successive *direct* transitions which we denote by underlining: $(\underline{2\ 0}\ 1\ 3)$, $(2\ \underline{0\ 1}\ 3)$, $(2\ 0\ \underline{1\ 3})$. In the flow table, the direct transition $(\underline{2\ 0}\ 1\ 3)$ leads us from column 2 to the cells $(c,0)$ and $(d,0)$ both containing a GOTO-f command. These f-commands are entered into the verification table in rows $(c,0)$ and $(d,0)$ of column (2 0 1 3). Now return our attention to the flow table, the follow-up direct transition, $(2\ \underline{0\ 1}\ 3)$, leads from stable f to stable e via cell $(f,1)$ which contains the GOTO-e command. This GOTO-e command is written to the verification table's cell in row $(f,1)$ and column (2 0 1 3). Again returning to the flow table, the last of the succession of direct transitions, $(2\ 0\ \underline{1\ 3})$, leads via the GOTO-a command in cell $(e,3)$. This transition is documented in the verification table by entering the GOTO-a command into the cell in row $(e,3)$ and column (2 0 1 3). The above two examples hopefully suffice to explain how to enter the GOTO-commands into a verification table.

The verification table shows us, to which level we need to develop (each branch of) the verification tree to be able to account for all GOTO-commands of the unstable cells of the flow table. For our example of the passed-sample problem this is visualised in Fig. 18.6. The empty and emphasized rows mark those cells of the flow table that cannot be accounted for by input sequences specifying the columns of a respective sub-figure. One sees quite clearly, that we need to employ longer and

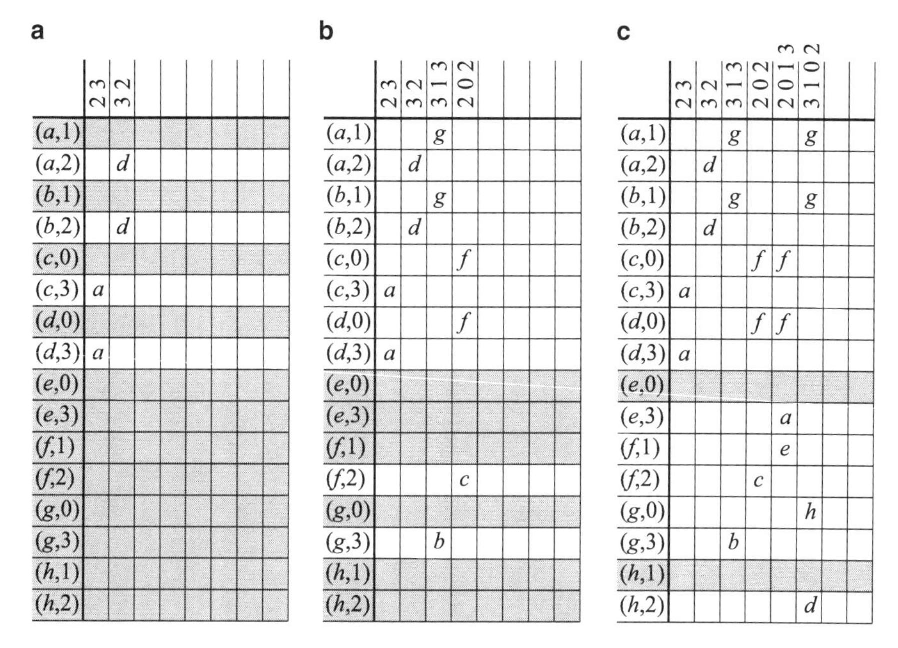

a

	2 3	3 2						
(a,1)								
(a,2)		d						
(b,1)								
(b,2)		d						
(c,0)								
(c,3)	a							
(d,0)								
(d,3)	a							
(e,0)								
(e,3)								
(f,1)								
(f,2)								
(g,0)								
(g,3)								
(h,1)								
(h,2)								

b

	2 3	3 2	3 1 3	2 0 2				
(a,1)			g					
(a,2)		d						
(b,1)			g					
(b,2)		d						
(c,0)				f				
(c,3)	a							
(d,0)				f				
(d,3)	a							
(e,0)								
(e,3)								
(f,1)								
(f,2)				c				
(g,0)								
(g,3)			b					
(h,1)								
(h,2)								

c

	2 3	3 2	3 1 3	2 0 2	2 0 1 3	3 1 0 2		
(a,1)			g			g		
(a,2)		d						
(b,1)			g			g		
(b,2)		d						
(c,0)				f	f			
(c,3)	a							
(d,0)				f	f			
(d,3)	a							
(e,0)								
(e,3)					a			
(f,1)					e			
(f,2)				c				
(g,0)						h		
(g,3)			b					
(h,1)								
(h,2)						d		

Fig. 18.6 Passed-Sample Problem: interpreting the verification table. (**a**) Emphasizing those cells in the flow table that cannot be specified by direct input transitions. (**b**) Emphasizing those cells that can neither be specified by direct nor by double input transitions. (**c**) Emphasizing those cells that can neither be specified by direct, nor by double, nor by triple input transitions

longer input sequences of the verification sub-tree until all rows of the verification table has at least one GOTO-command entered into it. You will always need to develop each branch of a verification tree to cover two periods to achieve this. Vice versa, you will not have to develop any branch of a verification tree further than its second period.

18.4 Verification Graph for the JK-Flipflop

All aspects of the JK-flipflop to be discussed are summarised in Fig. 18.7. The word-recognition tree shows that the JK-flipflop has quite a complex behaviour, see sub-figure (a): When the present input event is 0, the output value is identical to that caused by the previous input event; when the present input event is 3, the output value is identical to the *inverted* value caused by the previous input event. Unexpectedly, the verification tree—sub-figure (b)—is quite simple, as it, of its own design, only grows to the second level needing no pruning. It follows that the IO-sequences are few, short, and fully defined. Furthermore, connecting them to a verification graph is trivial, sub-figure (d) depicting a possible solution. (The K-map of sub-figure (c) is only used to visualise the chosen connection between the input variables J and K, and the values for the input events x.)

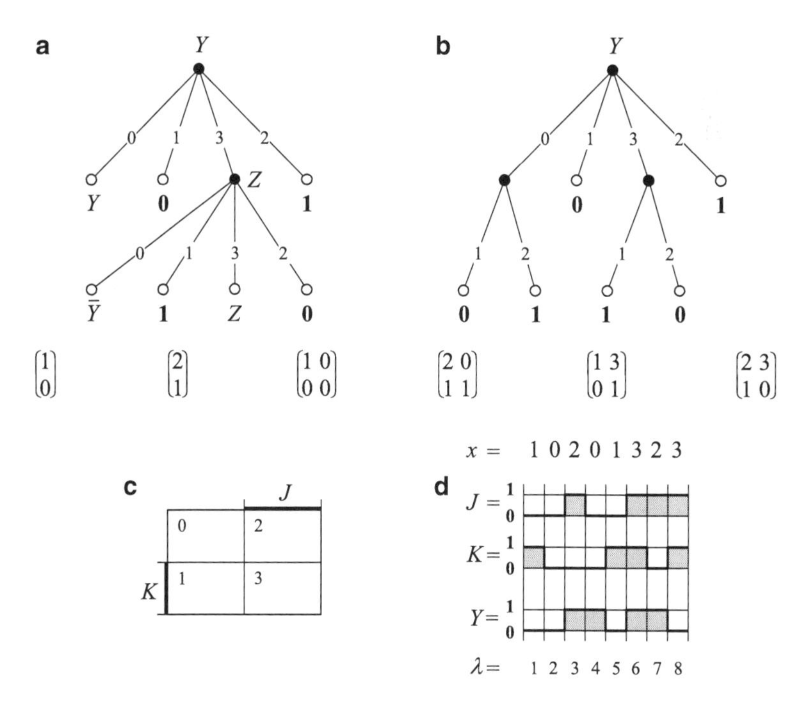

Fig. 18.7 JK-Flipflop: developing a verification graph

18.5 Verification Graph for the D-Latch

Latches are the simplest sequential circuits so that it is no wonder that this simplicity is also mirrored in their verification graphs. Let us study the D-latch, all aspects concerning its verification being summarised in Fig. 18.8. The word-recognition tree of the D-latch, sub-figure (a), is recognisably taken from the K-map of the D-latch in Fig. 11.9. Grafting to the first level provides the verification tree of sub-figure (b). Further grafting, until two periods of sub-trees are completed, leads to the verification tree of sub-figure (c). The IO-sequences of the end-nodes are read from and written to the right of the tree. The normally undefined, past output-values of an IO-sequence is taken from the IO-sequence a level lower. The result being that the IO-sequences of the highest level incorporate all IO-sequences of lower

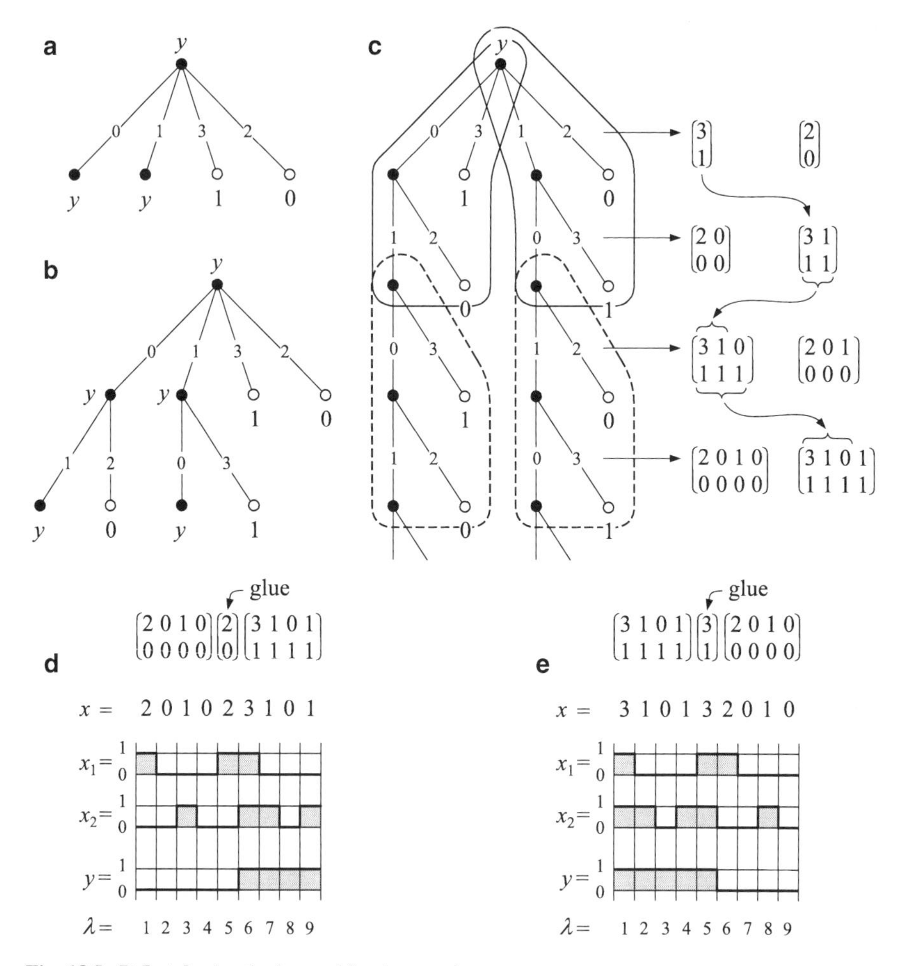

Fig. 18.8 D-Latch: developing verification graphs

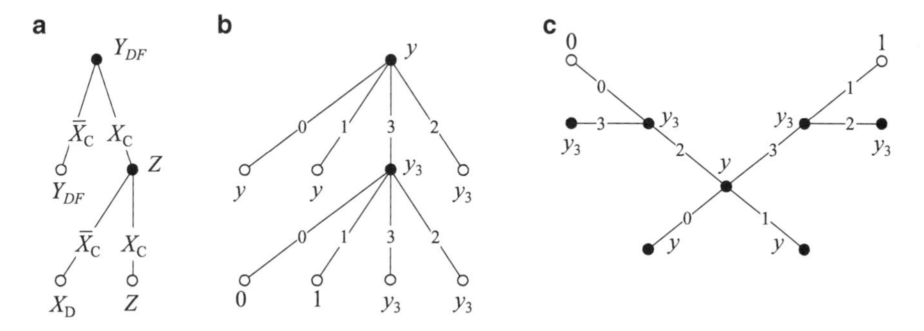

Fig. 18.9 D-flipflop: towards the verification tree

level enabling us to restrict our interest to the IO-sequences of the highest level. To combine these, so that only a single input variable changes its value, we need to insert an appropriate, redundant IO-sequence we call **glue**. This and the actual verification sequence is shown in sub-figure (d), a second one in sub-figure (e).

18.6 Verification Graph for the D-Flipflop

The original word-recognition tree of the D-flipflop (repeated in Fig. 18.9a, and taken from Fig. 13.6a) is deceptively simple. When it is expanded by grafting to a full grown verification tree which shows all periods of the sub-trees, it becomes quite formidable. Furthermore it is quite a task to draw and organise the verification tree in such a way as to be able to easily visualise, recognise, and emphasize the periods.

We start as pictured in Fig. 18.9 by switching from the condensed representation of sub-figure (a) to the standard representation of sub-figure (b) for two input variables, the numbering of the input events taken from the D-latch in Fig. 11.9. When we start drawing the verification tree, we rearrange the edges to emanate stellately from the root rotating them anti-clockwise as in sub-figure (c). This allows us to utilize the page better when developing the verification tree.

The fully developed verification tree is repeated successively—in Figs. 18.10–18.12—each of these figures emphasizing different periods of sub-trees. But first, I urge you to take paper and pencil and meticulously develop the verification tree; in principle, you should get a tree topologically equal to the tree shown in one of the above mentioned figures. Now let us look at these figures in detail.

Figure 18.10. Let us refer to the sub-trees emanating from the edges 2 and 3 as the branches 2 and 3. Each node in one of these branches is marked y_3 ensuring that these branches grow ad infinitum unless pruned. Pruning is always done after the second period of sub-trees. The periods are emphasized and thus easily recognisable. The prime period is marked within a box with a heavy line, the second period has a dashed-line box.

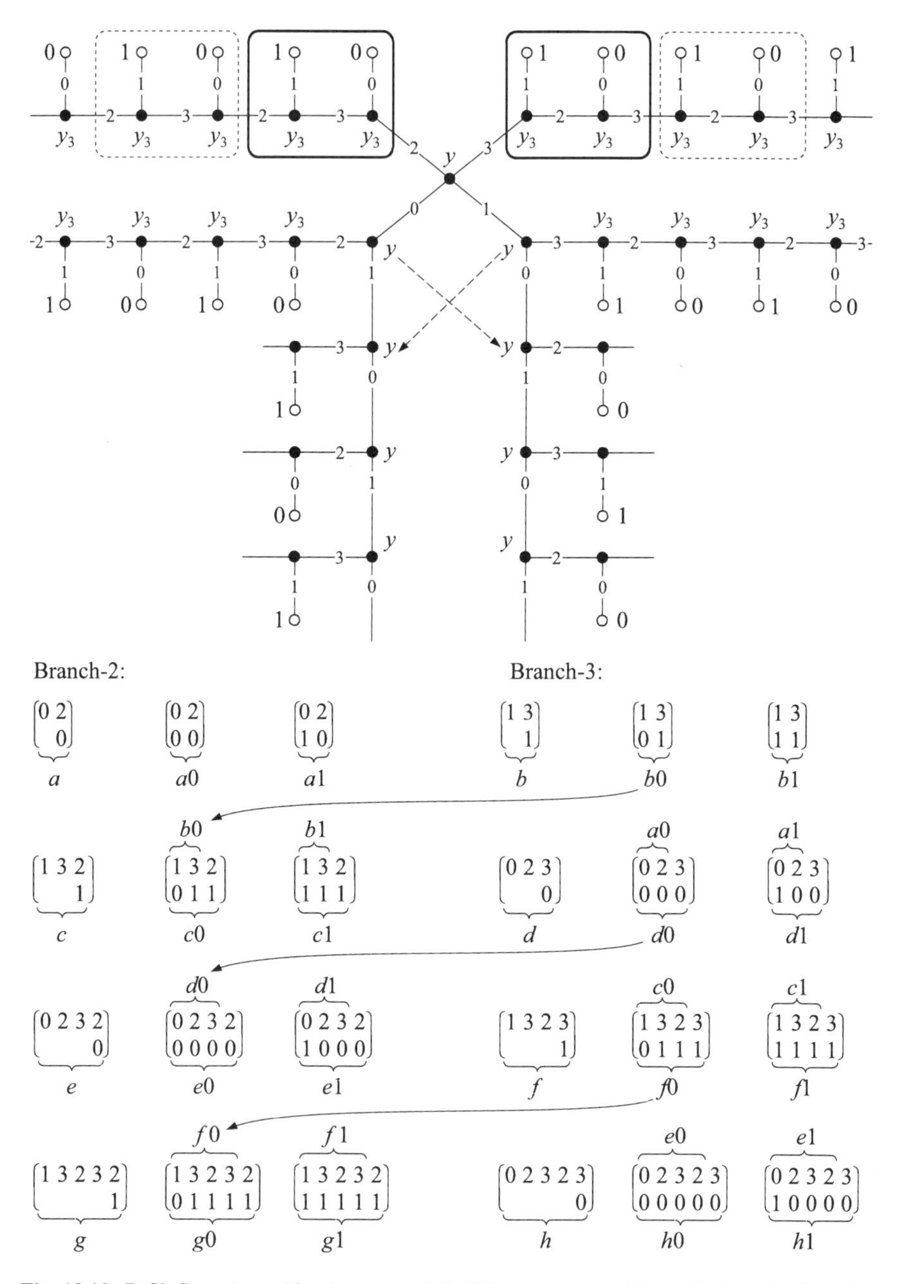

Fig. 18.10 D-flipflop: the verification tree and the IO-sequences read from the branches 2 and 3

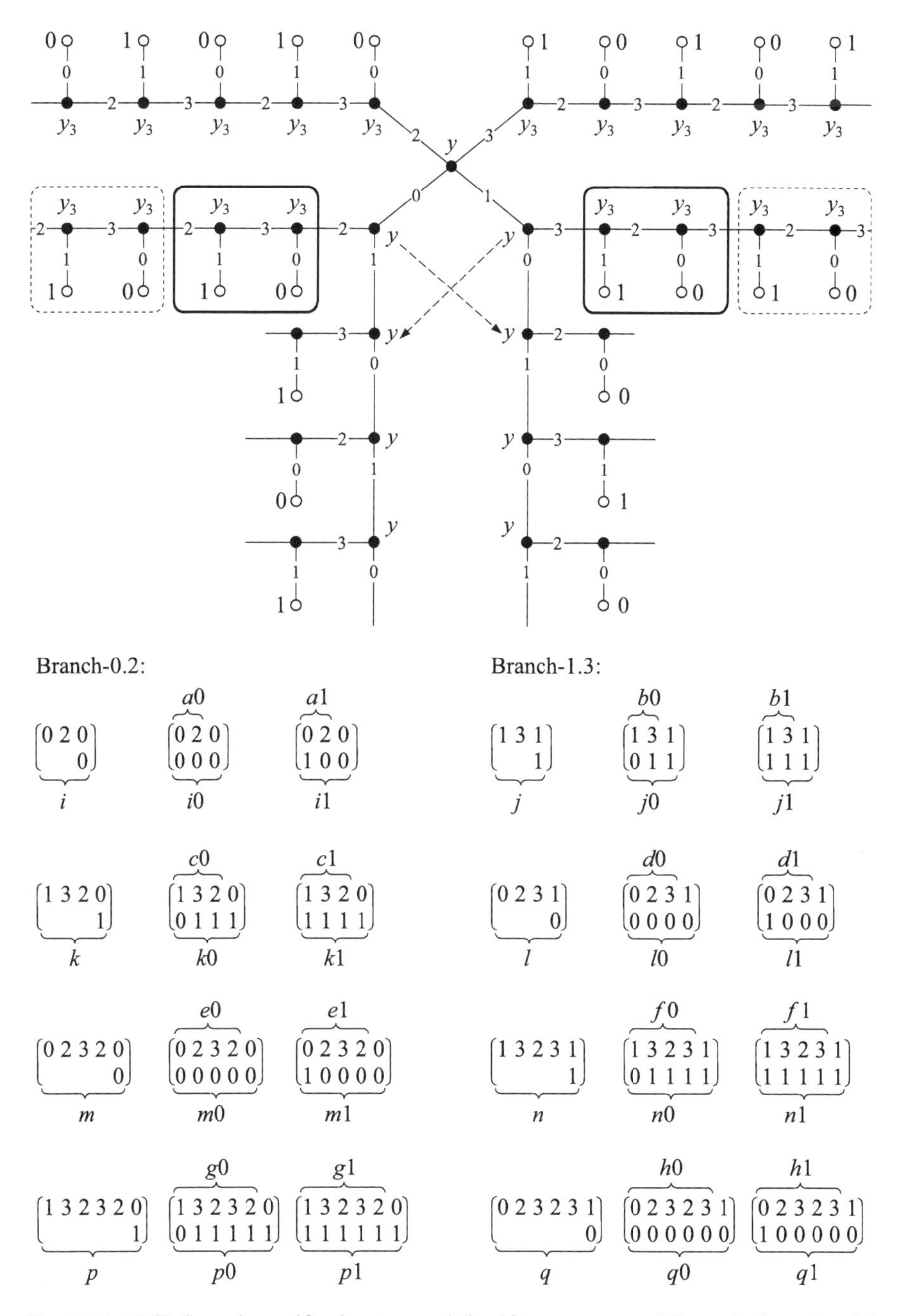

Fig. 18.11 D-flipflop: the verification tree and the IO-sequences read from the branches 0.2 and 1.3

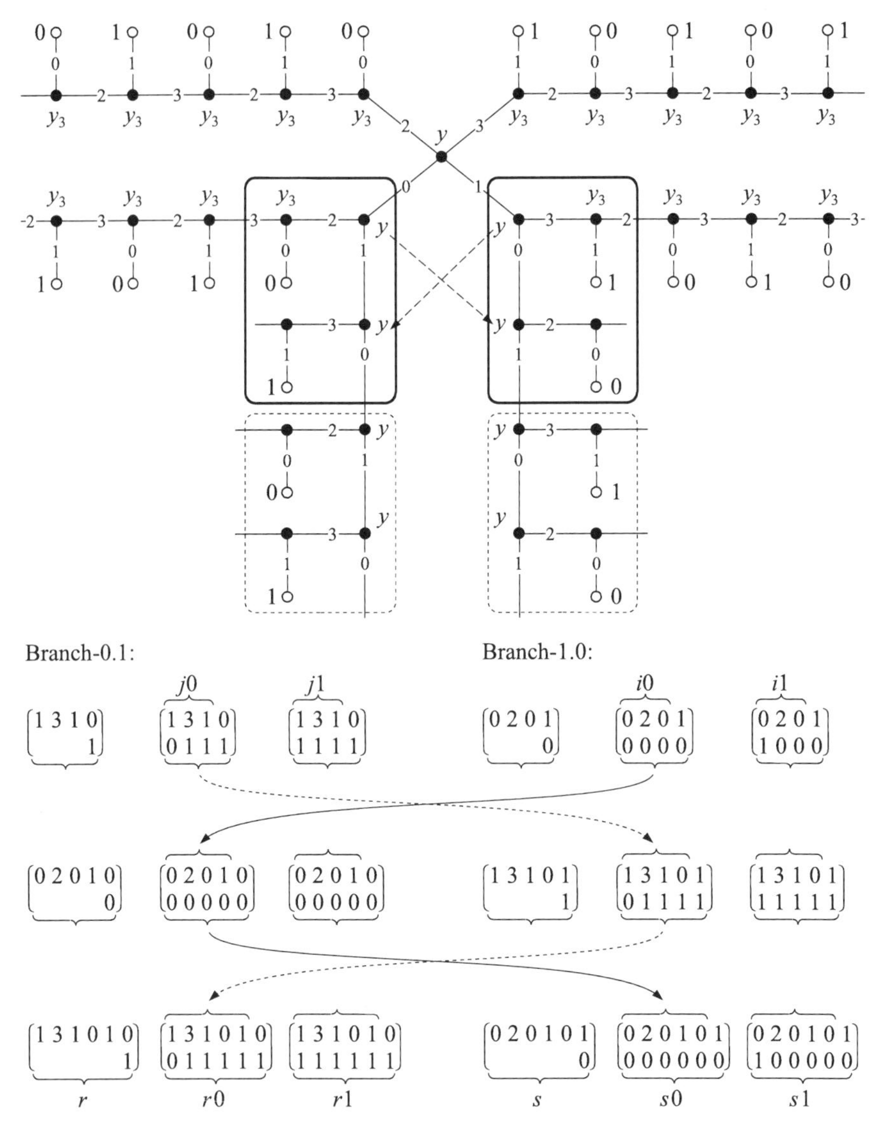

Fig. 18.12 **D-flipflop**: the verification tree and the IO-sequences read from the branches 0.1 and 1.0

An end-node, or leaf, is always defined by the succession of edges from the root to the that leaf. While the phrase *input word* refers to this succession of edges from the root to a certain end-node, we use the phrase *input sequence* for the reverse succession of edges, i.e., for the edges from a certain leaf to the root. The input sequence is always used together with the output value it causes and is then called a (partial) IO-sequence. The IO-sequence is written to a two-row matrix, its first or top row containing the input sequence, while the output value caused is written to the right-most position of the bottom row. The reason for using *input sequences* (instead of *input words*) in connection with verification trees clearly is that the input sequences read from the end-nodes correspond to partial *events-graphs* which, in turn, are catenated to comprise a full *verification graph.*

We now take a detailed look at the input sequences listed in Fig. 18.10. The verification tree shows that the input sequence $(0, 2)$ causes a 0-output (present simultaneously with the input event 2) but that the prior output value is undefined—this is pictured in IO-sequence a. The prior output could be either 0 or 1 which is documented as IO-sequences $a0$ and $a1$, respectively. Similar considerations hold for the first leaf of branch 3 allowing us to write the IO-sequences b, $b0$, and $b1$. Going one level higher in branch 2 we get IO-sequence c whose first two output values are undefined. But, do note that they can be taken from either $b0$ or $b1$ providing us with the fully specified IO-sequences $c0$ and $c1$. I trust, you will have no difficulty continuing iteratively for the remaining IO-sequences of the figure.

As a basic rule, all partial IO-sequences must be used in catenating a verification graph. But in this example only the final IO-sequences $g0$ and $g1$, and $h0$ and $h1$ need be considered, as these sequences contain all other IO-sequences listed in the figure.

Figure 18.11. With the help of the emphasizing boxes along the branches 0.2 and 1.3 it is easy to recognise the respective periods. The lowest level end-nodes of the periods mark the input sequences $(0, 2, 0)$ and $(1, 3, 1)$ with the respective output values 0 and 1 as documented by the IO-sequences i and j. The undefined output values of these IO-sequences can be taken from the IO-sequences $a0$ and $a1$, and $b0$ and $b1$ of the previous figure, providing us with the IO-sequences $i0$ and $i1$, and $j0$ and $j1$. Please do take the trouble to verify the remaining IO-sequences up to $p0$ and $p1$, and $q0$ and $q1$ noting that *all* undefined output values are taken from the previous figure.

Importantly, as in the previous figure, the only IO-sequences we need to consider in catenating a verification graph are the final sequences—$p0$ and $p1$, and $q0$ and $q1$—as these contain all shorter IO-sequences of the branches 0.2 and 1.3.

Figure 18.12. Periods that are slightly more complex and possibly easier to overlook are those of the branches 0.1 and 1.0. As above, the IO-sequences are carefully documented in the figure, and presumably need no further commenting (after the detailed explanations of the previous figure). Again, we need only take the final IO-sequences—$r0$ and $r1$, and $s0$ and $s1$—into account.

Figure 18.12. The last step is to catenate the IO-sequences—$g0$ and $g1$, and $h0$ and $h1$, then $p0$ and $p1$, and $q0$ and $q1$, as well as $r0$ and $r1$, and $s0$ and $s1$—to create a verification graph. One possible catenation is shown in the figure. There are two

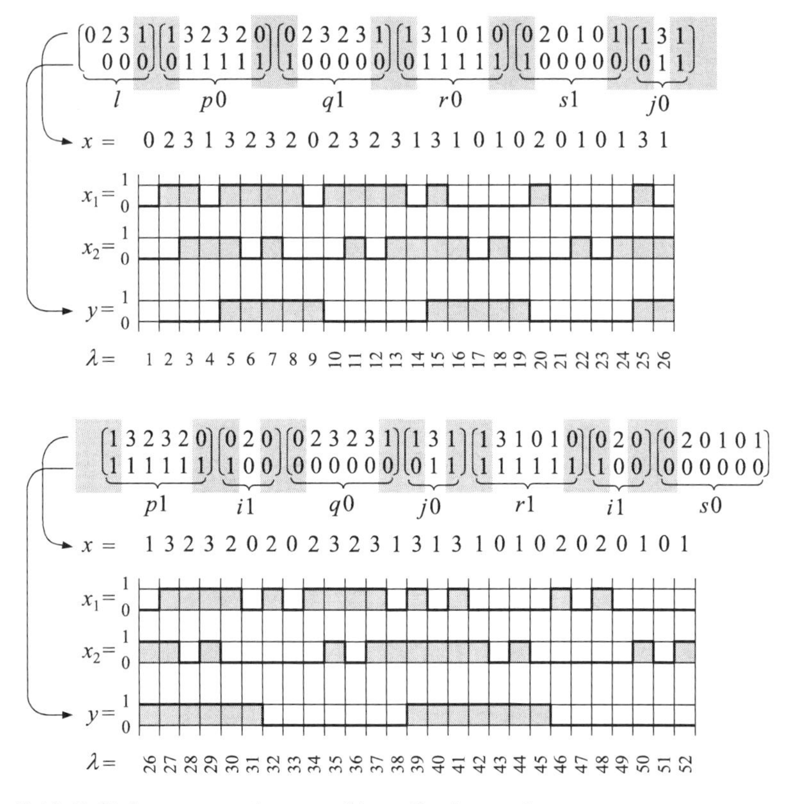

Fig. 18.13 D-flipflop: constructing a possible verification graph

things to point out. Firstly, you will always need a start-up sequence (in our case, the IO-sequence l). Secondly, you will always need a *glue* sequence as junction from one IO-sequence to the next. The glue sequences in our example are $j0$ and $i1$.

This final verification example will hopefully persuade you that developing verification graphs, even for complicated sequential functions, is realistic and provides technically useful results.

Glossary

In the main, the explanations for the words listed give their informal meaning. If you follow the list of contents or the index, you will be led to more in-depth explanations.

Adjacency originally refers to two cells (i.e., elementary K-sets) of a K-map being next to each other (i.e., adjacent). Adjacency can be extended to non-elementary K-sets: Two K-sets are said to be adjacent if they have no elements in common, and if they have, at least partially, a border in common.

Automaton in connection with sequential circuits, is an abbreviation for *sequential automaton*, this being a tape-reading device which takes the values of the input variables of the circuit as inputs, and outputs a 0 or 1, according to the input values, and the program written onto the tape it reads. The program the automaton reads is called a flow table.

Catenation is the chaining together of functions. It is used when composing a combinational circuit out of given sub-functions, or composing a latch using a chosen, basic hinge-function, or employing it repeatedly (iteratively) to obtain an encoded transition table.

Combinational circuit is a switching circuit without a memorising ability, or, formally, a *function* that assigns to each n-tuple of binary variables, 0 or 1, either the value 0, or the value 1.

Composition of combinational circuits is a method of designing a combinational circuit by using chosen combinational sub-functions as building blocks. This is opposed to designing a combinational circuit using, say, AND, OR, and NOT gates.

Composition of memory functions allows you to design a given memory function on the basis of a chosen memory function (the so-called hinge function) by developing a specific combinational circuit (the pre-logic) that adapts the hinge function to the given memory function.

Condition if an automaton is expressed by its state q and its input event x, and is represented by a cell of a flow table of a transition table.

S.P. Vingron, *Logic Circuit Design*, DOI 10.1007/978-3-642-27657-6,

Consensus of two adjacent Karnaugh sets is the largest K-set that can be constructed exclusively of elements of both the given K-sets.

Cover A cover is a set of K-sets of input events. To cover a chosen or given set of input events refers to finding or choosing K-sets of the input events such that their union equals the set of given input events.

Duality is a symmetry property of logic formulas. Duality reduces the burden of proof, for if you can prove a logic formula to be true, you know its dual formula also to be true.

Equivalent states are represented by identical rows of a transition table.

Evaluation formulas are formulas especially designed to evaluate combinational circuits and latches specified by K-maps. Their advantage over canonical normal forms is that they enable the direct derivation of minimised formulas.

Event or input event, is an ordered n-tuple of binary values, 0 or 1, of the n-tuple $(x_1, \ldots, x_n)$ of input variables. In the standard literature an input event is usually referred to as an input combination.

Events graph depicts a successive sequence of input events together with the output values they cause. Each input event can occur multiple times. Most importantly, an output value is only allowed to change if an input event changes.

Events table consists of columns specified as a sequence of unique input events together with the output values they cause. In the literature, events tables frequently and incorrectly are called truth tables.

Feedback is an output signal fed back to an input of the circuit causing the output.

Flipflop refers to a category of sequential circuits for which there is no generally accepted definition. There are four circuits honoured with the name flipflop: 7474 D-flipfop, D-, T-, and JK-flipflop. (This text always uses the spelling 'flipflop', instead of the standard 'flip-flop' so as to be able to write 'D-flipflop' and not have to use the disconnected name 'D flip-flop' or over-connected 'D-flip-flop'.) The only legitimate way to say what a flipflop is, is to specify a certain class of word-recognition trees whose characteristic we *agree* to be that of a flipflop. Taking a lead from the above mentioned 7474 D-flipfop, D-, T-, and JK-flipflops, you will find it quite difficult to adequately formulate the characteristics of a word-recognition tree of what you might accept to be a flipflop.

Let me attempt it: A flipflop is a circuit whose word-recognition tree

- Has two levels,
- The root sub-tree is a latch
- At least one higher-order sub-tree is dependant on the output of some other sub-tree (even be it the root).

This definition fits the above mentioned flipflop circuits while successfully excludes, for instance, the two-hand safety circuits.

Flow table is the term used for the program written to a tape a sequential automaton can read. Writing a flow table adheres to special rules.

Gates are graphic symbols that stand for logic connectives, and as such depict an idealised switching behaviour (e.g., they have no delays). They are not the real-world switching devices that emulate the behaviour of the logic connectives.

Hazard refers to an erroneous output signal that may occur.

Instance is the actual and therefore measured value of a (binary) variable, or input event.

Karnaugh map (or K-map) is a Venn diagram. Like any Venn diagram one first has to state what elements it depicts. The elements depicted in a K-map are the 2^n input events $(x_1, \ldots, x_n)$ of a circuit. The elements, or cells, are always arranged in a rectangle the number of its rows being 2^p, where p is chosen such that $0 < p < n$, while the number of its columns are 2^{n-p}. The 2^p rows are labelled using p binary input variables. Their values are listed according to a Gray code. The values of the remaining $n - p$ variables are used to label the 2^{n-p} columns—also according to a Gray code.

Latch is a real-world switching device or circuit that has the shortest binary memory possible. Defining its memorising capability is a different matter. This is done by way of a so-called memory function or memory. It is (alas) quite common to speak of a latch when one means memory.

An **Elementary latch** has two inputs and two mutually inverted outputs. There are tree types of elementary latches:

- **PSR-latches** are Predominantly Setting and Resetting latches.
- **PQ-latches** are Predominantly memorising latches (the symbol Q is usually used to denote the internal state, i.e., the internal feedback signal, of a latch and is thus also used to denote memorisation).
- **Eccles-Jordan latches** are the original latches invented 1919 by Eccles and Jordan.

Logic equivalence $\Leftrightarrow$, is the symbol used to state that two logic expressions, or logic variables, A and B, are the same, i.e., have the same truth value, $A \Leftrightarrow B$.

Logic variable is a variable that stands for either of the truth values *true* (for which we use the symbol **1**, reminiscent of the integer 1) or *false* (for which we use the symbol **0**, reminiscent of the integer 0). In this book I tried very hard to abide by the convention to use upper-case (Latin or Greek) letters for logic variables.

Maxterm is the logic expression of the union of all but one cells of a K-map. Except for the full area of a K-map, the maxtern expresses the largest area of a K-map.

Memory function or simply memory, is a function (in a mathematical sense) that describes the idealised behaviour of a latch. For details please refer to the book's text.

Mealy automaton (or Mealy machine) is not an automaton or a machine, rather, it is a system of two formulas said to describe a sequential circuit. Let $\mathcal{E}$ and $\mathcal{Q}$ be the set of input events, and of (internal) states, respectively. Then the Mealy automaton encompasses the functions $\sigma : \mathcal{E} \times \mathcal{Q} \mapsto \mathcal{Q}$, called the state transition function, and the function $\tau : \mathcal{E} \times \mathcal{Q} \mapsto \{0, 1\}$, called the Mealy output function.

Merging refers to dropping all identical rows but one of a transition table.

Minterm is the logic expression of a single cell of a K-map. A cell being the *smallest* unit of a K-map gives the minterm its name.

Moore automaton (or Moore machine) is not an automaton or a machine, rather, it is a system of two formulas said to describe a sequential circuit. Let $\mathcal{E}$ and $\mathcal{Q}$ be the set of input events, and of (internal) states, respectively. Then the Moore automaton encompasses the functions $\sigma : \mathcal{E} \times \mathcal{Q} \mapsto \mathcal{Q}$, called the state transition function, and the function $\tau : \mathcal{Q} \mapsto \{0, 1\}$, called the Moore output function.

Normal forms are equations that enable you to calculate unique formulas for logic circuits when given the circuit's events table. There are two types of normal forms, canonical and non-canonical. The canonical normal forms (the AND-to-OR and OR-to-AND formulas) have coefficients, each consisting of *all the* input variables (the minterms in the first case, the maxterms in the second) and are thus called canonical or well ordered. The non-canonical normal-forms are the Zhegalkin normal forms. The number of input variables of the coefficients of a Zhegalkin normal form can vary from 0 to the full number n of input variables. A canonical normal-form has a multiple OR or a multiple AND as output gate. A non-canonical normal-form has a multiple XOR or a multiple EQU as output gate.

Numeric variable represents a number, in our usage, usually an integer. In connection with logic circuits, numeric variables contrast with logic variable which stand for truth values (**0** and **1**). In our usage, a binary numeric-variable stands for the integers 0 and 1.

Proposition 'is a statement that either affirms or denies something'. Or, equivalently: 'A proposition is a sentence expressing an assertion that is either true (**1**) or false (**0**)'. Do note that it is not necessary to be able to prove the truth or falsity of the statement or assertion to call it a proposition. For instance, the unproven Goldbach conjecture '*Every even number greater than 2 can be represented by the sum of two primes*' is a proposition because something is being claimed that in principle can only be either true or false.

Propositional form stands for a class of propositions by splitting propositions into subjects and predicates. Take for instance the proposition $2 > 3$ in which the numbers 2 and 3 are its *subjects*, and the relation $>$ is the *predicate*. But writing $x > 3$, where x is a numeric variable, does not constitute a proposition as, without being given a specific value for x, it is impossible to say that $x > 3$ has a truth value ('*a number you choose is greater than* 3' is not very specific). But it does have the *form of a proposition* (thus the name) and, most importantly, substituting a value for the subject x the propositional form becomes a proposition. $x > 3$ stands for a whole class of propositions in which one subject has been replaced by a subject variable. In general, any number of subjects can be replaced subject variables thus obtaining propositional forms of higher degree. Importantly, within a logic expression, you can use a propositional form in the same way as you can use a propositional variable (a variable that stands for a proposition).

Sampling A so-called data input x_D is sampled, i.e., its value is passed on to the output during usually brief time intervals which are defined by a so-called control or sampling input x_C being 1; between sampling, i.e., while x_C is 0, the output retains the value of x_D last sampled.

Sequential automaton see automaton, above.

Signal is the numeric information conveyed by a quantifiable physical entity.

Stability is a principle and required property of a flow table. Assume a sequential automaton is momentarily reading line q of the flow table (i.e., is in state q), and that the input event present (i.e., the present input event) is x. The cell (q, x), consisting of the *present state* q and the *present input* x, is called the *present condition* of the automaton. The state q^+, inscribed into the present cell (or condition) is the *next state*, this naming the state or row of the flow table to be read next, i.e., the tape is moved so that row q^+ is moved under the reading head.

This is where **stability** comes into the picture: We *always* assume the tape to remain stationary (unmoving) in the next state. That is, the next condition (q^+, x) always has the next state q^+, the same q^+ that defines the condition (q^+, x) itself, inscribed into its cell, thereby ensuring that the tape is moved to, i.e., stays in row q^+. This forced ending of all tape movement when changing to the next state from a present one is what we call stability.

State refers to the rows of a flow table and a transition table.

State assignment is a binary encoding of a flow table by which all identical state lables (whether present or next) are assigned the same binary code. The thus encoded flow table is called a transition table.

In the standard Huffman theory there are practically know restrictions as to the choice of code. The encoding procedure of iterative catenation, on the other hand, is deterministic, not influenced by choice.

State encoding is the same as state assignment.

State reduction or merging refers to dropping, at most, all states but one of any set of identical states of a flow table or transition table.

Tautology is either a logical equivalence or a logical implication such as

$$X_1 \Leftrightarrow X_1\overline{X}_2 \vee X_1X_2, \qquad \text{or} \qquad (X_1 \vee X_2) \wedge \overline{X}_1 \Rightarrow X_2.$$

A tautology can be likened to what is sometimes called an **identity** in conventional algebra, e.g.

$$(a+b)^2 \equiv a^2 + 2ab + b^2.$$

Tautologies and identities have the characteristic property of being true for *all* instances of their variables.

Timing diagram depicts the time-dependant behaviour of the input and output signals of a circuit. The signals can be depicted in analogue or binary form. A binary timing diagramm is to be distinguished from an events graph.

Truth table lists the instances of the variables of a logical formula and the results obtained from all its connectives. If the final connective to be evaluated always provides the truth value **1**, the logic formula is a tautology.

Verification graph is an events graph whose input sequence (a) is if finite length, (b) is as short as possible, (c) successive input events must differ in the value of one and only one input variable, and (d) the input sequence enables a complete test of the circuit.

Verification tree is a word-recognition tree in which successive input events differ by the value of one and only one input variable. Sub-trees are always drawn (**grafted**), and each branch is **pruned** as soon as a period of sub-trees is established.

Well-behaved circuit is a sequential circuit whose latches are well-behaved. A memorising input event of a latch can produce a 1 or a 0 output. The output value must always equal that produced by the prior input event, and must remain unchanged as long as the present input event remains constant.

Word-recognition tree is a tree of input words. The tree's end-nodes state the output values caused by the input word associated with the leaf. The word-recognition tree is the formal representation of a circuit *per se*.

Bibliography

I have taken the liberty to present only a very short selection of books and papers so as to focus further reading. Of course the selection, especially on books, is highly subjective and not to be taken as a rating. The papers are chosen for their historical relevance. In these times of the internet I believe this selective presentation of literature justifiable and, possibly, advantageous.

Selected Books

Brzozowski, J.A., Yoeli, M.: Digital Networks. Prentice Hall,Englewood Cliffs (1976)

Brzozowski, J.A., Seger, C.-J.H.: Asynchronous Circuits. Springer, New York (1995)

Caldwell, S.H.: Switching Circuits and Logic Design. Wiley, New York (1958)

Clare, C.R.: Designing Logic Systems using State Machines. McGrew-Hill Book Company, New York (1973)

Dietmeyer, D.L.: Logic Design of Digital Systems. Allyn and Bacon, Boston (1971)

Eilenberg, S., Elgot, C.C.: Recursiveness. Academic Press, New York (1970)

Fasol, K.H., Vingron, P.: Synthese Industrieller Steuerungen. R. Oldenbourg Verlag, München (1975)

Harrison, M.A.: Introduction to Switching and Automata Theory. McGraw-Hill Book Company, New York (1965)

Kohavi, Z.: Switching and Finite Automata Theory. McGraw-Hill Book Company, New York (1970)

Krieger, M.: Basic Switching Circuit Theory. McGraw-Hill Book Company, New York (1969)

McCluskey, E.J.: Logic Design Principles. Prentice Hall, London (1986)

Mead, C., Conway, L.: Introduction to VLSI Systems. Addison-Wesley, Reading Massachusetts (1980)

Muroga, S.: Logic Design and Switching Theory. Wiley, New York (1979)

Pessen, D.W.: Ìndustrial Automation. Wiley, New York (1989)

Unger, S.H.: Asynchronous Sequential Switching Circuits. Wiley, New York (1969)

Vingron, S.P.: Switching Theory. Springer, Berlin (2004)

Zander, H.-J.: Entwurf von Folgeschaltungen. VEB Verlag Technik, Berlin (1974)

Selected Papers

Ashenhurst, R.A.: The decomposition of switching functions. In: Proceedings of an InternationalSymposium on the Theory of Switching, April 2–5, 1957. Annals of the Compotation Laboratory of Harvard University, vol. 29, pp. 74–116. Harvard University Press (1959)

S.P. Vingron, *Logic Circuit Design*, DOI 10.1007/978-3-642-27657-6,

Huffman, D.A.: The synthesis of sequential switching circuits. J. Frankl. Inst. **257**:161–190 (1954) and 275–303 (1954)

Huffman, D.A.: A study of memory requirements of sequential switching circuits. Massachusetts Institute of Technology, Research Laboratory of Electronics, Technical Report No. 293, April 1955

Huffman, D.A.: The design and use of Hazard-Free switching networks. J. Assoc. Comput. Mach. **4**:47–62 (1957)

Karnaugh, M.: The map method for synthesis of combinational logic circuits. Trans. AIEE, Part I, **72**(9):593–599 (1953)

McCluskey, E.J.: Minimisation of boolean functions. Bell Syst. Tech. J. **35**(6):1417–1445 (1956)

Mealy, G.H.: A method for synthesizing sequential circuits. Bell Syst. Tech. J. **34**(5):1045–1079 (1955)

Medvedev, I.T.: On a class of events representable in a finite automaton, pp. 385–401. Avtomaty, Moscow (1956). English translation in MIT Lincoln Laboratory Group Report, pp. 34–73, June 1958

Moore, E.F.: Gedankenexperiments on sequential machines. In: Shannon, C.E., McCarthy, J., Ashby, W.R. (eds.)Automata Studies. Princeton University Press, Princeton (1956)

Quine, W.V.: The problem of simplifying truth functions. Am. Math. Mon. **59**:521–531 (1952)

Quine, W.V.: A way to simplify truth functions. Am. Math. Mon. **63**:627–631 (1955)

Shannon, C.E.: A symbolic analysis of relays and switching circuits. Trans. AIEE **57**:713–723 (1938)

Tracey, J.H.: Internal state assignment for asynchronous sequential circuits. IEEE Trans. Electron. Comput. **EC-15**:551–560 (1966)

Unger, S.H.: Hazards and delays in asynchronous sequential circuits. IRE Trans. Circuit Theory **CT-6**:12–25 (1959)

Veitch, E.W.: A chart method for simplifying truth functions. Proceedings of Pittsburgh Association for Computing Machinery, University of Pittsburgh, May 1952

Vingron, P.: Coherent design of sequential circuits. IEE Proc. **130E**:190–201 (1983)

Zhegalkin, I.I.: The French title of the Russian original is Gégalkin I. I.: 'Sur le calcul des propositions dans la logique symbolique'. Mat. Sbornik **34**:9–28 (1927)

Index

S.P. Vingron, *Logic Circuit Design*, DOI 10.1007/978-3-642-27657-6,

M

N

O

P

Q

R

W

X

Z